Stochastic PDEs and Modelling of Multiscale Complex System

INTERDISCIPLINARY MATHEMATICAL SCIENCES*

Published

Vol. 20 *Stochastic PDEs and Modelling of Multiscale Complex System*
eds. Xiaopeng Chen, Yan Lv & Wei Wang

Vol. 19 *Kernel-based Approximation Methods using MATLAB*
by Gregory Fasshauer & Michael McCourt

Vol. 18 *Global Attractors of Non-Autonomous Dynamical and Control Systems (2nd Edition)*
by David N Cheban

Vol. 17 *Festschrift Masatoshi Fukushima: In Honor of Masatoshi Fukushima's Sanju*
eds. Zhen-Qing Chen, Niels Jacob, Masayoshi Takeda & Toshihiro Uemura

Vol. 16 *Hilbert–Huang Transform and Its Applications (2nd Edition)*
eds. Norden E Huang & Samuel S P Shen

Vol. 15 *Recent Advances in Applied Nonlinear Dynamics with Numerical Analysis: Fractional Dynamics, Network Dynamics, Classical Dynamics and Fractal Dynamics with Their Numerical Simulations*
eds. Changpin Li, Yujiang Wu & Ruisong Ye

Vol. 14 *Recent Developments in Computational Finance: Foundations, Algorithms and Applications*
eds. Thomas Gerstner & Peter Kloeden

Vol. 13 *Stochastic Analysis and Applications to Finance: Essays in Honour of Jia-an Yan*
eds. Tusheng Zhang & Xunyu Zhou

Vol. 12 *New Trends in Stochastic Analysis and Related Topics: A Volume in Honour of Professor K D Elworthy*
eds. Huaizhong Zhao & Aubrey Truman

Vol. 11 *Advances in Interdisciplinary Applied Discrete Mathematics*
eds. Hemanshu Kaul & Henry Martyn Mulder

*For the complete list of titles in this series, please go to
http://www.worldscientific.com/series/ims

Interdisciplinary Mathematical Sciences – Vol. 20

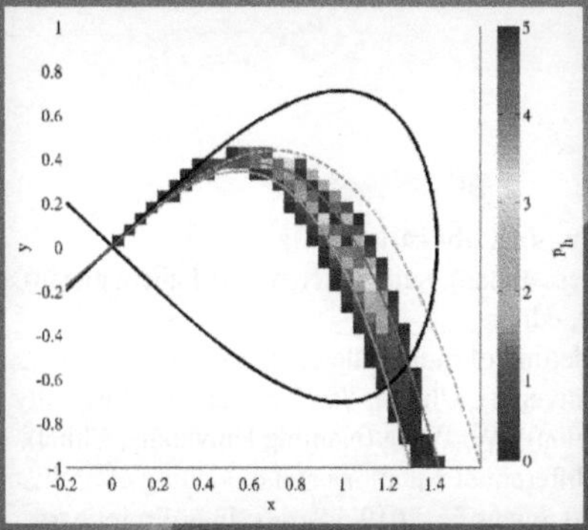

Stochastic PDEs and Modelling of Multiscale Complex System

Xiaopeng Chen
Shantou University, China

Yan Lv
Nanjing University of Sciences and Technology, China

Wei Wang
Nanjing University, China

World Scientific

NEW JERSEY • LONDON • SINGAPORE • BEIJING • SHANGHAI • HONG KONG • TAIPEI • CHENNAI • TOKYO

Published by

World Scientific Publishing Co. Pte. Ltd.

5 Toh Tuck Link, Singapore 596224

USA office: 27 Warren Street, Suite 401-402, Hackensack, NJ 07601

UK office: 57 Shelton Street, Covent Garden, London WC2H 9HE

Library of Congress Cataloging-in-Publication Data

Names: Chen, Xiaopeng (Mathematician), editor. | Lv, Yan (Mathematician), editor. | Wang, Wei (Mathematician), editor.

Title: Stochastic PDEs and modelling of multiscale complex system / editors, Xiaopeng Chen (Shantou University, China), Yan Lv (Nanjing University of Science and Technology, China), Wei Wang (Nanjing University, China).

Other titles: Stochastic partial differential equations and modelling of multiscale complex system

Description: New Jersey : World Scientific, 2019. | Series: Interdisciplinary mathematical sciences ; 20 | Includes bibliographical references.

Identifiers: LCCN 2019008184 | ISBN 9789811200342 (hard cover : alk. paper)

Subjects: LCSH: Stochastic partial differential equations. | Differential equations, Partial. | Multiscale modeling.

Classification: LCC QA274.23 .S8645 2019 | DDC 519.2/2--dc23

LC record available at https://lccn.loc.gov/2019008184

British Library Cataloguing-in-Publication Data

A catalogue record for this book is available from the British Library.

For any available supplementary material, please visit
https://www.worldscientific.com/worldscibooks/10.1142/11280#t=suppl

Desk Editor: Liu Yumeng

Preface

This volume is dedicated to Professor Anthony Roberts, on the occasion of his 60th birthday, to celebrate his remarkable contribution and significant influence on modern dynamical system theory in conjunction with new computer algebra algorithms in scientific computing in deriving and interpreting mathematical and computational models of complex multiscale systems, both deterministic and stochastic. Mathematical models for spatial-temporal physical, chemical and biological systems under random influences are often described as stochastic partial differential equations (SPDEs) with different scale in both time and space. Many mathematical tools are developed to understand qualitatively and simulate the complex system described by SPDEs.

This volume contains 12 review/new articles from leading authorities as well as active young researchers in the fields of stochastic partial differential equations and related fields, which covers stochastic model reduction, multiscale modelling, multiscale stochastic simulation, stochastic averaging for SPDEs, small parameter approximation of SPDEs, stochastic nonlocal equations and optimal control.

We are grateful to all the authors for their contributions to this volume. We also thank all the referees for the valuable comments under an even tighter time frame. We would like to thank Prof Jinqiao Duan for his invitation for such a volume.

Xiaopeng Chen, Shantou University
Yan Lv, Nanjing University of Science and Technology
Wei Wang, Nanjing University

A Biographical Note and Tribute to Anthony Roberts on His 60th Birthday

Professor Anthony Roberts was born in Nigeria in 1957, and his family moved to Adelaide, Australia, when he was three. He was awarded a scholarship to Pulteney Grammar School, where he studied 1962–73, and in 1973 won a prize in the state-wide IBM Mathematics Competition. While studying Mathematical Science at Adelaide University 1974–77, he met his wife Barbara Ralph. They married in 1979 before going on to study for a PhD in the University of Cambridge with the supervision of Prof Herbert Huppert. They currently have three children (two of whom also earned PhDs from the University of Cambridge), and three grandchildren.

As a leader in developing and applying a branch of modern dynamical system theory, in conjunction with new computer algebra algorithms in scientific computing, Professor Roberts derives and interprets mathematical and computational models of complex multiscale systems, both deterministic and stochastic. His development of the systematic centre manifold approach gives a powerful new paradigm for the construction of dynamic models, complete with initial and boundary conditions, and sensitivity projections. He develops applications of this methodology mainly to free surface fluid dynamics in the flow of thin fluid layers, water waves, and on to turbulent floods and tsunamis. Research interests include the modelling of complex multiscale dynamical systems, stochastic differential equations, centre manifold theory and applications, free-surface hydrodynamics, thin film flows, turbulent floods, multifractal geometry, scientific computing, computer algebra algorithms, and multiscale modelling of emergent dynamics in complex systems.

Anthony Roberts was awarded a Knight Prize by the University of Cambridge in 1981 and was elected a fellow of Trinity College in 1982. He was also awarded for excellence research at USQ in 1997. He was a member of the board of the Australian Mathematical Sciences Institute and was appointed Electronic Editor of the Australian Mathematical Society from 1997. He was served on the founding executive committee of the Australasian Fluid Mechanics Society from 2008 to 2011.

Anthony Roberts is not just a brilliant mathematician but also a great director in mathematics. He was the foundation professor of applied mathematics at the University of Southern Queensland and deputy director of the Computational Engineering and Sciences Research Centre at the University of Southern Queensland. Anthony Roberts is also a kindly teacher. He always has good ideas to solve the problems from his students and published a series of articles in the Gazette describing some simple techniques to improve writing.

Anthony Roberts has published over one hundred refereed journal articles and three books, and he also has edited fifteen proceedings. Even now he is still working hard on the mathematics he loves and writing research papers. His contributions to mathematics is immense and his spirit always encourages us. It is our great pleasure to be able to edit this volume and dedicate it to Anthony Roberts.

Xiaopeng Chen, Yan Lv and Wei Wang

December 2018

Contents

Chapter 1

Geometric Methods for Stochastic Dynamical Systems

Jinqiao Duan[1,*] and Hui Wang[2,†]

[1]*Department of Applied Mathematics, Illinois Institute of Technology, Chicago, IL 60561, USA*

[2]*School of Mathematics and Statistics, Zhengzhou University, Zhengzhou 450001, China*

[*]*duan@iit.edu*; [†]*huiwang2018@zzu.edu.cn*

1. Introduction

Noisy fluctuations are ubiquitous in complex systems. They play a crucial or delicate role in the dynamical evolution of gene regulation, signal transduction, biochemical reactions, among other systems. Therefore, it is essential to consider the effects of noise on dynamical systems. It has been a challenging topic to have better understanding of the impact of the noise on the dynamical behaviors of complex systems. See [1, 2] for more information.

1.1. *Stochastic dynamical systems*

A dynamical system may be thought as evolution of mechanical 'particles', and is then described by a system of ordinary differential equations

$$\frac{dx}{dt} = f(x), \tag{1}$$

where f is often called a vector field. In natural science and applied science, the dynamical systems are used to describe the evolution of complex phenomena [3–5]. However, dynamical systems are often influenced by random factors in the environment, such as in the systems of the propagation of waves through random media, stochastic particle acceleration, signal detection, and optimal control with fluctuating constraints. Indeed, a small random disturbance may have an unexpected effect on the whole dynamical system. A stochastic dynamical system is a dynamical system with noisy components or under random influences [1, 2] and thus may be modeled by a stochastic differential equation

$$\frac{dx}{dt} = f(x) + \text{noise}. \tag{2}$$

Usually, we use two kinds of noise. One is Gaussian noise, and the other is non-Gaussian noise. Gaussian noise is modeled by Brownian motion B_t, while non-Gaussian noise is expressed via Lévy process (especially α-stable Lévy motion $L_t^{\alpha,\beta}$) [6, 7]. Dynamical systems driven by Gaussian noise have been widely studied, but in some complex systems, the random influences or stochastic processes are non-Gaussian. For example, during the regulation of gene expression, transcriptions of DNA from genes and translations into proteins take place in a bursty, intermittent, unpredictable manner [8–13]. It is more suitable to model these processes by dynamical systems with non-Gaussian Lévy noise.

In order to better describe the impact of noise on the dynamical systems, we consider some geometric methods. These include most probable phase portraits, mean phase portraits, invariant manifolds, and slow manifolds. These geometric tools may help us characterize the impact of noise on the dynamical systems vividly from the geometric perspective.

2. Stochastic dynamial systems

Consider a stochastic dynamical system in n-dimensional Euclidean space $\mathbb{R}^n$, either with (Gaussian) Brownian noise

$$dX_t = f(X_t)dt + \sigma(X_t)B_t, \quad X_0 = x_0, \tag{3}$$

or (non-Gaussian) Lévy noise

$$dX_t = f(X_t)dt + \sigma(X_t)L_t^{\alpha,\beta}, \quad X_0 = x_0, \tag{4}$$

where σ is noise intensity. Let us briefly recall the definition of Brownian motion and Lévy motion.

2.1. *Brownian motion*

Definition ([14]) A Brownian motion B_t is a stochastic process defined on a probability space Ω equipped with probability $\mathbb{P}$, with the following properties:
(i) $B_0 = 0$, almost surely;
(ii) B_t has independent increments;
(iii) B_t has stationary increments with normal distribution: $B_t - B_s \sim \mathcal{N}(t-s, 0)$, for $t > s$;
(iv) B_t has continuous sample paths, almost surely.

2.2. *Lévy motion*

Definition ([2]) On a sample space Ω equipped with probability $\mathbb{P}$, a scalar asymmetric stable Lévy motion $L_t^{\alpha,\beta}$, with the non-Gaussianity index $\alpha \in (0, 2)$ and the skewness index $\beta \in [-1, 1]$, is a stochastic process with the following properties:
(i) $L_0^{\alpha,\beta} = 0$, almost surely;
(ii) $L_t^{\alpha,\beta}$ has independent increments;

(iii) $L_t^{\alpha,\beta}$ has stationary increments with stable distribution: $L_t^{\alpha,\beta} - L_s^{\alpha,\beta} \sim S_\alpha((t-s)^{\frac{1}{\alpha}}, \beta, 0)$, for $t > s$;
(iv) $L_t^{\alpha,\beta}$ has stochastically continuous sample paths, i.e., for every s, $L_t^{\alpha,\beta} \to L_s^{\alpha,\beta}$ in probability (i.e., for all $\delta > 0$, $\mathbb{P}(|L_t^{\alpha,\beta} - L_s^{\alpha,\beta}| > \delta) \to 0$), as $t \to s$.

The jump measure, which describes jump intensity and size for sample paths, for the asymmetric Lévy motion $L_t^{\alpha,\beta}$ is [2, 6],

$$\nu_{\alpha,\beta}(dy) = \frac{C_1 I_{\{0<y<\infty\}}(y) + C_2 I_{\{-\infty<y<0\}}(y)}{\mid y \mid^{1+\alpha}} dy, \tag{5}$$

with $C_1 = \frac{H_\alpha(1+\beta)}{2}$ and $C_2 = \frac{H_\alpha(1-\beta)}{2}$. When $\alpha = 1$, $H_\alpha = \frac{2}{\pi}$; when $\alpha \neq 1$, $H_\alpha = \frac{\alpha(1-\alpha)}{\Gamma(2-\alpha)\cos(\frac{\pi\alpha}{2})}$.

Especially for $\beta = 0$, this is the symmetric stable Lévy motion, which is usually denoted by $L_t^\alpha \triangleq L_t^{\alpha,0}$. More especially, for $\alpha = 2, \beta = 0$, this is the well-known Brownian motion B_t.

3. Geometric methods for stochastic dynamical systems

As in the geometrical approaches for deterministic dynamical systems [3, 15, 16], we consider geometric concepts and methods for stochastic dynamical systems qualitatively. These include phase portraits and invariant manifolds.

We now introduce some geometric methods for stochastic dynamical systems in this section.

3.1. *Most probable phase portraits*

The Fokker–Planck equation for the stochastic differential equation (5) or (4) describes the time evolution of the probability density $p(x,t) \triangleq p(x,t|x_0,0)$ for the solution process X_t with initial condition $X_0 = x_0$. It is a linear deterministic partial differential equation [2]

$$p_t = A^* p, \;\; p(x,0) = \delta(x - x_0), \tag{6}$$

where A^* is the adjoint operator of the generator A for this stochastic differential equation, and δ is the Dirac delta function.

As the solution of the Fokker–Planck equation, the probability density function $p(x,t)$ is a surface in the (x,t,p)-space. At a given time instant t, the maximizer $x_m(t)$ for $p(x,t)$ indicates the most probable (i.e., maximal likely) location of this orbit at time t. The orbit traced out by $x_m(t)$ is called a most probable orbit starting at x_0. Thus, the deterministic orbit $x_m(t)$ — also denoted by $x_m(x_0,t)$ — follows the top ridge of the surface in the (x,t,p)-space as time goes on.

Definition 1.1. (Most probable equilibrium point) A most probable equilibrium point (state) is a point (state) which either attracts or repels all nearby points (states). When it attracts all nearby points (states), it is called a most probable stable equilibrium point (state), while if it repels all nearby points (states), it is called a most probable unstable equilibrium point (state).

The most probable phase portrait [2, 17] for a stochastic dynamical system is the state space with representative most probable orbits including equilibrium states. It is a deterministic geometric object, which describes which sample orbit is most probable (maximal likely).

3.2. *Mean phase portraits*

Given the probability density function $p(x,t) \triangleq p(x,t|x_0,0)$ as the solution of the Fokker–Planck equation (6). The mean orbit starting at an initial point x_0 in state space is defined as

$$\bar{x}(x_0,t) = \int_{\mathbb{R}} \xi\, p(\xi,t|x_0,0)d\xi. \tag{7}$$

We also give the definition of a mean equilibrium point (state).

Definition 1.2. (Mean equilibrium point) A mean equilibrium point (state) is a point (state) which either attracts or repels all nearby points (states). When it attracts all nearby points (states), it is called a mean stable equilibrium point (state), while if it repels all nearby points (states), it is called a mean unstable equilibrium point (state).

The mean probable phase portrait is the state space with representative mean orbits including mean equilibrium states. It is also a deterministic geometric object, which describes the orbits in the mean sense, starting in state space.

3.3. *Invariant manifolds*

Invariant geometry structures play an important role in our understanding of dynamical systems. Invariant manifolds to nonlinear systems just as eigenspace to linear systems. With a better understanding of stochastic invariant manifolds, we may have a new insight to understand stochastic dynamical systems.

We recall some definitions.

Definition 1.3. (Random set) A random set for a random dynamical φ in $\mathbb{R}^n$ is a collection $M = M(\omega)$, $\omega \in \Omega$, satisfies the following property:
(i) $M(\omega)$ is a nonempty closed set, $M(\omega) \subset \mathbb{R}^n$, $\forall\, \omega \in \Omega$;
(ii) $V_x : \Omega \to \mathbb{R}^1$, and $V_x(\omega) \triangleq inf_{y\in M(\omega)} d(x,y)$ is a scale random variable $\forall\, x \in \mathbb{R}^n$.

Definition 1.4. (Random invariant set) An invariant set for a random dynamical φ is a random set M satisfies: $\varphi(t,\omega,M(\omega)) = M(\theta_t\omega)$, $\forall\, t \in \mathbb{R}$ and $\omega \in \Omega$.

Definition 1.5. (Random invariant manifold) If a random invariant set M for a random dynamical φ can be represented by a graph of a Lipschitz mapping: $\gamma^*(\omega,\cdot) : H^+ \to H^-$, with direct sum decomposition $H^{\oplus}H^- = \mathbb{R}^n$, satisfies that $M(\omega) = (x^+, \gamma^*(\omega,x^+)), x^+ \in H^+$, then M is called a Lipschitz invariant manifold.

For more information about random invariant manifolds, see [1, 18, 19].

3.4. *Slow manifolds*

When a system has a slow component and a fast component, we call it slow–fast system. For example, temperature and molecular motion, climate and weather, are slow and fast components in a body and in the climate system, respectively. In this kind of systems, there are two time scales, and usually, we are more interested in the slow dynamics. But we still like to take the fast dynamics into account, even when the fast evolution is less important to our modeling purpose. We thus represent fast components in terms of slow components as slow manifolds. For more information about slow manifolds, see [19–23].

Consider a stochastic slow–fast system:

$$\begin{cases} \dot{x} &= Ax + f(x,y), x(0) = x_0 \in \mathbb{R}^n, \\ \dot{y} &= \frac{1}{\varepsilon}By + \frac{1}{\varepsilon}g(x,y) + \frac{\sigma}{\sqrt{\varepsilon}}\dot{W}_t, \end{cases} \tag{8}$$

where A and B are matrices, ε is a small positive parameter measuring slow and fast scale separation, f and g are nonlinear Lipschitz continuous functions with Lipschitz constant L_f and L_g respectively, σ is a noise intensity constant, and $W_t : t \in \mathbb{R}$ is a two-sided $\mathbb{R}^m$-*valued* Brownian motion on a probability space $(\Omega, \mathcal{F}, \mathbb{P})$.

Under the exponential dichotomy and gap condition, there exists a random slow manifold $\tilde{\mathcal{M}}^\varepsilon(\omega) = (\xi, \tilde{h}^\varepsilon(\xi,\omega)) : \xi \in \mathbb{R}^n$ with $\tilde{h}^\varepsilon$ can be expressed by Liapunov-Perron equation:

$$\tilde{h}^\varepsilon(\xi,\omega) = \frac{1}{\varepsilon}\int_{-\infty}^{0} e^{-\frac{B}{\varepsilon}s} g(X(s,\omega,\xi), Y(s,\omega,\xi) + \sigma\eta^\varepsilon(\theta_s\omega))ds \tag{9}$$

Then the slow system is

$$\dot{x} = Ax + f(x, \tilde{h}^\varepsilon(x,\theta_t\omega) + \sigma\eta(\psi_\varepsilon\omega)) \tag{10}$$

Moreover, $\hat{h}^\varepsilon(\xi,\omega) = \tilde{h}_d^\varepsilon(\xi,\omega) + \varepsilon\tilde{h}_1^\varepsilon(\xi,\omega)$ is an approximation or a first order truncation of $\tilde{h}^\varepsilon(\xi,\omega)$. Thus, we have an approximate slow system

$$\dot{x} = Ax + f(x, \hat{h}^\varepsilon(x,\theta_t\omega) + \sigma\eta(\psi_\varepsilon\omega)) \tag{11}$$

Based on the reduced slow system, we can conduct parameter estimation [22], data assimilation [24, 25], and stochastic bifurcation [26].

4. Applications

In this section, we will discuss some applications associated with the geometric methods for stochastic dynamical systems.

4.1. *Transition phenomena*

In contrast to deterministic situation, it is natural to consider transitions among metastable states in stochastic dynamical systems. That is to say, a transition orbit is likely to occur between two stable equilibrium states, when a system is under random fluctuations.

Transition phenomena occur in gene regulation, tumor cell density, climate change, parametric oscillator, Briggs–Rauscher chemical reaction, and predator-prey systems. By the geometric methods for stochastic dynamical systems, we may qualitatively demonstrate these noise-induced transition phenomena.

In genetic regulatory systems, specific protein concentrations play an important role in cell life. The low concentration and high concentration of specific protein correspond to different cell activities. Recent studies [27–29] have recognized that Lévy motion can induce switches between different protein concentrations. Multiple phenotypic states often arise in a single cell with different gene expression states. The transitions between two phenotypic states due to stochastic fluctuations have been verified in [30].

In population dynamics, determining the amount of predation which a prey population can sustain without endangering its survival is an important problem [31]. The problem is to find out the best strategy for a good management of biological resources. We may also think in the opposite direction, driving a prey population to extinction by sufficient predation. In this case the extinction problem does not only find applications in pest control but also in the medical sciences.

4.2. *Stochastic bifurcation*

Although bifurcation studies for deterministic dynamical systems have a long history, the stochastic bifurcation investigation is still in its early stage. One reason for this slow development in stochastic bifurcation is due to the lack of appropriate phase portraits, in contrast to deterministic dynamical systems.

Stochastic bifurcations have been observed in a wide range of nonlinear systems in physical science and engineering [32–34]. Some works about stochastic bifurcation are analytical studies of invariant measures, together with their spectra and supports. We recently have used most probable phase portraits to detect stochastic bifurcation [35].

5. Conclusion

Stochastic dynamical systems widely arise as mathematical models in biology, physics, chemistry and engineering. In this chapter, we have reviewed several geometric methods for stochastic dynamical systems and their applications. By most probable phase portraits and mean phase portraits, we may better understand stochastic dynamical behaviors such as transition orbits and stochastic bifurcation. Based on the reduced systems on slow manifolds, we may estimate parameters, detect stochastic bifurcation, and conduct data assimilation on lower dimensional slow systems.

References

[1] Arnold, L. [2003]. Random Dynamical Systems, 2nd Ed. (Springer, New York, USA).
[2] Duan, J. [2015]. An Introduction to Stochastic Dynamics. (Cambridge University Press, New York, USA).
[3] Meiss, J. D. [2017]. Differential Dynamical Systems, Revised. (Society for Industrial and Applied, Philadelphia).
[4] Wiggins, S. [2003]. Introduction to Applied Nonlinear Dynamical Systems and Chaos, 2nd Ed. (Springer, New York, USA).
[5] Guckenheimer, J. and Holmes, P. [1983]. Nonlinear Oscillations, Dynamical Systems and Bifurcations of Vector Fields. (Springer, New York, USA).
[6] Applebaum, D. [2009]. Lévy Processes and Stochastic Calculus, 2nd Edition. (Cambridge University Press, New York, USA).
[7] Sato, K. [1999]. Lévy Processes and Infinitely Divisible Distributions. (Cambridge University Press, New York, USA).
[8] Holloway, D. M. and Spirov, A. V. [2017]. Transcriptional bursting in Drosophila development: Stochastic dynamics of evestripe 2 expression, Plos One **12 (4)**, e0176228.
[9] Kumar, N., Singh, A. and Kulkarni, R. V. [2015]. Transcriptional bursting in gene expression: analytical results for general stochastic models, Plos Computational Biology **11**, e1004292.
[10] Dar, R. D., Razooky, B. S., Singh, A., Trimeloni, T. V., McCollum, J. M., Cox, C. D., Simpson, M. L. and Weinberger, L. S. [2012]. Transcriptional burst frequency and burst size are equally modulated across the human genome, PNAS **109**, 17454–17459.
[11] Ozbudak, E. M., Thattai, M., Kurtser, I., Grossman, A. D. and Oudenaarden, A. V. [2002]. Regulation of noise in the expression of a single gene, Nature Genetics **31**, 69–73.
[12] Blake, W. J., Kærn, M., Cantor, C. R. and Collins, J. J. [2003]. Noise in eukaryotic gene expression, Nature **422**, 633–637.
[13] Sanchez, A. and Golding, I. [2013]. Genetic determinants and cellular constraints in noisy gene expression, Science **342 (6163)**, 1188–1193.
[14] Karatzas, I. and Shreve, S. E. [1991]. Brownian Motion and Stochastic Calculus, 2nd Edition. (Springer–Verlag, New York, USA).
[15] Arnold, V. I. [1988]. Geometrical Methods in the Theory of Ordinary Differential Equations, 2nd Ed. (Springer–Verlag, New York, USA).
[16] Arnold, L. [1989]. Mathematical Methods of Classical Mechanics, 2nd Ed. (Springer–Verlag, New York, USA).
[17] Cheng, Z., Duan, J. and Wang, L. [2016]. Most probable dynamics of some nonlinear systems under noisy fluctuatuons, Commun. Nonlinear Sci. Numer. Simulat. **30**, 108–114.
[18] Duan, J., Lu, K. and Schmalfuß, B. [2003]. Invariant manifolds for stochastic partial differential equations, Annals of Probability **31**, 2109–2135.
[19] Schmalfuß, B. and Schneider, R. [2008]. Invariant manifolds for random dynamical systems with slow and fast variables, Journal of Dynamics and Differential Equations **20**, 133–164.

[20] Fu, H., Liu, X. and Duan, J. [2013]. Slow manifolds for multi-time-scale stochastic evolutionary systems, Communications in Mathematical Sciences **11(1)**, 141–162.
[21] Ren, J., Duan, J. and Christopher, J. [2015]. Approximation of random slow manifolds and settling of inertial particles under uncertainty, Journal of Dynamics and Differential Equations **27**, 961–979.
[22] Ren, J., Duan, J. and Wang, X. [2015]. A parameter estimation method based on random slow manifolds, Applied Mathematical Modelling **39**, 3721–3732.
[23] Constable, G., McKane, A. and Rogers, T. [2013]. Stochastic dynamics on slow manifolds, Journal of Physics A: Mathematical and Theoretical **46 (295002)**.
[24] Zhang, Y., Cheng, Z., Zhang, X., Chen, X., Duan, J. and Li, X. [2017]. Data assimilation and parameter estimation for a multiscale stochastic system with α-stable Lévy noise, Journal of Statistical Mechanics: Theory and Experiment, 113401.
[25] Qiao, H., Zhang, Y. and Duan, J. [2018]. Effective filtering on a random slow manifold, Nonlinearity **31**, 4649–4666.
[26] He, Z., Cai, R., Duan, J. and Liu, X. [2018]. A role of random slow manifolds in detecting stochastic bifurcation. arxiv: 1805.04653v1.
[27] Xu, Y., Feng, J., Li, J. and Zhang, H. [2013]. Lévy noise induced switch in the gene transcriptional regulatory system, Chaos **23 (1)**, 013110.
[28] Zheng, Y., Larissa, S., Duan, J. and Jürgen, K. [2016]. Transitions in a genetic transcriptional regulatory system under Lévy motion, Scient. Rep. **6**, 29274.
[29] Wang, H., Cheng, X., Duan, J., Jürgen, K. and Li, X. [2018]. Likelihood for transcriptions in a genetic regulatory system under asymmetric stable Lévy noise, Chaos **28 (1)**, 013121.
[30] Ge, H., Qian, H. and Xie, X. S. [2015]. Stochastic phenotype transition of a single cell in an intermediate region of gene state switching, Phys. Rev. Lett. **114 (7)**, 078101.
[31] Horsthemke, W. and Lefever, R. [2006]. Noise-Induced Transitions. (Springer: Berlin Heidelberg).
[32] Deco, G. and Martí, D. [2007]. Deterministic analysis of stochastic bifurcations in multi-stable neurodynamical systems, Biological Cybernetics **96 (5)**, 487–496.
[33] Bashkirtseva, I., Nasyrova, V. and Ryashko, L. [2018]. Noise-induced bursting and chaos in the two-dimensional Rulkov model, Chaos Solitons & Fractals **110**, 76–81.
[34] Bogatenko, T. and Semenov, V. [2018]. Coherence resonance in an excitable potential well, Physics Letters A **382**, 2645–2649.
[35] Wang, H., Chen, X. and Duan, J. [2018]. A stochastic pitchfork bifurcation in most probable phase portraits, International Journal of Bifurcation and Chaos **28 (1)**.

Chapter 2

Stochastic 3D Navier–Stokes Equations with Nonlinear Damping: Martingale Solution, Strong Solution and Small Time LDP

Hongjun Gao* and Hui Liu

Jiangsu Provincial Key Laboratory for NSLSCS, School of Mathematical Sciences, Nanjing Normal University, Nanjing 210023, P.R. China

**gaohj@njnu.edu.cn*

In this paper, by using classical Faedo–Galerkin approximation and compactness method, the existence of martingale solutions for the stochastic 3D Navier–Stokes equations with nonlinear damping is obtained. The existence and uniqueness of strong solution are proved for $\beta > 3$ with any $\alpha > 0$ and $\alpha \geq \frac{1}{2}$ as $\beta = 3$. Meanwhile a small time large deviation principle (LDP) for the stochastic 3D Navier–Stokes equations with damping is obtained for $\beta > 3$ with any $\alpha > 0$ and $\alpha \geq \frac{1}{2}$ as $\beta = 3$.

Keywords Stochastic Navier–Stokes equations, martingale solution, nonlinear damping, strong solution, large deviations.

2010 Mathematics Subject Classification 60H15, 35Q30, 76M35, 60F10

1. Introduction

We are concerned with a class of stochastic Navier–Stokes equations with damping in this paper. The damping is from the resistance to the motion of the flow. It describes various physical situations such as drag or friction effects, and some dissipative mechanisms [2, 18]. Importance of such problems for climate modeling and physical fluid dynamics is well known [16, 31].

In this paper we consider the following stochastic three-dimensional Navier–Stokes equations with nonlinear damping

$$\begin{cases} \frac{\partial u}{\partial t} - \mu\Delta u + (u \cdot \nabla)u + \nabla p + \alpha|u|^{\beta-1}u = f(t) + G(t,u)dW(t), \\ \nabla \cdot u = 0, \\ u(t,x)|_{\partial D} = 0, \\ u(x,0) = u_0(x), \end{cases} \tag{1.1}$$

where $D \subset R^3$ is an open connected bounded subset with smooth boundary ∂D, $u = (u_1, u_2, u_3)$ is the velocity, p is the pressure, $\beta > 1$, $\alpha > 0$ and $t \in [0,T]$, u_0 is the initial velocity, f is the deterministic external force and $G(t,u)dW(t)$, where W is a cylindrical Wiener process, stands for the random forces.

The non-negative coefficients μ and α are called kinematic viscosity and sticky viscosity, respectively. For a fixed $\alpha > 0$, if $\mu > 0$, it is called stochastic damped Navier–Stokes equations, whereas if $\mu = 0$, it is called the stochastic damped Euler equations.

The deterministic Navier–Stokes equations with nonlinear damping has been extensively investigated. For instance, Cai and Jiu have studied the existence and regularity of solutions for three-dimensional Navier–Stokes equations with nonlinear damping [6], and they obtained the global weak solution for $\beta \geq 1$, the global strong solution for $\beta \geq \frac{7}{2}$ and the strong solution was unique for any $\frac{7}{2} \leq \beta \leq 5$. Based on it, Song and Hou considered the global attractor in [29] and [30]. In [19], the authors considered the L^2 decay of weak solutions with $\beta \geq \frac{10}{3}$, the optimal upper bounds of the higher-order derivative of the strong solution for $\frac{7}{2} \leq \beta < 5$ and the asymptotic stability of the large solution to the system with $\beta \geq \frac{7}{2}$. In [33], the authors considered the regularity criterion of the 3D Navier–Stokes equations with nonlinear damping. In [35], for $\alpha = 1$, Zhou proved that the strong solution exists globally for $\beta \geq 3$ and strong-weak uniqueness for $\beta \geq 1$, and established two regularity criteria as $1 \leq \beta \leq 3$. Oliveira has studied the existence of weak solutions for the generalized Navier–Stokes equations with damping [24] for non-Newtonian fluids.

In a previous work, the existence of martingale solution of stochastic Euler equations was studied by many authors, for instance, Capiński and Cutland [7], Bessaih [3], and Brzeźniak and Peszat [5]. Flandoli and Gatarek [13] obtained the martingale and stationary solutions for n-dimensional stochastic Navier–Stokes equations. In the paper [14, 15], the existence of martingale solutions and Markov selections of stochastic 3D Navier–Stokes equations were proved. The existence of martingale solutions has been proved for the stochastic 3D Navier–Stokes equations with jump under proper assumptions [12].

Large deviation has applications in many areas, such as in thermodynamics, stochastic mechanics, information theory, risk management, etc. (please refer to [32]). A small time large deviation principle for two-dimensional stochastic Navier–Stokes equations driven by multiplicative noise was established in [34]. Using weak convergence method, a large deviation principle of Freidlin–Wentzell type for stochastic tamed 3D Navier–Stokes equations driven by multiplicative noise was proved in [27]. The small time large deviation principle for the stochastic 3D tamed Navier–Stokes equations was obtained in [26].

In this paper, for simplicity, we set $\mu = 1$. Using Galerkin approximation, Prokhorov's theorem and Skorokhod's embedding theorem, the existence of martingale solutions for stochastic 2D and 3D Navier–Stokes equations on unbounded domain was obtained [4]. The presence of the nonlinear damping $|u|^{\beta-1}u$ in (2.1) is beneficial to the regularity of the martingale solution. Using the nonlinear structure and delicated analysis we have overcome some difficulties (such as the estimate of $\int_0^t \int_D |u \cdot \nabla u|^2 dxds$). The existence and uniqueness of strong solution for problem

(2.1) can be proved for $\beta > 3$ with any $\alpha > 0$ and $\alpha \geq \frac{1}{2}$ as $\beta = 3$. If we use the method in [35] with further delicate analysis, we still need the condition $\alpha \geq \frac{1}{2}$ for $\beta \geq 3$, so our result improved the early results even for deterministic case.

As for the existence of local solution for stochastic 3D Navier–Stokes equations with damping, we could use the monotonicity method as in [20]. Our purpose is to prove the global existence of strong solution which depends on the various estimates of solution, the LDP and ergodicity also depend on such estimates of solution, so we use the Galerkin approximation, a priori estimate and weak convergence here.

The paper is organized as follows. In section 2, we recall basic definitions and introduce some auxiliary operators. Martingale solutions will be obtained in section 3. In section 4, we prove the existence and uniqueness of strong solution for $\beta > 3$ with any $\alpha > 0$ and $\alpha \geq \frac{1}{2}$ as $\beta = 3$. In section 5, we prove the small time large deviation principle for the stochastic 3D Navier–Stokes equations with damping for $\beta > 3$ with any $\alpha > 0$ and $\alpha \geq \frac{1}{2}$ as $\beta = 3$.

2. Preliminaries

In this section, we introduce some basic definitions, important lemmas which will be needed in this paper.

Let $D \subset R^3$ be an open subset with smooth boundary ∂D. The norm in $L^p(D, R^3)$ is defined by

$$||u||_{L^p} = (\textstyle\int_D |u(x)|^p dx)^{\frac{1}{p}}, \quad u \in L^p(D, R^3),\ p > 1.$$

Let $H^1(D, R^3)$ be the Sobolev space of all $u \in L^2(D, R^3)$ and its weak derivatives $\frac{\partial u}{\partial x_i} \in L^2(D, R^3)$, $i = 1, 2, 3$. It is a Hilbert space with the scalar product defined by

$$(u, v)_{H^1} = (u, v)_{L^2} + ((u, v)), \quad u,\ v \in H^1(D, R^3),$$

where

$$((u, v)) = (\nabla u, \nabla v)_{L^2} = \sum_{i=1}^{3} \int_D \frac{\partial u}{\partial x_i} \frac{\partial v}{\partial x_i} dx, \quad u,\ v \in H^1(D, R^3).$$

Let $C_c^\infty(D, R^3)$ be the space of all R^3-valued functions of class $\mathcal{C}^\infty$ with compact supports included in D, and denote

$$\begin{aligned} \mathcal{V} &= \{u \in \mathcal{C}_c^\infty(D, R^3) : divu = 0\}, \\ H &= \textit{ the closure of } \mathcal{V} \textit{ in } L^2(D, R^3), \\ V &= \textit{ the closure of } \mathcal{V} \textit{ in } H^1(D, R^3). \end{aligned}$$

For equation (2.1), we consider the abstract form as a stochastic evolution equation

$$du(t) + \mathcal{A}u(t)dt + B(u(t))dt + g(u(t))dt = f(t)dt + G(u(t))dW(t) \qquad (2.1)$$

with $u(0) = u_0$.
We define the bilinear operator $B(u, v)$, as $\langle B(u, v), z\rangle = b(u, v, z) = \int_D (u \cdot \nabla v) z dx$ for all $u, v, z \in V$, and also define the linear operator $\mathcal{A}$ as $\mathcal{A}u = -\Delta u$ and $g(u) = \alpha |u|^{\beta-1} u$.

By the incompressibility condition, we have (Temam [31])

$$b(u,w,v) = -b(u,v,w), \qquad u,v,w \in V.$$

In particular,

$$b(u,v,v) = 0, \quad u,v \in V.$$

Let $(X, |\cdot|_X)$, $(Y, |\cdot|_Y)$ be two real normed spaces. The symbol $\mathcal{L}(X,Y)$ stands for the space of all bounded linear operators from X to Y. The symbol ${}_{X'}\langle\cdot,\cdot\rangle_X$ denotes the standard duality pair. If both spaces X and Y are separable Hilbert, $\mathcal{T}_2(Y,X)$ stands for the Hilbert space of all Hilbert–Schmidt operators from Y to X endowed with the standard norm.

For all $s > 0$, the following standard scale of Hilbert spaces is defined by

$$V_s = the\ closure\ of\ \mathcal{V}\ in\ H^s(D, R^3).$$

Consider the natural embedding $j : V \hookrightarrow H$ and its adjoint $j^* : H \to V$, and suppose that $s > \frac{5}{2}$, we get that V_s is dense in V and the embedding $j_s : V_s \hookrightarrow V$ is continuous. By Lemma 2.5 in [17], then there exists a Hilbert space U such that $U \subset V_s$, U is dense in V_s and the natural embedding $\iota_s : U \hookrightarrow V_s$ is compact.
We get

$$U \hookrightarrow V_s \hookrightarrow V \hookrightarrow H \cong H' \hookrightarrow V' \hookrightarrow V_s' \hookrightarrow U'.$$

Fixing $n \in N$ and let $H_n = \text{span}\{e_1, e_2, \cdots e_n\}$ be the linear space spanned by the vectors $e_1, e_2, \cdots, e_n$. Let P_n be the operator from U' to $\text{span}\{e_1, e_2, \cdots e_n\}$ given by

$$P_n u^* = \sum_{i=1}^{n} {}_{U'}\langle u^*, e_i\rangle_U e_i, \quad u^* \in U'. \tag{2.2}$$

We consider the following functional spaces, similarly to those considered in [23]:

$\mathcal{C}([0,T];U') :=$ the space of continuous functions $u : [0,T] \to U'$ with the topology $\mathcal{T}_1$ induced by the norm $||u||_{\mathcal{C}([0,T],U')} = \sup_{t\in[0,T]} |u(t)|_{U'}$,

$L^2_w(0,T;V) :=$ the space $L^2(0,T;V)$ with the weak topology $\mathcal{T}_2$,

$L^2(0,T;H) :=$ the space of measurable functions $u : [0,T] \to H$ such that

$$q_T(u) := ||u||_{L^2(0,T;H)} = \left(\int_0^T \int_D |u(t,x)|^2 dx dt\right)^{\frac{1}{2}} < \infty,$$

with the topology $\mathcal{T}_3$ generated by the seminorms (q_T).
Let H_w be the Hilbert space H endowed with the weak topology. Let

$\mathcal{C}([0,T];H_w) :=$ the space of weakly continuous functions $u : [0,T] \to H$ endowed with the weakest topology $\mathcal{T}_4$, such that for any $h \in H$, the mappings $\mathcal{C}([0,T];H_w) \ni u \to (u(\cdot),h)_H \in \mathcal{C}([0,T];R)$ are continuous.
We introduce the ball

$$\mathbb{B} = \{x \in H : |x|_H \le r\}.$$

Let $\mathbb{B}_w$ be the ball $\mathbb{B}$ endowed with the weak topology and q be the metric compatible with the weak topology on $\mathbb{B}$. We will introduce the following subspace of the space $\mathcal{C}([0,T];H_w)$

$$\mathcal{C}([0,T];\mathbb{B}_w) = \{u \in \mathcal{C}([0,T];H_w) : \sup_{t\in[0,T]} |u(t)|_H \leq r\}.$$

The space $\mathcal{C}([0,T];\mathbb{B}_w)$ is metrizable with

$$\varrho(u,v) = \sup_{t\in[0,T]} q(u(t),v(t)).$$

Lemma 2.1. ([23, Lemma 2.7]) Let

$$\mathcal{Z} = \mathcal{C}([0,T];U') \cap L^2_w(0,T;V) \cap L^2(0,T;H) \cap \mathcal{C}([0,T];H_w)$$

and $\mathcal{T}$ be the supremum of the corresponding topologies. Then a set $\mathcal{K} \subset \mathcal{Z}$ is $\mathcal{T}$-relatively compact if the following three conditions hold

(a) $sup_{u\in\mathcal{K}} sup_{s\in[0,T]} |u(s)|_H < \infty$,

(b) $sup_{u\in\mathcal{K}} \int_0^T ||u(s)||^2 ds < \infty$,

(c) $lim_{\delta\to 0} sup_{u\in\mathcal{K}} sup_{s,t\in[0,T]} |u(t)-u(s)|_{U'} = 0, \quad \forall\, |t-s| \leq \delta.$

Let $(\mathbb{S},\varrho)$ be a separable and complete metric space.

Definition 2.1. Let $u \in \mathcal{C}([0,T],\mathbb{S})$. The modulus of continuity of u on $[0,T]$ is given by

$$m(u,\delta) = \sup_{s,t\in[0,T],|t-s|\leq\delta} \varrho(u(t),u(s)), \quad \delta > 0.$$

Let $(\Omega,\mathcal{F},\mathbb{P})$ be a probability space with filtration $\mathbb{F} = (\mathcal{F})_{t\in[0,T]}$ satisfying the ordinary condition [22], let $(X_n)_{n\in N}$ be a sequence of continuous $\mathbb{F}$-adapted $\mathbb{S}$-valued processes.

Definition 2.2. ([21]) A sequence $(X_n)_{n\in N}$ satisfies condition [**A**] iff $\forall\, \varepsilon > 0$, $\forall\, \eta > 0$, $\exists\, \delta > 0$ such that for any sequence $(\tau_n)_{n\in N}$ of $\mathbb{F}$-stopping times with $\tau_n \leq T$, we have

$$\sup_{n\in\mathbb{N}} \sup_{0\leq\theta\leq\delta} \mathbb{P}\{\varrho(X_n(\tau_n+\theta), X_n(\tau_n)) \geq \eta\} \leq \varepsilon.$$

Applying the deterministic compactness criterion formulated in Lemma 2.1, we get the following Lemma 2.2.

Lemma 2.2. (Tightness criterion [21]) Assume that $(X_n)_{n\in N}$ is a sequence of continuous $\mathbb{F}$-adapted U'-valued processes such that

(a) there exists a positive constant C_1 such that

$$\sup_{n\in\mathbb{N}} \mathbb{E}[\sup_{s\in[0,T]} |X_n(s)|^2_H] \leq C_1,$$

(b) there exists a positive constant C_2 such that

$$\sup_{n\in\mathbb{N}} \mathbb{E}[\int_0^T ||X_n(s)||^2 ds] \leq C_2,$$

(c) $(X_n)_{n\in N}$ satisfies the Aldous condition [**A**] in U'.

Let $\tilde{\mathbb{P}}_n$ be the law of X_n on $\mathcal{Z}$. Then there exists a compact subset K_ε of $\mathcal{Z}$ such that for any $\varepsilon > 0$

$$\sup_{n\in\mathbb{N}} \tilde{\mathbb{P}}_n(K_\varepsilon) \geq 1 - \varepsilon.$$

Lemma 2.3. ([29, Lemma 2.4]) Let $g(u) = \alpha|u|^{\beta-1}u$, then

(1) g is continuously differentiable in R^3 and the Jacobian matrix is defined by, for any $u = (u_1, u_2, u_3)$ in R^3 :

$$g'(u) = \alpha|u|^{\beta-3} \begin{pmatrix} (\beta-1)u_1^2 + |u|^2 & (\beta-1)u_1u_2 & (\beta-1)u_1u_3 \\ (\beta-1)u_1u_2 & (\beta-1)u_2^2 + |u|^2 & (\beta-1)u_2u_3 \\ (\beta-1)u_1u_3 & (\beta-1)u_2u_3 & (\beta-1)u_3^2 + |u|^2 \end{pmatrix}.$$

Moreover, $g'(u)$ is positive definite and for any $u, v, w \in R^3$:

$$|(g'(u)v) \cdot w| \leq c|u|^{\beta-1}|v||w|,$$

where c is a positive constant depending on β and α.

(2) g is monotonic in R^3, i.e., for any $u, v \in R^3$:

$$(g(u) - g(v), u - v) \geq 0.$$

Assumption A. There exist positive constants k_1, k_2, k_3 such that

$$||\nabla G_n(u_n(r))P_n||^2_{\mathcal{T}_2(Y,H)} \leq k_1||u||^2 + k_2|u|^2_H + k_3.$$

3. Existence of martingale solutions

We shall consider the following stochastic evolution equation

$$\begin{cases} du(t) + \mathcal{A}u(t)dt + B(u(t), u(t))dt + g(u)dt = f(t)dt + G(u(t))dW(t), \\ u(0) = u_0. \end{cases} \tag{3.1}$$

Definition 2.3. We say that there exists a martingale solution of the equation (3.1) iff there exists a stochastic basis $(\hat{\Omega}, \hat{\mathcal{F}}, \hat{\mathbb{F}}, \hat{\mathbb{P}})$ with filtration $\hat{\mathbb{F}} = \{\hat{\mathcal{F}}_t\}_{t\geq 0}$, a cylindrical Wiener process $\hat{W}$ on the space Y and a progressively measurable process $u : [0,T] \times \hat{\Omega} \to H$, with $\hat{\mathbb{P}}$-*a.s.* paths

$$u(\cdot, \omega) \in C([0,T]; H_\omega) \cap L^2(0,T; V) \cap L^{\beta+1}(0,T; L^{\beta+1})$$

such that $\hat{\mathbb{P}}$-*a.s.*, the identity

$$(u(t), v)_H + \int_0^t \langle \mathcal{A}u(s), v\rangle ds + \int_0^t \langle B(u(s), u(s)), v\rangle ds + \int_0^t \langle g(u(s)), v\rangle ds$$

$$= (u_0, v)_H + \int_0^t \langle f(s), v\rangle ds + \langle \int_0^t G(u(s))d\hat{W}(s), v\rangle \tag{3.2}$$

holds true for any $t \in [0,T]$ and all $v \in \mathcal{V}$.

Theorem 2.1. *Let following assumptions (A.1)–(A.3) be satisfied,*

(A.1) $W(t)$ is a cylindrical Wiener process in a separable Hilbert space Y defined on the stochastic basis $(\Omega, \mathcal{F}, \mathbb{F}, \mathbb{P})$ with a filtration $\mathbb{F} = \{\mathcal{F}_t\}_{t\geq 0}$ and $H \subset Y$,

(A.2) $u_0 \in H, f \in L^2(0,T;V')$,

(A.3) the mapping $G : V \to \mathcal{T}_2(Y,H)$ is Lipschitz continuous and

$$2\langle \mathcal{A}u, u\rangle - ||G(u)||^2_{\mathcal{T}_2(Y,H)} \geq \eta||u||^2 - \lambda_0|u|^2_H - \rho, \tag{3.3}$$

where $u \in V$, for some constants λ_0, ρ and $\eta \in (0,2]$.
Then there exists a martingale solution $(\hat{\Omega}, \hat{\mathcal{F}}, \hat{\mathbb{F}}, \hat{\mathbb{P}}, u)$ of problem (2.4) such that

$$\hat{\mathbb{E}}[\sup_{t\in[0,T]} |u(t)|^2_H + \textstyle\int_0^T ||u(t)||^2 dt] < \infty. \tag{3.4}$$

However, G could extend to a mapping $G : H \to \mathcal{T}_2(Y,V')$ such that

$$||G(u)||^2_{\mathcal{T}_2(Y,V')} \leq C(1 + |u|^2_H), \quad u \in H. \tag{3.5}$$

Let $H_n = span\{e_i, 1 \leq i \leq n\}$ be the orthogonal projection of H on H_n, and $P_n : U' \to H_n$ be defined by (2.2).

Consider the following classical Faedo–Galerkin approximation in the space H_n :

$$\begin{cases} du_n(t) = -[P_n\mathcal{A}u_n(t) + B_n(u_n(t)) + P_n g(u_n) - P_n f(t)]dt + P_n G(u_n(t))dW(t), \\ u_n(0) = P_n u_0. \end{cases} \tag{3.6}$$

Using the Itô formula and the Burkholder–Davis–Gundy inequality [8], we will prove the following Lemma 2.4. Let p satisfy the following conditions

$$\begin{cases} p \in [2, 2 + \frac{\eta}{2-\eta}) \ \ if \ \eta \in (0,2), \\ p \in [2, \infty) \ \ if \ \eta = 2. \end{cases} \tag{3.7}$$

Since $||u|| := ||\nabla u||_{L^2}$ and $\langle \mathcal{A}u, u\rangle = ((u,u)) = (\nabla u, \nabla u)_{L^2}$,

$$2\langle \mathcal{A}u, u\rangle - \eta||u||^2 = (2-\eta)||u||^2$$

for some constants $\eta \in (0,2]$, $\lambda_0 \geq 0$ and $\rho \in R$. So inequality (2.6) could be written equivalently in the following form

$$||G(u)||^2_{\mathcal{T}_2(Y,H)} \leq (2-\eta)||u||^2 + \lambda_0|u|^2_H + \rho. \tag{3.8}$$

The following estimate and Lemma 2.5 are based on a similar method used by Brzeźniak and Motyl in [4].

Lemma 2.4. There exist three positive constants $C_1(p)$, $C_2(p)$, C_3, for each $p \geq 2$, such that

$$\sup_{n\geq 1} \mathbb{E}(\sup_{0\leq s\leq t} |u_n(s)|^p_H) \leq C_1(p), \tag{3.9}$$

$$\sup_{n\geq 1} \mathbb{E}\int_0^t |u_n(s)|^{p-2}_H ||u_n(s)||^2 ds \leq C_2(p), \tag{3.10}$$

$$\sup_{n\geq 1}\mathbb{E}\int_0^t |u_n(s)|_{\beta+1}^{\beta+1}ds \leq C_3. \tag{3.11}$$

For any $n \in N$, the solution u_n of the Galerkin equations defines a measure $\mathcal{L}(u_n)$ on $(\mathcal{Z},\mathcal{T})$. Using Lemma 2.2 and (3.5), we will prove the tightness of this set of measures.

Lemma 2.5. The family $\mathcal{L}(u_n)$ is tight on $(\mathcal{Z},\mathcal{T})$.

Proof of Theorem 2.1.

By Lemma 2.5, we have the set of measures $\{\mathcal{L}(u_n), n \in N\}$ tight on the space $(\mathcal{Z},\mathcal{T})$, and by Corollary 3.12 in [4], then there exists a subsequence $(n_k)_k$, a probability space $(\tilde{\Omega},\tilde{\mathcal{F}},\tilde{\mathbb{P}})$, and $\mathcal{Z}$-valued random variables $\tilde{u}$, $\tilde{u}_{n_k}$, $k \geq 1$ such that

$\tilde{u}_{n_k}$ has the same law as u_{n_k} on $\mathcal{Z}$ and $\tilde{u}_{n_k} \to u_{n_k}$ in $\mathcal{Z}$, $\tilde{\mathbb{P}}$-a.s.

We denote the subsequence $(\tilde{u}_{n_k})_k$ by $(\tilde{u}_n)_n$.

Let us define the process $\tilde{M}_n(t)$ with trajectories in $\mathcal{C}([0,T];H)$

$$\tilde{M}_n(t) = \tilde{u}_n(t) - P_n\tilde{u}_0 + \int_0^t P_n\mathcal{A}\tilde{u}_n(s)ds + \int_0^t B_n(\tilde{u}_n(s))ds$$
$$+ \int_0^t P_n g(\tilde{u}_n(s))ds - \int_0^t P_n f(s)ds, \quad t\in[0,T], \quad n\geq 1. \tag{3.12}$$

The following proof is based on a similar method used by Brzeźniak and Motyl [4]. Since $\tilde{u}_n$ and u_n obey the same laws, for all $r,t \in [0,T]$, $r \leq t$, all functions h bounded continuously on $\mathcal{C}([0,s];U')$, and all $\psi,\zeta \in U$, we have

Lemma 2.6. (Lemma 5.5 [4]) For all $r,t\in[0,T]$ such that $r \leq t$ and all $\psi \in U$, we have

(a) $\lim_{n\to\infty}(\tilde{u}_n, P_n\psi)_H = (\tilde{u},\psi)_H$, $\tilde{\mathbb{P}}$-*a.s.*,

(b) $\lim_{n\to\infty}\int_r^t\langle\mathcal{A}\tilde{u}_n(\sigma), P_n\psi\rangle d\sigma = \int_r^t\langle\mathcal{A}\tilde{u}(\sigma),\psi\rangle d\sigma$, $\tilde{\mathbb{P}}$-*a.s.*,

(c) $\lim_{n\to\infty}\int_r^t\langle B(\tilde{u}_n(\sigma)), P_n\psi\rangle d\sigma = \int_r^t\langle B(\tilde{u}(\sigma)),\psi\rangle d\sigma$, $\tilde{\mathbb{P}}$-*a.s.*.

Lemma 2.7. For all $r,t\in[0,T]$ such that $r\leq t$ and all $\psi\in U$, we have

$$lim_{n\to\infty}\int_r^t\langle g(\tilde{u}_n(\sigma)), P_n\psi\rangle d\sigma = \int_r^t\langle g(\tilde{u}(\sigma)),\psi\rangle d\sigma, \qquad \tilde{\mathbb{P}}\text{-}a.s.$$

Proof. Let us fix $r,t\in[0,T]$, $r\leq t$ and $\psi\in U\subset V_s\subset L^\infty$ since $s>\frac{5}{2}$. By Lemma 2.4, we get that $g(\tilde{u}_n)\rightharpoonup g(\tilde{u})$ in $L^{\frac{\beta+1}{\beta}}(0,T;L^{\frac{\beta+1}{\beta}})$, $\tilde{\mathbb{P}}$-a.s. and by (2.3), we get $P_n\psi\to\psi$ in L^2, then

$$\begin{aligned}
&\langle g(\tilde{u}_n(\sigma)), P_n\psi\rangle - \langle g(\tilde{u}(\sigma)),\psi\rangle\\
&= \langle g(\tilde{u}_n(\sigma)), P_n\psi-\psi\rangle + \langle g(\tilde{u}_n(\sigma)) - g(\tilde{u}(\sigma)),\psi\rangle\\
&\leq |\tilde{u}_n|_{\beta+1}^{\beta}|P_n\psi-\psi|_{\frac{\beta+1}{\beta}} + \langle g(\tilde{u}_n(\sigma)) - g(\tilde{u}(\sigma)),\psi\rangle\\
&\leq C|P_n\psi-\psi|_2 + \langle g(\tilde{u}_n(\sigma)) - g(\tilde{u}(\sigma)),\psi\rangle\\
&\to 0, \quad as\ n\to\infty.
\end{aligned}$$

This completes the proof of Lemma 2.7.

Lemma 2.8. (Lemma 5.6 [4]) For all $r, t \in [0,T]$ such that $r \leq t$ and all $\psi \in U$, we have

$$\lim_{n\to\infty} \mathbb{E}[\langle \tilde{M}_n(t) - \tilde{M}_n(r), \psi\rangle h(\tilde{u}_{n|[0,r]})] = \mathbb{E}[\langle \tilde{M}(t) - \tilde{M}(r), \psi\rangle h(\tilde{u}_{|[0,r]})].$$

Lemma 2.9. (Lemma 5.7 [4]) For all $r, t \in [0,T]$ such that $r \leq t$ and all $\psi, \zeta \in U$, we have

$$\begin{aligned}\lim\nolimits_{n\to\infty} \mathbb{E}[(\langle \tilde{M}_n(t), \psi\rangle\langle \tilde{M}_n(t), \zeta\rangle - \langle \tilde{M}_n(s), \psi\rangle\langle \tilde{M}_n(s), \zeta\rangle) h(\tilde{u}_{n|[0,r]})] \\ = \mathbb{E}[(\langle \tilde{M}(t), \psi\rangle\langle \tilde{M}(t), \zeta\rangle - \langle \tilde{M}(r), \psi\rangle\langle \tilde{M}(r), \zeta\rangle) h(\tilde{u}_{|[0,r]})].\end{aligned}$$

Lemma 2.10. (Lemma 5.8 [4]) For all $r, t \in [0,T]$ such that $r \leq t$ and all $\psi, \zeta \in U$, we have

$$\begin{aligned}\lim\nolimits_{n\to\infty} \mathbb{E}[(\textstyle\int_r^t (G(\tilde{u}_n(\sigma))^* P_n\psi, G(\tilde{u}_n(\sigma))^* P_n\zeta)_Y d\sigma) \cdot h(\tilde{u}_{n|[0,r]})] \\ = \mathbb{E}[(\textstyle\int_r^t (G(\tilde{u}(\sigma))^*\psi, G(\tilde{u}(\sigma))^*\zeta)_Y d\sigma) \cdot h(\tilde{u}_{|[0,r]})].\end{aligned}$$

Now, we apply the idea analogous to those used by Da Prato and Zabczyk ([9], Section 8.3), [17] and [4]. By the martingale representation theorem [9], we will get the conclusion. This completes the proof of Theorem 2.1.

4. Existence and uniqueness of strong solution

In order to get the strong solution for the stochastic Navier–Stokes equation with damping, we need the following Galerkin method.

We first introduce the Galerkin systems associated to the original equation and establish some uniform a priori estimates. Recall that

$$\{e_1, e_2, \cdots\} \subset V$$

is an orthonormal basis of H. For any $n \geq 1$, let $H_n = span\{e_1, e_2, \cdots, e_n\}$ and $P_n : V' \to H_n$ be defined by

$$P_n u = \sum_{i=1}^{n} {}_{V'}\langle u, e_i\rangle_V e_i, \quad u \in V'. \tag{4.1}$$

Hence, $P_n|_H$ denotes the orthogonal projection onto H_n in H.

We consider the classical Faedo–Galerkin approximation in the space H_n defined by

$$du_n(t) = -[P_n\mathcal{A}u_n(t) + B_n(u_n(t)) + P_n g(u_n) - P_n f(t)]dt + P_n G(u_n(t))dW(t) \tag{4.2}$$

for $u_n(0) = P_n u(0)$.

The following lemma is the main preliminary step in the proof of Theorem 2.3.

Lemma 2.11. Suppose that $\beta > 3$ with any $\alpha > 0$ and $\alpha \geq \frac{1}{2}$ as $\beta = 3$, $u_n(0) \in V \cap L^{\beta+1}(D)$, and $\nabla f \in L^2(\Omega; L^2(0,T;H))$, then there exists a positive constant C, such that

$$\begin{aligned}\sup_{n\geq 1} \mathbb{E}(\sup_{0\leq t\leq T} |\nabla u_n(t)|_2^2 + \int_0^T ||\nabla u_n(r)||^2 dr + \int_0^T \int_D |u_n|^{\beta-1}|\nabla u_n|^2 ds dr \\ + \int_0^T |\nabla |u_n|^{\frac{\beta+1}{2}}|_2^2 dr) \leq C(\mathbb{E}|\nabla u_n(0)|_2^2 + 1).\end{aligned}$$

Proof. We apply Itô formula to $|\nabla u_n|_2^2$ for $t \in [0, T]$,

$$\begin{aligned}
&|\nabla u_n(t)|_2^2 + 2\int_0^t ||\nabla u_n(r)||^2 dr \\
&+ 2\alpha \int_0^t \int_D |u_n|^{\beta-1} |\nabla u_n|^2 ds dr + \frac{8\alpha(\beta-1)}{(\beta+1)^2} \int_0^t |\nabla |u_n|^{\frac{\beta+1}{2}}|_2^2 dr \\
&\leq |P_n \nabla u_n(0)|_2^2 + 2\int_0^t |\langle \Delta u_n, B(u_n)\rangle| dr + 2\int_0^t |\langle \Delta u_n, f\rangle| dr \\
&+ 2|\int_0^t \langle G_n(u_n(r)) dW(r), \Delta u_n(r)\rangle| + \int_0^t ||\nabla G_n(u_n(r)) P_n||_{\mathcal{T}_2(Y,H)}^2 dr \\
&= |P_n \nabla u_n(0)|_2^2 + \sum_{i=1}^4 J_i(t). \qquad (4.3)
\end{aligned}$$

Since

$$0 < \frac{2}{\beta - 1} < 1, \qquad for \ \ \beta > 3, \qquad (4.4)$$

using Young's inequality and (4.4) to estimate $J_1(t)$, we deduce

$$\begin{aligned}
&J_1(t) \\
&\leq 2\int_0^t (\frac{1}{2}|\Delta u_n|_2^2 + \frac{1}{2}|u_n \cdot \nabla u_n|_2^2) dr \\
&\leq \int_0^t |\Delta u_n|_2^2 dr + \int_0^t \int_D |\nabla u_n|^2 |u_n|^2 ds dr \\
&= \int_0^t ||\nabla u_n||^2 dr + \int_0^t \int_D |u_n|^2 |\nabla u_n|^{\frac{4}{\beta-1}} |\nabla u_n|^{2-\frac{4}{\beta-1}} ds dr \\
&\leq \int_0^t ||\nabla u_n||^2 dr + \int_0^t [\int_D (|u_n|^2 |\nabla u_n|^{\frac{4}{\beta-1}})^{\frac{\beta-1}{2}} ds]^{\frac{2}{\beta-1}} [\int_D (|\nabla u_n|^{2-\frac{4}{\beta-1}})^{\frac{\beta-1}{\beta-3}} ds]^{\frac{\beta-3}{\beta-1}} dr \\
&\leq \int_0^t ||\nabla u_n||^2 dr + \varepsilon \int_0^t \int_D |\nabla u_n|^2 |u_n|^{\beta-1} ds dr + C(\varepsilon) \int_0^t |\nabla u_n(r)|_2^2 dr. \qquad (4.5)
\end{aligned}$$

The Hölder's inequality and Young's inequality imply that

$$J_2(t) \leq \frac{1}{4}\int_0^t ||\nabla u_n(r)||^2 dr + 4(|\nabla f|_{L^2(0,t;H)}^2). \qquad (4.6)$$

Taking the supremum and expectation over the interval $[0, t]$ on the equality (4.3), we estimate the last two items of (4.3). Now, applying Burkholder–Davis–Gundy inequality, Young's inequality and the Assumption A, we have

$$\begin{aligned}
&\mathbb{E}(\sup_{0\leq s\leq t} |J_3(s)|) \leq 2\sqrt{2}\mathbb{E}\{\int_0^t |\nabla u_n(r)|_2^2 |\nabla G_n(u_n(r)) P_n|_{\mathcal{T}_2(Y,H)}^2 dr\}^{\frac{1}{2}} \\
&\leq \frac{1}{2}\mathbb{E} \sup_{0\leq s\leq t} |\nabla u_n|_2^2 + 4\mathbb{E}\int_0^t ||\nabla G_n(u_n(r))||_{\mathcal{T}_2(Y,H)}^2 dr \\
&\leq \frac{1}{2}\mathbb{E} \sup_{0\leq s\leq t} |\nabla u_n|_2^2 + 4\mathbb{E}\int_0^t (k_1||u_n||^2 + k_2|u_n|_H^2 + k_3) dr. \qquad (4.7)
\end{aligned}$$

Similarly, we have

$$\mathbb{E}(\sup_{0\le s\le t}|J_4(s)|)\le \mathbb{E}\int_0^t (k_1||u_n||^2+k_2|u_n|_H^2+k_3)dr. \tag{4.8}$$

Substituting (4.5)–(4.8) into (4.3), we have

$$\begin{aligned}
&\mathbb{E}\sup_{0\le s\le t}|\nabla u_n(s)|_2^2+\frac{3}{4}\mathbb{E}\int_0^t ||\nabla u_n(r)||^2dr\\
&+2\alpha\mathbb{E}\int_0^t\int_D |u_n|^{\beta-1}|\nabla u_n|^2dsdr+\frac{8\alpha(\beta-1)}{(\beta+1)^2}\mathbb{E}\int_0^t |\nabla|u_n|^{\frac{\beta+1}{2}}|_2^2dr\\
&\le \mathbb{E}|P_n\nabla u_n(0)|_2^2+\varepsilon\mathbb{E}\int_0^t\int_D|\nabla u_n|^2|u_n|^{\beta-1}dsdr+C(\varepsilon)\mathbb{E}\int_0^t|\nabla u_n(r)|_2^2dr\\
&+\frac{1}{2}\mathbb{E}\sup_{0\le s\le t}|\nabla u_n|_2^2+5\mathbb{E}\int_0^t(k_1||u_n||^2+k_2|u_n|_H^2+k_3)dr+C|\nabla f|^2_{L^2(0,t;H)}.
\end{aligned} \tag{4.9}$$

So, we could have

$$\begin{aligned}
&\mathbb{E}\sup_{0\le s\le t}|\nabla u_n(s)|_2^2+\frac{3}{2}\mathbb{E}\int_0^t ||\nabla u_n(r)||^2dr\\
&+2(2\alpha-\varepsilon)\mathbb{E}\int_0^t\int_D |u_n|^{\beta-1}|\nabla u_n|^2dsdr+\frac{16\alpha(\beta-1)}{(\beta+1)^2}\mathbb{E}\int_0^t |\nabla|u_n|^{\frac{\beta+1}{2}}|_2^2dr\\
&\le 2\mathbb{E}|P_n\nabla u_n(0)|_2^2+C\mathbb{E}\int_0^t|\nabla u_n(r)|_2^2dr\\
&+10\mathbb{E}\int_0^t(k_1||u_n||^2+k_2|u_n|_H^2+k_3)dr+C|\nabla f|^2_{L^2(0,t;H)}\\
&\le 2\mathbb{E}|P_n\nabla u_n(0)|_2^2+C\int_0^t\mathbb{E}\sup_{0\le\tau\le r}|\nabla u_n(\tau)|_2^2dr+C(t)(1+|\nabla f|^2_{L^2(0,t;H)}).
\end{aligned} \tag{4.10}$$

Choosing a sufficiently small ε and Gronwall's inequality, we deduce

$$\begin{aligned}
\sup_{n\ge1}\mathbb{E}(\sup_{0\le t\le T}&|\nabla u_n(t)|_2^2+\int_0^T||\nabla u_n(r)||^2dr+\int_0^T\int_D|u_n|^{\beta-1}|\nabla u_n|^2dsdr\\
&+\int_0^T|\nabla|u_n|^{\frac{\beta+1}{2}}|_2^2dr)\le C(\mathbb{E}|\nabla u_n(0)|_2^2+1).
\end{aligned} \tag{4.11}$$

For $\alpha\ge\frac{1}{2}$ as $\beta=3$, we can get the above estimate easily. This completes the proof of Lemma 2.11.

Now we will study the weak convergence of approximating sequences. Using the priori estimate in the Lemma 2.4 and Lemma 2.11, we deduce the existence of a subsequence of Galerkin elements u_n and process $u\in L^2([0,T]\times\Omega,V)\cap L^{\beta+1}([0,T]\times\Omega,L^{\beta+1})\cap L^2(\Omega,L^\infty([0,T],H))$, $\nabla u\in L^2([0,T]\times\Omega,V)\cap L^2(\Omega,L^\infty([0,T],H))$, $-\mathcal{A}u-B(u)+f\in L^2([0,T]\times\Omega,V')$, $g\in L^s([0,T]\times\Omega,L^s(D))$, $s=(\beta+1)^*=\frac{\beta+1}{\beta}$ and $G\in L^2([0,T]\times\Omega,\mathcal{T}_2)$ for which the following limits hold:

(i) $u_n\to u$ weakly in $L^2([0,T]\times\Omega,V)$, $\nabla u_n\to\nabla u$ weakly in $L^2([0,T]\times\Omega,V)$,

(ii) $u_n\to u$ weakly in $L^{\beta+1}([0,T]\times\Omega,L^{\beta+1})$,

(iii) u_n is weak star converging to u in $L^2(\Omega, L^\infty([0,T],H))$, ∇u_n is weak star converging to ∇u in $L^2(\Omega, L^\infty([0,T],H))$,
(iv) $-\mathcal{A}u_n - B(u_n) + f \to -\mathcal{A}u - B(u) + f$ weakly in $L^2([0,T]\times\Omega, V')$,
(v) $G_n(u_n)P_n \to G$ weakly in $L^2([0,T]\times\Omega, \mathcal{T}_2)$,
(vi) $P_n g(u_n) \to g$ weakly in $g \in L^s([0,T]\times\Omega, L^s(D))$.

4.1. *Uniqueness of strong solution*

Let us recall that by assumption (A.3) the mapping $G : V \to \mathcal{T}_2(Y,H)$ is Lipschitz continuous, i.e. for some $L > 0$, the following inequality holds

$$||G(u_1(s)) - G(u_2(s))||^2_{\mathcal{T}_2(Y,H)} \leq L||u_1(s) - u_2(s)||, \quad s \in [0,T]. \tag{4.12}$$

Definition 2.4. (Pathwise uniqueness) We say that the pathwise uniqueness holds for Eq. (3.1) if whenever we are given two weak solutions of Eq. (3.1) defined on the same probability space together with the same Brownian motion

$$(\Omega, \mathcal{F}, P; (\mathcal{F})_{t\geq 0}; W; u_1),$$

$$(\Omega, \mathcal{F}, P; (\mathcal{F})_{t\geq 0}; W; u_2),$$

the condition $P\{u_1(0) = u_2(0)\} = 1$ implies $P\{\omega : u_1(t,\omega) = u_2(t,\omega), \forall t \geq 0\} = 1$.

The following Yamada–Watanabe theorem holds in this case ([25, 28]).

Theorem 2.2. *Under the conditions of Lemma 2.11, the existence of martingale solutions plus pathwise uniqueness implies the existence of a unique strong solution.*

Theorem 2.3. *Under the conditions of Lemma 2.11, and suppose that assumption (A.1)–(A.3) are satisfied with $\beta > 3$ with any $\alpha > 0$ and $\alpha \geq \frac{1}{2}$ as $\beta = 3$ and $L < 2$, then pathwise uniqueness holds for (3.1).*

Proof. Let u_1 and u_2 be two solutions of Eq. (3.1) defined on the same probability space together with the same Brownian motion and starting from the same initial value.
Set

$$U = u_1 - u_2.$$

Then U satisfies the following equation

$$dU(t) + [\mathcal{A}U(t) + (B(u_1(t)) - B(u_2(t))) + (g(u_1(t)) - g(u_2(t)))]dt$$

$$= [G(u_1(t)) - G(u_2(t))]dW(t).$$

Let $r(t) = a\int_0^t ||u_2||^4 ds$, $t \in [0,T]$, where a is a positive constant. Let

$$F(t,\chi) = e^{-r(t)}|\chi|^2_H, \qquad (t,\chi) \in [0,T]\times H.$$

Then by Itô's formula, we have

$$
\begin{aligned}
& e^{-r(t)}|U(t)|_H^2 \\
&= \int_0^t e^{-r(s)}\{-r'(s)|U(s)|_H^2 \\
&\quad -2\langle \mathcal{A}U(s) + (B(u_1(s)) - B(u_2(s))) + (g(u_1(s)) - g(u_2(s))), U(s)\rangle\} ds \\
&\quad +\frac{1}{2}\int_0^t Tr[(G(u_1(s)) - G(u_2(s)))\frac{\partial^2 F}{\partial \chi^2}(G(u_1(s)) - G(u_2(s)))^*] ds \\
&\quad +2\int_0^t e^{-r(s)}\langle G(u_1(s)) - G(u_2(s)), U(s)dW(s)\rangle \\
&\le \int_0^t e^{-r(s)}[-r'(s)|U(s)|_H^2 - 2||U(s)||^2 \\
&\quad -2\langle B(u_1(s)) - B(u_2(s)), U(s)\rangle - 2\langle g(u_1(s)) - g(u_2(s)), U(s)\rangle] ds \\
&\quad +\int_0^t e^{-r(s)}||G(u_1(s)) - G(u_2(s))||^2_{\mathcal{T}_2(Y,H)} ds \\
&\quad +2\int_0^t e^{-r(s)}\langle G(u_1(s)) - G(u_2(s)), U(s)dW(s)\rangle. \qquad (4.13)
\end{aligned}
$$

Since

$$B(u_1(s)) - B(u_2(s)) = B(u_1(s), U(s)) + B(U(s), u_2(s)), \quad s \in [0, T],$$

so, we have

$$\langle B(u_1(s)) - B(u_2(s)), U(s)\rangle = \langle B(U(s), u_2(s)), U(s)\rangle,$$

and hence

$$|2\langle B(u_1(s)) - B(u_2(s)), U(s)\rangle| \le C|U|_H^{\frac{1}{2}}||U||^{\frac{3}{2}}||u_2||, \quad s \in [0, T].$$

Therefore, for every $\varepsilon > 0$, there exists $C_\varepsilon > 0$ such that for $s \in [0, T]$,

$$|2\langle B(u_1(s)) - B(u_2(s)), U(s)\rangle| \le \varepsilon||U||^2 + C_\varepsilon|U|_H^2||u_2||^4. \qquad (4.14)$$

By Lemma 2.3, we have

$$2\langle g(u_1(s)) - g(u_2(s)), U(s)\rangle \ge 0. \qquad (4.15)$$

Let $a = C_\varepsilon$, we get

$$-r'(s)|U(s)|_H^2 + C_\varepsilon|U|_H^2||u_2||^4 = -a|U|_H^2||u_2||^4 + C_\varepsilon|U|_H^2||u_2||^4 = 0.$$

By (4.13)–(4.15), we have

$$
\begin{aligned}
& e^{-r(t)}|U(t)|_H^2 + \int_0^t e^{-r(s)}(2 - \varepsilon - L)||U(s)||^2 ds \\
&\le 2\int_0^t e^{-r(s)}\langle G(u_1(s)) - G(u_2(s)), U(s)dW(s)\rangle. \qquad (4.16)
\end{aligned}
$$

Let us choose $\varepsilon > 0$ such that $2 - \varepsilon - L > 0$, we have

$$e^{-r(t)}|U(t)|_H^2 \le 2\int_0^t e^{-r(s)}\langle G(u_1(s)) - G(u_2(s)), U(s)dW(s)\rangle. \qquad (4.17)$$

Taking expectations for (4.17), we get

$$\mathbb{E}(e^{-r(t)}|U(t)|_H^2) \leq 0. \tag{4.18}$$

By Lemma 2.11, we have $\mathbb{E}[\int_0^T ||u_2||^4 dt] < \infty$ for $\beta > 3$ with any $\alpha > 0$ and $\alpha \geq \frac{1}{2}$ as $\beta = 3$, with (4.18), then we get

$$\mathbb{E}(|U(t)|_H^2) \leq 0. \tag{4.19}$$

This completes the proof of Theorem 2.3.

Remark If $W(\cdot)$ is an H-valued Brownian motion on $(\Omega, \mathcal{F}, P)$ with the covariance operator Q, which is a positive, symmetric, trace class operator on H and $G(t,u)$ satisfies the following hypotheses (A.1)–(A.4), using a similar method, we could get the existence and uniqueness of strong solution on the space $u \in L^2(\Omega; C([0,T]; H)) \cap L^2(\Omega \times [0,T]; V) \cap L^{\beta+1}(\Omega \times [0,T]; L^{\beta+1})$ and $\nabla u \in L^2([0,T] \times \Omega; V) \cap L^2(\Omega, C([0,T]; H))$ for $\beta > 3$ with any $\alpha > 0$ and $\alpha \geq \frac{1}{2}$ as $\beta = 3$.

5. Large deviation principle

Let $W(\cdot)$ be an H-valued Brownian motion on $(\Omega, \mathcal{F}, P)$ with the covariance operator Q, which is a positive, symmetric, trace class operator on H. Assume that $L_Q(H; H)$ denotes the space of linear operator S such that $SQ^{\frac{1}{2}}$ is a Hilbert–Schmidt operator from H to H. Define the norm on the space $L_Q(H; H)$ by $|S|_{L_Q} = \sqrt{Tr(SQS^*)}$. Assume that L_Q^V denotes the class of linear operators $\tilde{S}$ such that $\tilde{S}Q^{\frac{1}{2}}$ is a Hilbert–Schmidt operator from H to V, endowed with the norm $|\tilde{S}|_{L_Q^V}^2 = Tr(\tilde{S}Q\tilde{S}^*)$. Let $f(t) = 0$ for simplicity.

In order to obtain the small time large deviation principle, we need the following hypotheses

(A.1) $\mathbb{E}(|\xi|^4) < \infty$.

(A.2) There exists a constant L such that $|G(t,u)|_{L_Q}^2 \leq L(1+|u|^2)$, for all $t \in (0,T)$, and all $u \in H$.

(A.3) There exists a constant $\hat{L}$ such that $|G(t,u)|_{L_Q^V}^2 \leq \hat{L}(1+||u||^2)$, for all $t \in (0,T)$, and all $u \in V$.

(A.4) There exists a constant K such that $|G(t,u) - G(t,v)|_{L_Q}^2 \leq K|u-v|^2$, for all $t \in (0,T)$, and all $u, v \in H$.

(A.5) There exists a constant $\hat{K}$ such that $|G(t,u) - G(t,v)|_{L_Q^V}^2 \leq \hat{K}||u-v||^2$, for all $t \in (0,T)$, and all $u, v \in V$.

Under the assumptions (A.1)–(A.4), we consider the following 3D stochastic Navier–Stokes equation with damping

$$du(t) + [Au(t) + B(u(t)) + g(u(t))]dt = G(t, u(t))dW(t), \tag{5.1}$$

$$u(0,x) = \xi, \tag{5.2}$$

which has a unique solution $u \in L^2(\Omega; C([0,T]; H)) \cap L^2(\Omega \times [0,T]; V) \cap L^{\beta+1}(\Omega \times [0,T]; L^{\beta+1})$ and $\nabla u \in L^2([0,T] \times \Omega; V) \cap L^2(\Omega, C([0,T]; H))$ for $\beta > 3$ with any $\alpha > 0$ and $\alpha \geq \frac{1}{2}$ as $\beta = 3$ and we have

$$u(t) = \xi - \int_0^t Au(s)ds - \int_0^t B(u(s))ds - \int_0^t g(u(s))ds + \int_0^t G(s, u(s))dW(s). \tag{5.3}$$

Consider the small time process $u(\varepsilon t)$, we have the following equation:

$$u^\varepsilon(t) = \xi - \varepsilon \int_0^t Au^\varepsilon ds - \varepsilon \int_0^t B(u^\varepsilon(s))ds - \varepsilon \int_0^t g(u^\varepsilon(s))ds + \sqrt{\varepsilon} \int_0^t G(\varepsilon s)dW(s). \tag{5.4}$$

Let μ_ξ^ε be the law of u^ε and ∇u^ε on $C([0,1]; H)$. Define the following functional $I(g)$ by

$$I(g) = \inf_{h \in \Gamma_g} \left\{ \frac{1}{2} \int_0^1 |\dot{h}(t)|_{H_0}^2 dt \right\},$$

where $H_0 = Q^{\frac{1}{2}} H$ endowed with the norm $|h|_{H_0}^2 = |Q^{-\frac{1}{2}} h|^2$, and

$$\Gamma_g = \Big\{ h = (h_1, h_2, \cdots, h_k, \cdots) \in C([0,T]; H) : \text{h is absolutely continuous}$$
$$\text{and such that } g(t) = \xi + \int_0^t G(s, g(s))\dot{h}(s)ds, 0 \leq t \leq 1 \Big\}.$$

Theorem 2.4. *Assume (A.1)–(A.5) are satisfied, $\xi \in H$, then there exists a sequence $\{\xi_n\}_{n=1}^\infty \subset V$ such that $\xi_n \to \xi$, for $\beta > 3$ with any $\alpha > 0$ and $\alpha \geq \frac{1}{2}$ as $\beta = 3$, then μ_ξ^ε satisfies a large deviation with rate function $I(\cdot)$, i.e.,*
(i) for every closed subset $F \subset C([0,1]; H)$,

$$\lim_{\varepsilon \to 0} \sup_{\xi_n \to \xi} \varepsilon log \mu_{\xi_n}^\varepsilon (F) \leq - \inf_{g \in F} (I(g)),$$

(ii) for every open subset $E \subset C([0,1]; H)$,

$$\lim_{\varepsilon \to 0} \inf_{\xi_n \to \xi} \varepsilon log \mu_{\xi_n}^\varepsilon (E) \geq - \inf_{g \in E} (I(g)).$$

Proof. This section is devoted to the proof of Theorem 2.4, which will be split into a number of lemmas. Let $v^\varepsilon(\cdot)$ be the solution of the stochastic equation

$$v^\varepsilon(t) = \xi + \sqrt{\varepsilon} \int_0^t G(\varepsilon s, v^\varepsilon(s))dW(s), \tag{5.5}$$

and ν^ε be the law of $v^\varepsilon(\cdot)$ on $C([0,1]; H)$. By [9], we obtain that ν^ε satisfies a large deviation principle with the rate function $I(\cdot)$. First, we will prove that the two families of probability measure μ^ε and ν^ε are exponentially equivalent, that is, for any $\delta > 0$, we deduce

$$\lim_{\varepsilon \to 0} \varepsilon log P(\sup_{0 \leq t \leq 1} |u^\varepsilon(t) - v^\varepsilon(t)|^2 > \delta) = -\infty. \tag{5.6}$$

Then Theorem 2.4 follows from (5.6) and Theorem 4.2.13 in [11] for ξ. Theorem 4.2.13 in [11] satisfies that if one of the two exponentially equivalent families satisfies a large deviation principle, so does the other.
The following result is a useful estimate of the probability that the solution of (5.4) leaves an energy ball. It will play an important role in the rest of the paper.

Lemma 2.12. Let $u^\varepsilon(t)$ be the solution of (5.4), then we have

$$\lim_{M\to\infty}\sup_{0<\varepsilon\leq 1}\varepsilon log P((|u^\varepsilon|_V^H(1))^2 > M) = -\infty, \tag{5.7}$$

where $(|u^\varepsilon|_V^H(1))^2 = \sup_{0\leq t\leq 1}|u^\varepsilon(t)|^2 + 2\varepsilon\int_0^1 ||u^\varepsilon(t)||^2 dt$.

Proof. Applying Itô's formula to $|u^\varepsilon(t)|^2$, we deduce

$$\begin{aligned}|u^\varepsilon(t)|^2 &= |\xi|^2 - 2\varepsilon\int_0^t \langle u^\varepsilon(s), Au^\varepsilon(s)\rangle ds - 2\varepsilon\int_0^t \langle u^\varepsilon(s), B(u^\varepsilon(s))\rangle ds\\ &\quad - 2\varepsilon\int_0^t \langle u^\varepsilon(s), g(u^\varepsilon(s))\rangle ds\\ &\quad + 2\sqrt{\varepsilon}\int_0^t (u^\varepsilon(s), G(\varepsilon s, u^\varepsilon(s))dW(s)) + \varepsilon\int_0^t |G(\varepsilon s, u^\varepsilon(s))|_{L_Q}^2 ds,\end{aligned}$$

that is,

$$\begin{aligned}&|u^\varepsilon(t)|^2 + 2\varepsilon\int_0^t (u^\varepsilon(s), Au^\varepsilon(s))ds + 2\varepsilon\int_0^t (u^\varepsilon(s), g(u^\varepsilon(s)))ds\\ &= |\xi|^2 - 2\varepsilon\int_0^t (u^\varepsilon(s), B(u^\varepsilon(s)))ds\\ &+ 2\sqrt{\varepsilon}\int_0^t (u^\varepsilon(s), G(\varepsilon s, u^\varepsilon(s))dW(s)) + \varepsilon\int_0^t |G(\varepsilon s, u^\varepsilon(s))|_{L_Q}^2 ds\\ &= |\xi|^2 + I_1 + I_2 + I_3.\end{aligned} \tag{5.8}$$

For the first term, we deduce

$$I_1(t) = 0. \tag{5.9}$$

By virtue of (A.2), we have

$$|I_3(t)| \leq \varepsilon L\int_0^t (1 + |u^\varepsilon(s)|^2)ds. \tag{5.10}$$

Substituting (5.9) and (5.10) into (5.8), we deduce

$$\begin{aligned}&|u^\varepsilon(t)|^2 + 2\varepsilon\int_0^t ||u^\varepsilon(s)||^2 ds + 2\varepsilon\int_0^t |u^\varepsilon(s)|_{\beta+1}^{\beta+1} ds\\ &\leq (|\xi|^2 + \varepsilon Lt) + \varepsilon L\int_0^t |u^\varepsilon(s)|^2 ds\\ &+ 2\sqrt{\varepsilon}|\int_0^t (u^\varepsilon(s), G(\varepsilon s, u^\varepsilon(s))dW(s))|.\end{aligned}$$

Hence, we have the following inequality

$$
\begin{aligned}
&(|u^\varepsilon|_V^H(T))^2 + 2\varepsilon \int_0^T |u^\varepsilon(s)|_{\beta+1}^{\beta+1} ds \\
&\leq 2(|\xi|^2 + \varepsilon LT) + 4\varepsilon L \int_0^T (|u^\varepsilon(s)|_V^H(s))^2 ds \\
&+ 4\sqrt{\varepsilon} \sup_{0\leq t\leq T} |\int_0^t (u^\varepsilon(s), G(\varepsilon s, u^\varepsilon(s)) dW(s))|.
\end{aligned}
$$

For $p \geq 2$, there exists a constant c such that

$$
\begin{aligned}
&(\mathbb{E}(|u^\varepsilon|_V^H(T))^{2p})^{\frac{1}{p}} \\
&\leq 2(|\xi|^2 + \varepsilon LT) + c(\mathbb{E}(\int_0^T (|u^\varepsilon(s)|_V^H(s))^2 ds)^p)^{\frac{1}{p}} \\
&+ 4\sqrt{\varepsilon}(\mathbb{E}(\sup_{0\leq t\leq T} |\int_0^t (u^\varepsilon(s), G(\varepsilon s, u^\varepsilon(s)) dW(s))|^p))^{\frac{1}{p}}.
\end{aligned}
\tag{5.11}
$$

For estimating the stochastic integral term, we will use the following result from [1, 10], and there exists a positive constant C such that, for any $p \geq 2$ and for any continuous martingale (M_t) with $M_0 = 0$, one has

$$
||M_t^*||_p \leq C p^{\frac{1}{2}} ||\langle M \rangle_t^{\frac{1}{2}}||_p, \tag{5.12}
$$

where $M_t^* = \sup_{0\leq s\leq t} |M_s|$ and $||\cdot||_p$ stands for the L^p-norm.

$$
\begin{aligned}
&4\sqrt{\varepsilon}(\mathbb{E}(\sup_{0\leq t\leq T} |\int_0^t (u^\varepsilon(s), G(\varepsilon s, u^\varepsilon(s)) dW(s))|^p))^{\frac{1}{p}} \\
&\leq 4C\sqrt{p\varepsilon}((\mathbb{E}(\int_0^T (1 + |u^\varepsilon(s)|^2)^2 ds)^{\frac{p}{2}})^{\frac{2}{p}})^{\frac{1}{2}} \\
&\leq 4C\sqrt{p\varepsilon}(\int_0^T 1 + (\mathbb{E}|u^\varepsilon(s)|^{2p})^{\frac{2}{p}} ds)^{\frac{1}{2}},
\end{aligned}
\tag{5.13}
$$

where (A.2) has been used. In other words,

$$
c(\mathbb{E}(\int_0^T (|u^\varepsilon|_V^H(s))^2 ds)^p)^{\frac{1}{p}} \leq c\int_0^T (\mathbb{E}(|u^\varepsilon|_V^H(s))^{2p})^{\frac{1}{p}} ds. \tag{5.14}
$$

Combining (5.11), (5.13) and (5.14), we obtain

$$
\begin{aligned}
&(\mathbb{E}(|u^\varepsilon|_V^H(T))^{2p})^{\frac{2}{p}} \\
&\leq 8(|\xi|^2 + \varepsilon LT)^2 + 2c^2 T \int_0^T (\mathbb{E}(|u^\varepsilon(s)|_V^H(s))^{2p})^{\frac{2}{p}} ds \\
&+ 32C^2 p\varepsilon T + 32C^2 p\varepsilon \int_0^T (\mathbb{E}(|u^\varepsilon(s)|_V^H(s))^{2p})^{\frac{2}{p}} ds.
\end{aligned}
\tag{5.15}
$$

Applying the Gronwall's inequality, we get

$$
(\mathbb{E}(|u^\varepsilon|_V^H(1))^{2p})^{\frac{2}{p}} \leq C[8(|\xi|^2 + \varepsilon L)^2 + 32C^2 p\varepsilon] exp(2c^2 + 32C^2 p\varepsilon). \tag{5.16}
$$

Since $P((|u^\varepsilon|_V^H(1))^2 > M) \le M^{-p}E(|u^\varepsilon|_V^H(1))^{2p}$. Let $p = \frac{2}{\varepsilon}$ in (2.50), we obtain

$$\begin{aligned}&\varepsilon log P((|u^\varepsilon|_V^H(1))^2 > M)\\ &\le -logM + log(E(|u^\varepsilon|_V^H(1))^{2p})^{\frac{1}{p}}\\ &\le -logM + log\sqrt{C[8(|\xi|^2+\varepsilon L)^2 + 32C^2]} + c^2 + 32C^2.\end{aligned}$$

Hence, we get the following inequality

$$\begin{aligned}&\sup_{0<\varepsilon\le 1} \varepsilon log P((|u^\varepsilon|_V^H(1))^2 > M)\\ &\le -logM + log\sqrt{C[8(|\xi|^2+L)^2 + 32C^2]} + c^2 + 32C^2.\end{aligned}$$

Letting $M \to \infty$ on both sides of the above inequality, this completes the proof of Lemma 2.12.

For $\xi \in H$, since V is dense in H, then there exists a sequence $\{\xi_n\}_{n=1}^\infty \subset V$ such that

$$\lim_{n\to+\infty} |\xi_n - \xi| = 0.$$

Let $u_n^\varepsilon(\cdot)$ be the solution of (5.4) with the initial value ξ_n. By the proof of Lemma 2.12, it is easy to see that we get the following equality

$$\lim_{M\to+\infty} \sup_n \sup_{0<\varepsilon\le 1} \varepsilon log P((|u_n^\varepsilon|_V^H(1))^2 > M) = -\infty. \tag{5.17}$$

Let $v_n^\varepsilon(\cdot)$ be the solution of (5.5) with the initial value ξ_n, and we have the following result whose proof is similar to that of the Lemma 2.12.

Lemma 2.13. For any fixed $n \in Z^+$,

$$\lim_{M\to+\infty} \sup_{0<\varepsilon\le 1} \varepsilon log P(\sup_{0\le t\le 1} ||v_n^\varepsilon(t)||^2 > M) = -\infty,$$

$$\lim_{M\to+\infty} \sup_{0<\varepsilon\le 1} \varepsilon log P(\sup_{0\le t\le 1} |v_n^\varepsilon(t)|_{\beta+1}^{\beta+1} > M) = -\infty.$$

Proof. Applying Itô's formula to $||v_n^\varepsilon(t)||^2$, we deduce

$$||v_n^\varepsilon(t)||^2 = ||\xi_n||^2 + 2\sqrt{\varepsilon}\int_0^t ((v_n^\varepsilon(s), G(\varepsilon s, v_n^\varepsilon(s))dW(s))) + \varepsilon\int_0^t |G(\varepsilon s, v_n^\varepsilon(s))|_{L_Q^V}^2 ds.$$

By (A.3) and inequality (5.12), we obtain

$$\begin{aligned}(\mathbb{E}[\sup_{0\le t\le r} ||v_n^\varepsilon(t)||^{2p}])^{\frac{2}{p}} &\le 2||\xi_n||^4 + 8C\varepsilon p(\mathbb{E}[\int_0^r ||v_n^\varepsilon(s)||^2 |G(\varepsilon s, v_n^\varepsilon(s))|_{L_Q^V}^2 ds]^{\frac{p}{2}})^{\frac{2}{p}}\\ &\quad + 4\varepsilon^2\hat{L}^2 r(r + \int_0^r (\mathbb{E}[\sup_{0\le l\le s} ||v_n^\varepsilon(l)||^{2p}])^{\frac{2}{p}} ds)\\ &\le 2||\xi_n||^4 + 16C\varepsilon p\hat{L}(r + \int_0^r (\mathbb{E}[\sup_{0\le l\le s} ||v_n^\varepsilon(l)||^{2p}])^{\frac{2}{p}} ds)\\ &\quad + 4\varepsilon^2\hat{L}^2 r(r + \int_0^r (\mathbb{E}[\sup_{0\le l\le s} ||v_n^\varepsilon(l)||^{2p}])^{\frac{2}{p}} ds).\end{aligned} \tag{5.18}$$

By Gronwall's inequality,

$$(\mathbb{E}[\sup_{0\le t\le 1}||v_n^\varepsilon(t)||^{2p}])^{\frac{2}{p}} \le (2||\xi_n||^4 + 16C\varepsilon p\hat{L} + 4\varepsilon^2\hat{L}^2)e^{16C\varepsilon p\hat{L}+4\varepsilon^2\hat{L}^2}. \quad (5.19)$$

According to the proof of Lemma 2.12, we have

$$\lim_{M\to+\infty}\sup_{0<\varepsilon\le 1}\varepsilon log P(\sup_{0\le t\le 1}||v_n^\varepsilon(t)||^2 > M) = -\infty.$$

Similarly, we could get the following equality

$$\lim_{M\to+\infty}\sup_{0<\varepsilon\le 1}\varepsilon log P(\sup_{0\le t\le 1}|v_n^\varepsilon(t)|_{\beta+1}^{\beta+1} > M) = -\infty.$$

Lemma 2.14. For any $\delta > 0$ and $\beta > 3$ with any $\alpha > 0$ and $\alpha \ge \frac{1}{2}$ as $\beta = 3$,

$$\lim_{n\to\infty}\sup_{0<\varepsilon\le 1}\varepsilon log P(\sup_{0\le t\le 1}|u^\varepsilon(t) - u_n^\varepsilon(t)|^2 > \delta) = -\infty. \quad (5.20)$$

Proof. For any $M > 0$, we define a stopping time

$$\tau_{\varepsilon,M} = inf\{t : \varepsilon\int_0^t ||u^\varepsilon(r)||^2 dr > M,\ or\ |u^\varepsilon(t)|^2 > M, or\ ||u^\varepsilon(t)||^2 > M\},$$

then, we deduce

$$\begin{aligned}
&P(\sup_{0\le t\le 1}|u^\varepsilon(t) - u_n^\varepsilon(t)|^2 > \delta, (|u^\varepsilon|_V^H(1))^2 \le M, ||u^\varepsilon(t)||^2 \le M)\\
&\le P(\sup_{0\le t\le 1}|u^\varepsilon(t) - u_n^\varepsilon(t)|^2 > \delta, \tau_{\varepsilon,M} \ge 1)\\
&\le P(\sup_{0\le t\le 1}|u^\varepsilon(t) - u_n^\varepsilon(t)|^2 > \delta). \qquad (5.21)
\end{aligned}$$

Let k be a positive constant. Applying Itô's formula to

$$e^{-k\varepsilon\int_0^{t\wedge\tau_{\varepsilon,M}}||u^\varepsilon(s)||^4 ds}|u^\varepsilon(t\wedge\tau_{\varepsilon,M}) - u_n^\varepsilon(t\wedge\tau_{\varepsilon,M})|^2,$$

we deduce

$$\begin{aligned}
&e^{-k\varepsilon\int_0^{t\wedge\tau_{\varepsilon,M}}||u^\varepsilon||^4 ds}|u^\varepsilon(t\wedge\tau_{\varepsilon,M}) - u_n^\varepsilon(t\wedge\tau_{\varepsilon,M})|^2\\
&+ 2\varepsilon\int_0^{t\wedge\tau_{\varepsilon,M}} e^{-k\varepsilon\int_0^s ||u^\varepsilon||^4 dr}||u^\varepsilon(s) - u_n^\varepsilon(s)||^2 ds\\
=\ & |\xi - \xi_n|^2 - k\varepsilon\int_0^{t\wedge\tau_{\varepsilon,M}} e^{-k\varepsilon\int_0^s ||u^\varepsilon(r)||^4 dr}||u^\varepsilon(s)||^4|u^\varepsilon(s) - u_n^\varepsilon(s)|^2 ds\\
&- 2\varepsilon\int_0^{t\wedge\tau_{\varepsilon,M}} e^{-k\varepsilon\int_0^s ||u^\varepsilon(r)||^4 dr}(b(u^\varepsilon(s), u^\varepsilon(s), u^\varepsilon(s) - u_n^\varepsilon(s))\\
&- b(u_n^\varepsilon(s), u_n^\varepsilon(s), u^\varepsilon(s) - u_n^\varepsilon(s)))ds\\
&- 2\varepsilon\int_0^{t\wedge\tau_{\varepsilon,M}} e^{-k\varepsilon\int_0^s ||u^\varepsilon(r)||^4 dr}(g(u^\varepsilon(s)) - g(u_n^\varepsilon(s)), u^\varepsilon(s) - u_n^\varepsilon(s))\\
&+ 2\sqrt{\varepsilon}\int_0^{t\wedge\tau_{\varepsilon,M}} e^{-k\varepsilon\int_0^s ||u^\varepsilon(r)||^4 dr}(u^\varepsilon(s) - u_n^\varepsilon(s),\\
&\qquad (G(\varepsilon s, u^\varepsilon(s)) - G(\varepsilon s, u_n^\varepsilon(s)))dW(s))\\
&+ \varepsilon\int_0^{t\wedge\tau_{\varepsilon,M}} e^{-k\varepsilon\int_0^s ||u^\varepsilon(r)||^4 dr}|G(\varepsilon s, u^\varepsilon(s)) - G(\varepsilon s, u_n^\varepsilon(s))|_{L_Q}^2 ds. \qquad (5.22)
\end{aligned}$$

Since $b(u_n^\varepsilon(t), u_n^\varepsilon(t), u^\varepsilon(t) - u_n^\varepsilon(t)) = b(u_n^\varepsilon(t), u^\varepsilon(t), u^\varepsilon(t) - u_n^\varepsilon(t))$, then we have

$$\begin{aligned}
&|b(u^\varepsilon(t), u^\varepsilon(t), u^\varepsilon(t) - u_n^\varepsilon(t)) - b(u_n^\varepsilon(t), u_n^\varepsilon(t), u^\varepsilon(t) - u_n^\varepsilon(t))| \\
&= |b(u^\varepsilon(t) - u_n^\varepsilon(t), u^\varepsilon(t), u^\varepsilon(t) - u_n^\varepsilon(t))| \\
&\le c|u^\varepsilon(t) - u_n^\varepsilon(t)|^{\frac{1}{2}} \cdot ||u^\varepsilon(t)|| \cdot ||u^\varepsilon(t) - u_n^\varepsilon(t)||^{\frac{3}{2}}.
\end{aligned}$$

Combining with (5.22), we deduce

$$\begin{aligned}
& e^{-k\varepsilon \int_0^{t\wedge\tau_{\varepsilon,M}} ||u^\varepsilon(s)||^4 ds} |u^\varepsilon(t\wedge\tau_{\varepsilon,M}) - u_n^\varepsilon(t\wedge\tau_{\varepsilon,M})|^2 \\
&\quad + 2\varepsilon \int_0^{t\wedge\tau_{\varepsilon,M}} e^{-k\varepsilon \int_0^s ||u^\varepsilon(r)||^4 dr} ||u^\varepsilon(s) - u_n^\varepsilon(s)||^2 ds \\
&\le |\xi - \xi_n|^2 - k\varepsilon \int_0^{t\wedge\tau_{\varepsilon,M}} e^{-k\varepsilon \int_0^s ||u^\varepsilon(r)||^4 dr} ||u^\varepsilon(s)||^4 |u^\varepsilon(s) - u_n^\varepsilon(s)|^2 ds \\
&\quad + 2c\varepsilon \int_0^{t\wedge\tau_{\varepsilon,M}} e^{-k\varepsilon \int_0^s ||u^\varepsilon(r)||^4 dr} |u^\varepsilon(t) - u_n^\varepsilon(t)|^{\frac{1}{2}} \cdot ||u^\varepsilon(t)|| \cdot ||u^\varepsilon(t) - u_n^\varepsilon(t)||^{\frac{3}{2}} ds \\
&\quad + 2\sqrt{\varepsilon} \left| \int_0^{t\wedge\tau_{\varepsilon,M}} e^{-k\varepsilon \int_0^s ||u^\varepsilon(r)||^4 dr} (u^\varepsilon(s) - u_n^\varepsilon(s), \right. \\
&\qquad\qquad \left. (G(\varepsilon s, u^\varepsilon(s)) - G(\varepsilon s, u_n^\varepsilon(s))) dW(s)) \right| \\
&\quad + K\varepsilon \int_0^{t\wedge\tau_{\varepsilon,M}} e^{-k\varepsilon \int_0^s ||u^\varepsilon(r)||^4 dr} |u^\varepsilon(s) - u_n^\varepsilon(s)|^2 ds \\
&\le |\xi - \xi_n|^2 - k\varepsilon \int_0^{t\wedge\tau_{\varepsilon,M}} e^{-k\varepsilon \int_0^s ||u^\varepsilon(r)||^4 dr} ||u^\varepsilon(s)||^4 |u^\varepsilon(s) - u_n^\varepsilon(s)|^2 ds \\
&\quad + C\varepsilon \int_0^{t\wedge\tau_{\varepsilon,M}} e^{-k\varepsilon \int_0^s ||u^\varepsilon(r)||^4 dr} |u^\varepsilon(s) - u_n^\varepsilon(s)|^2 \cdot ||u_n^\varepsilon(s)||^4 ds \\
&\quad + \varepsilon \int_0^{t\wedge\tau_{\varepsilon,M}} e^{-k\varepsilon \int_0^s ||u^\varepsilon(r)||^4 dr} ||u^\varepsilon(s) - u_n^\varepsilon(s)||^2 ds \\
&\quad + 2\sqrt{\varepsilon} \left| \int_0^{t\wedge\tau_{\varepsilon,M}} e^{-k\varepsilon \int_0^s ||u^\varepsilon(r)||^4 dr} (u^\varepsilon(s) - u_n^\varepsilon(s), \right. \\
&\qquad\qquad \left. (G(\varepsilon s, u^\varepsilon(s)) - G(\varepsilon s, u_n^\varepsilon(s))) dW(s)) \right| \\
&\quad + K\varepsilon \int_0^{t\wedge\tau_{\varepsilon,M}} e^{-k\varepsilon \int_0^s ||u^\varepsilon(r)||^4 dr} |u^\varepsilon(s) - u_n^\varepsilon(s)|^2 ds. \qquad (5.23)
\end{aligned}$$

Choosing $k > C$ and using (5.12), we deduce

$$
\begin{aligned}
&(\mathbb{E}[\sup_{0\leq s\leq t\wedge\tau_{\varepsilon,M}}(e^{-k\varepsilon\int_0^s\|u^\varepsilon(r)\|^4dr}|u^\varepsilon(s)-u_n^\varepsilon(s)|^2)]^p)^{\frac{2}{p}}\\
&\leq 2|\xi-\xi_n|^4+2\varepsilon^2K^2\int_0^t(\mathbb{E}[(\sup_{0\leq r\leq s\wedge\tau_{\varepsilon,M}}e^{-k\varepsilon\int_0^r\|u^\varepsilon(l)\|^4dl}|u^\varepsilon(r)-u_n^\varepsilon(r)|^2)^p])^{\frac{2}{p}}ds\\
&+8C\varepsilon pK^2\int_0^t(\mathbb{E}[(\sup_{0\leq r\leq s\wedge\tau_{\varepsilon,M}}e^{-2k\varepsilon\int_0^r\|u^\varepsilon(l)\|^4dl}|u^\varepsilon(r)-u_n^\varepsilon(r)|^4)^{\frac{p}{2}}])^{\frac{2}{p}}ds\\
&\leq 2|\xi-\xi_n|^4+2\varepsilon^2K^2\int_0^t(\mathbb{E}[(\sup_{0\leq r\leq s\wedge\tau_{\varepsilon,M}}e^{-k\varepsilon\int_0^r\|u^\varepsilon(l)\|^4dl}|u^\varepsilon(r)-u_n^\varepsilon(r)|^2)^p])^{\frac{2}{p}}ds\\
&+8C\varepsilon pK^2\int_0^t(\mathbb{E}[(\sup_{0\leq r\leq s\wedge\tau_{\varepsilon,M}}e^{-k\varepsilon\int_0^r\|u^\varepsilon(l)\|^4dl}|u^\varepsilon(r)-u_n^\varepsilon(r)|^2)^p])^{\frac{2}{p}}ds. \qquad (5.24)
\end{aligned}
$$

Applying Gronwall's inequality, we have the following inequality

$$
\begin{aligned}
&(\mathbb{E}[\sup_{0\leq t\leq 1\wedge\tau_{\varepsilon,M}}(e^{-k\varepsilon\int_0^t\|u^\varepsilon(s)\|^4ds}|u^\varepsilon(t)-u_n^\varepsilon(t)|^2)^p])^{\frac{2}{p}}\\
&\leq 2|\xi-\xi_n|^4e^{2\varepsilon^2K^2+8C\varepsilon pK^2}. \qquad (5.25)
\end{aligned}
$$

Hence, we have

$$
\begin{aligned}
&(\mathbb{E}[\sup_{0\leq t\leq 1\wedge\tau_{\varepsilon,M}}(|u^\varepsilon(t)-u_n^\varepsilon(t)|^2)^p])^{\frac{2}{p}}\\
&\leq (\mathbb{E}[\sup_{0\leq t\leq 1\wedge\tau_{\varepsilon,M}}(e^{-k\varepsilon\int_0^t\|u^\varepsilon(s)\|^4ds}|u^\varepsilon(t)-u_n^\varepsilon(t)|^2)^pe^{kp\varepsilon\int_0^{1\wedge\tau_{\varepsilon,M}}\|u^\varepsilon(s)\|^4ds}])^{\frac{2}{p}}\\
&\leq e^{2kM}(\mathbb{E}[\sup_{0\leq t\leq 1\wedge\tau_{\varepsilon,M}}(e^{-k\varepsilon\int_0^t\|u^\varepsilon(s)\|^4ds}|u^\varepsilon(t)-u_n^\varepsilon(t)|^2)^p])^{\frac{2}{p}}\\
&\leq 2e^{2kM}|\xi-\xi_n|^4e^{2\varepsilon^2K^2+8C\varepsilon pK^2}. \qquad (5.26)
\end{aligned}
$$

Fix a positive constant M and $p = \frac{2}{\varepsilon}$, we have the following inequalities

$$
\begin{aligned}
&\sup_{0<\varepsilon\leq 1}\varepsilon log P(\sup_{0\leq t\leq 1\wedge\tau_{\varepsilon,M}}(|u^\varepsilon(t)-u_n^\varepsilon(t)|^2>\delta)\\
&\leq \sup_{0<\varepsilon\leq 1}\varepsilon log\frac{\mathbb{E}[\sup_{0\leq t\leq 1\wedge\tau_{\varepsilon,M}}|u^\varepsilon(t)-u_n^\varepsilon(t)|^{2p}]}{\delta^p}\\
&\leq 2kM+2K^2+16CK^2-2log\delta+log2|\xi-\xi_n|^4\\
&\to -\infty,\quad as\ n\to+\infty. \qquad (5.27)
\end{aligned}
$$

For any given $R > 0$, by Lemma 2.12, there exists a positive constant M such that for any $\varepsilon \in (0, 1]$, we get the following inequality

$$
P((|u^\varepsilon|_V^H(1))^2 > M) \leq e^{-\frac{R}{\varepsilon}}. \qquad (5.28)
$$

For any M, (5.22) and (5.27) imply that there exists a positive integer N, such that for any $n \geq N$,

$$
\begin{aligned}
&\sup_{0<\varepsilon\leq 1}\varepsilon log P(\sup_{0\leq t\leq 1}|u^\varepsilon(t)-u_n^\varepsilon(t)|^2>\delta,(|u^\varepsilon|_V^H(1))^2 \qquad (5.29)\\
&\leq M,\|u^\varepsilon(t)\|^2\leq M)\leq -R.
\end{aligned}
$$

Combining (5.28) and (5.29) together, we can get a positive number N, such that for any $n \geq N$, $\varepsilon \in (0,1]$,

$$P(\sup_{0\leq t\leq 1}|u^\varepsilon(t)-u_n^\varepsilon(t)|^2>\delta)\leq 2e^{-\frac{R}{\varepsilon}}. \tag{5.30}$$

Since R is arbitrary, this completes the proof of Lemma 2.14.

The following lemma can be proved similarly as Lemma 2.14.

Lemma 2.15. For any $\delta > 0$,

$$\lim_{n\to\infty}\sup_{0<\varepsilon\leq 1}\varepsilon log P(\sup_{0\leq t\leq 1}|v^\varepsilon(t)-v_n^\varepsilon(t)|^2>\delta)=-\infty. \tag{5.31}$$

The following lemma shows that for any fixed integer n, the two families $\{u_n^\varepsilon, \varepsilon > 0\}$ and $\{v_n^\varepsilon, \varepsilon > 0\}$ are exponentially equivalent.

Lemma 2.16. For any $\delta > 0$, every positive integer n and $\beta > 3$ with any $\alpha > 0$ or $\alpha \geq \frac{1}{2}$ as $\beta = 3$,

$$\lim_{\varepsilon\to 0}\varepsilon log P(\sup_{0\leq t\leq 1}|u_n^\varepsilon(t)-v_n^\varepsilon(t)|^2>\delta)=-\infty. \tag{5.32}$$

Proof. For any $M > 0$, we define stopping times

$$\begin{aligned}
\tau_{\varepsilon,M}^{1,n} &= inf\{t\geq 0; \varepsilon\int_0^t ||u_n^\varepsilon(s)||^2ds>M,\ or\ |u_n^\varepsilon(t)|^2>M,\ or\ ||u_n^\varepsilon(s)||^2>M\},\\
\tau_{\varepsilon,M}^{2,n} &= inf\{t\geq 0; ||v_n^\varepsilon(t)||^2>M, |v_n^\varepsilon(t)|_{\beta+1}^{\beta+1}>M\},
\end{aligned}$$

and we have

$$\begin{aligned}
&P(\sup_{0\leq t\leq 1}|u_n^\varepsilon(t)-v_n^\varepsilon(t)|^2>\delta, (|u_n^\varepsilon|_V^H(1))^2\leq M, \sup_{0\leq t\leq 1}||u_n^\varepsilon(t)||^2\leq M,\\
&\sup_{0\leq t\leq 1}||v_n^\varepsilon(t)||^2\leq M, \sup_{0\leq t\leq 1}|v_n^\varepsilon(t)|_{\beta+1}^{\beta+1}\leq M)\\
&\leq P(\sup_{0\leq t\leq 1}|u_n^\varepsilon(t)-v_n^\varepsilon(t)|^2>\delta, 1\leq \tau_{\varepsilon,M}^{1,n}\wedge\tau_{\varepsilon,M}^{2,n})\\
&\leq P(\sup_{0\leq t\leq 1\wedge\tau_{\varepsilon,M}^{1,n}\wedge\tau_{\varepsilon,M}^{2,n}}|u_n^\varepsilon(t)-v_n^\varepsilon(t)|^2>\delta).
\end{aligned} \tag{5.33}$$

Let $\tau^n_{\varepsilon,M} = \tau^{1,n}_{\varepsilon,M} \wedge \tau^{2,n}_{\varepsilon,M}$. By Itô's formula, we deduce

$$
\begin{aligned}
&|u^\varepsilon_n(t\wedge\tau^n_{\varepsilon,M}) - v^\varepsilon_n(t\wedge\tau^n_{\varepsilon,M})|^2 + 2\varepsilon\int_0^{t\wedge\tau^n_{\varepsilon,M}} ||u^\varepsilon_n(s) - v^\varepsilon_n(s)||^2 ds\\
&= -2\varepsilon\int_0^{t\wedge\tau^n_{\varepsilon,M}} (u^\varepsilon_n(s) - v^\varepsilon_n(s), \mathcal{A}v^\varepsilon_n(s))ds\\
&\qquad\qquad - 2\varepsilon\int_0^{t\wedge\tau^n_{\varepsilon,M}} (u^\varepsilon_n(s) - v^\varepsilon_n(s), g(u^\varepsilon_n(s)))ds\\
&- 2\varepsilon\int_0^{t\wedge\tau^n_{\varepsilon,M}} (b(u^\varepsilon_n(s), u^\varepsilon_n(s)), u^\varepsilon_n(s) - v^\varepsilon_n(s))ds\\
&+ 2\sqrt{\varepsilon}\int_0^{t\wedge\tau^n_{\varepsilon,M}} (u^\varepsilon_n(s) - v^\varepsilon_n(s), (G(\varepsilon s, u^\varepsilon_n(s)) - G(\varepsilon s, v^\varepsilon_n(s)))dW(s))\\
&+ \varepsilon\int_0^{t\wedge\tau^n_{\varepsilon,M}} |G(\varepsilon s, u^\varepsilon_n(s)) - G(\varepsilon s, v^\varepsilon_n(s))|^2_{L_Q} ds\\
&= -2\varepsilon\int_0^{t\wedge\tau^n_{\varepsilon,M}} (u^\varepsilon_n(s) - v^\varepsilon_n(s), \mathcal{A}v^\varepsilon_n(s))ds\\
&\qquad\qquad - 2\varepsilon\int_0^{t\wedge\tau^n_{\varepsilon,M}} (u^\varepsilon_n(s) - v^\varepsilon_n(s), g(u^\varepsilon_n(s)))ds\\
&- 2\varepsilon\int_0^{t\wedge\tau^n_{\varepsilon,M}} b(u^\varepsilon_n(s) - v^\varepsilon_n(s), u^\varepsilon_n(s), u^\varepsilon_n(s) - v^\varepsilon_n(s))ds\\
&+ 2\varepsilon\int_0^{t\wedge\tau^n_{\varepsilon,M}} b(v^\varepsilon_n(s), u^\varepsilon_n(s) - v^\varepsilon_n(s), v^\varepsilon_n(s))ds\\
&+ 2\sqrt{\varepsilon}\int_0^{t\wedge\tau^n_{\varepsilon,M}} (u^\varepsilon_n(s) - v^\varepsilon_n(s), (G(\varepsilon s, u^\varepsilon_n(s)) - G(\varepsilon s, v^\varepsilon_n(s)))dW(s))\\
&+ \varepsilon\int_0^{t\wedge\tau^n_{\varepsilon,M}} |G(\varepsilon s, u^\varepsilon_n(s)) - G(\varepsilon s, v^\varepsilon_n(s))|^2_{L_Q} ds\\
&= I_1 + I_2 + I_3 + I_4 + I_5 + I_6. \qquad (5.34)
\end{aligned}
$$

For the first term, we have

$$
\begin{aligned}
I_1 &\le 2\varepsilon\int_0^{t\wedge\tau^n_{\varepsilon,M}} |(u^\varepsilon_n(s) - v^\varepsilon_n(s), \mathcal{A}v^\varepsilon_n(s))|ds\\
&\le 2\varepsilon\int_0^{t\wedge\tau^n_{\varepsilon,M}} ||u^\varepsilon_n(s) - v^\varepsilon_n(s)|| \cdot ||v^\varepsilon_n(s)||ds. \qquad (5.35)
\end{aligned}
$$

For the second term, using Young inequality, we deduce

$$\begin{aligned}
I_2 &= -2\varepsilon \int_0^{t\wedge\tau^n_{\varepsilon,M}} (u^\varepsilon_n(s) - v^\varepsilon_n(s), g(u^\varepsilon_n(s)))ds \\
&\le -2\varepsilon \int_0^{t\wedge\tau^n_{\varepsilon,M}} |u^\varepsilon_n(s)|^{\beta+1}_{\beta+1} ds + 2\varepsilon \int_0^{t\wedge\tau^n_{\varepsilon,M}} \int_D |u^\varepsilon_n(s)|^\beta |v^\varepsilon_n(s)| dx ds \\
&\le -2\varepsilon \int_0^{t\wedge\tau^n_{\varepsilon,M}} |u^\varepsilon_n(s)|^{\beta+1}_{\beta+1} ds + 2\varepsilon \int_0^{t\wedge\tau^n_{\varepsilon,M}} (\xi_1 |u^\varepsilon_n(s)|^{\beta+1}_{\beta+1} + C(\xi_1)|v^\varepsilon_n(s)|^{\beta+1}_{\beta+1}) ds \\
&\le -2\varepsilon(1-\xi_1) \int_0^{t\wedge\tau^n_{\varepsilon,M}} |u^\varepsilon_n(s)|^{\beta+1}_{\beta+1} ds + 2\varepsilon C \int_0^{t\wedge\tau^n_{\varepsilon,M}} |v^\varepsilon_n(s)|^{\beta+1}_{\beta+1} ds. \qquad (5.36)
\end{aligned}$$

Next, we have

$$\begin{aligned}
I_3 &\le 2\varepsilon \int_0^{t\wedge\tau^n_{\varepsilon,M}} |b(u^\varepsilon_n(s) - v^\varepsilon_n(s), u^\varepsilon_n(s), u^\varepsilon_n(s) - v^\varepsilon_n(s))| ds \\
&\le c\varepsilon \int_0^{t\wedge\tau^n_{\varepsilon,M}} |u^\varepsilon_n(s) - v^\varepsilon_n(s)|^{\frac{1}{2}} \cdot ||u^\varepsilon_n(s)|| \cdot ||u^\varepsilon(s) - v^\varepsilon_n(s)||^{\frac{3}{2}} ds. \qquad (5.37)
\end{aligned}$$

Similarly, we get

$$\begin{aligned}
I_4 &\le 2\varepsilon \int_0^{t\wedge\tau^n_{\varepsilon,M}} |b(v^\varepsilon_n(s), u^\varepsilon_n(s) - v^\varepsilon_n(s), v^\varepsilon_n(s))| ds \\
&\le c\varepsilon \int_0^{t\wedge\tau^n_{\varepsilon,M}} |v^\varepsilon_n(s)|^{\frac{1}{2}} \cdot ||u^\varepsilon_n(s) - v^\varepsilon_n(s)|| \cdot ||v^\varepsilon_n(s)||^{\frac{3}{2}} ds. \qquad (5.38)
\end{aligned}$$

By (A.4), we have

$$\begin{aligned}
I_6 &= \varepsilon \int_0^{t\wedge\tau^n_{\varepsilon,M}} |G(\varepsilon s, u^\varepsilon_n(s)) - G(\varepsilon s, v^\varepsilon_n(s))|^2_{L_Q} ds \\
&\le \varepsilon K \int_0^{t\wedge\tau^n_{\varepsilon,M}} |u^\varepsilon_n(s) - v^\varepsilon_n(s)|^2 ds. \qquad (5.39)
\end{aligned}$$

Using (5.35)–(5.39) in (5.34) and choosing ξ_1 small enough, then there exists a constant c such that

$$\begin{aligned}
&|u^\varepsilon_n(t\wedge\tau^n_{\varepsilon,M}) - v^\varepsilon_n(t\wedge\tau^n_{\varepsilon,M})|^2 + 2\varepsilon \int_0^{t\wedge\tau^n_{\varepsilon,M}} ||u^\varepsilon_n(s) - v^\varepsilon_n(s)||^2 ds \\
&\le \varepsilon \int_0^{t\wedge\tau^n_{\varepsilon,M}} ||u^\varepsilon_n(s) - v^\varepsilon_n(s)||^2 ds + 4\varepsilon \int_0^{t\wedge\tau^n_{\varepsilon,M}} ||v^\varepsilon_n(s)||^2 ds \\
&+ c\varepsilon \int_0^{t\wedge\tau^n_{\varepsilon,M}} |u^\varepsilon_n(s) - v^\varepsilon_n(s)|^2 \cdot ||u^\varepsilon_n(s)||^4 ds \\
&+ 2\varepsilon C \int_0^{t\wedge\tau^n_{\varepsilon,M}} |v^\varepsilon_n(s)|^{\beta+1}_{\beta+1} ds \\
&+ c\varepsilon \int_0^{t\wedge\tau^n_{\varepsilon,M}} (|v^\varepsilon_n(s)|^2 + ||v^\varepsilon_n(s)||^6) ds + \varepsilon K \int_0^{t\wedge\tau^n_{\varepsilon,M}} |u^\varepsilon_n(s) - v^\varepsilon_n(s)|^2 ds \\
&+ 2\sqrt{\varepsilon} \left| \int_0^{t\wedge\tau^n_{\varepsilon,M}} (u^\varepsilon_n(s) - v^\varepsilon_n(s), (G(\varepsilon s, u^\varepsilon_n(s)) - G(\varepsilon s, v^\varepsilon_n(s))) dW(s)) \right|. \qquad (5.40)
\end{aligned}$$

Applying Gronwall's inequality, we get

$$
\begin{aligned}
&|u_n^\varepsilon(t\wedge\tau_{\varepsilon,M}^n)-v_n^\varepsilon(t\wedge\tau_{\varepsilon,M}^n)|^2\\
&\leq (4\varepsilon\int_0^{t\wedge\tau_{\varepsilon,M}^n}||v_n^\varepsilon(s)||^2ds+c\varepsilon\int_0^{t\wedge\tau_{\varepsilon,M}^n}(|v_n^\varepsilon(s)|^2+||v_n^\varepsilon(s)||^6)ds\\
&+2\varepsilon C\int_0^{t\wedge\tau_{\varepsilon,M}^n}|v_n^\varepsilon(s)|_{\beta+1}^{\beta+1}ds\\
&+2\sqrt{\varepsilon}|\int_0^{t\wedge\tau_{\varepsilon,M}^n}(u_n^\varepsilon(s)-v_n^\varepsilon(s),(G(\varepsilon s,u_n^\varepsilon(s))-G(\varepsilon s,v_n^\varepsilon(s)))dW(s))|)\\
&\times e^{c\int_0^{t\wedge\tau_{\varepsilon,M}^n}||u_n^\varepsilon(t)||^4ds+\varepsilon Kt}.
\end{aligned}
\tag{5.41}
$$

Similarly to the proof of (5.24), it follows (5.41) that

$$
\begin{aligned}
&(\mathbb{E}\sup_{0\leq s\leq 1\wedge\tau_{\varepsilon,M}^n}|u_n^\varepsilon(s)-v_n^\varepsilon(s)|^{2p})^{\frac{2}{p}}\\
&\leq e^{(2cM^2+2\varepsilon K)}\cdot(2c^2\varepsilon^2(\mathbb{E}(\int_0^{t\wedge\tau_{\varepsilon,M}^n}|v_n^\varepsilon(s)|^2ds)^p)^{\frac{2}{p}}\\
&\qquad+2c^2\varepsilon^2(\mathbb{E}(\int_0^{t\wedge\tau_{\varepsilon,M}^n}||v_n^\varepsilon(s)||^6ds)^p)^{\frac{2}{p}}\\
&+8C^2\varepsilon^2(\mathbb{E}(\int_0^{t\wedge\tau_{\varepsilon,M}^n}|v_n^\varepsilon(s)|_{\beta+1}^{\beta+1}ds)^p)^{\frac{2}{p}}+32\varepsilon^2(\mathbb{E}(\int_0^{t\wedge\tau_{\varepsilon,M}^n}||v_n^\varepsilon(s)||^2ds)^p)^{\frac{2}{p}}\\
&+8C\varepsilon pK^2\int_0^t(\mathbb{E}\sup_{0\leq r\leq s\wedge\tau_{\varepsilon,M}^n}|u_n^\varepsilon(r)-v_n^\varepsilon(r)|^{2p})^{\frac{2}{p}}ds)\\
&\leq e^{(2cM^2+2\varepsilon K)}\cdot(2c^2M^2\varepsilon^2+2c^2M^6\varepsilon^2+8C^2\varepsilon^2M^2+32\varepsilon^2M^2)\\
&+e^{(2cM^2+2\varepsilon K)}\cdot 8C\varepsilon pK^2\int_0^t(\mathbb{E}\sup_{0\leq r\leq s\wedge\tau_{\varepsilon,M}^n}|u_n^\varepsilon(r)-v_n^\varepsilon(r)|^{2p})^{\frac{2}{p}}ds.
\end{aligned}
\tag{5.42}
$$

Let $C_{\varepsilon,M}=e^{(2cM^2+2\varepsilon K)}\cdot 8C\varepsilon pK^2$, we get

$$
\begin{aligned}
&(\mathbb{E}\sup_{0\leq s\leq 1\wedge\tau_{\varepsilon,M}^n}|u_n^\varepsilon(s)-v_n^\varepsilon(s)|^{2p})^{\frac{2}{p}}\\
&\leq e^{(2cM^2+2\varepsilon K)}\cdot(2c^2M^2\varepsilon^2+2c^2M^6\varepsilon^2+8C^2\varepsilon^2M^2+32\varepsilon^2M^2)\cdot e^{C_{\varepsilon,M}}.
\end{aligned}
\tag{5.43}
$$

From (5.17) and Lemma 2.12 and 2.13, for any $R>0$, there exists a positive constant M such that

$$\sup_{0<\varepsilon\leq 1}\varepsilon logP((|u_n^\varepsilon|_V^H(1))^2>M)\leq -R, \tag{5.44}$$

$$\sup_{0<\varepsilon\leq 1}\varepsilon logP(\sup_{0<\varepsilon\leq 1}||u_n^\varepsilon||^2>M)\leq -R, \tag{5.45}$$

$$\sup_{0<\varepsilon\leq 1}\varepsilon logP(\sup_{0<\varepsilon\leq 1}||v_n^\varepsilon||^2>M)\leq -R, \tag{5.46}$$

$$\sup_{0<\varepsilon\leq 1}\varepsilon logP(\sup_{0<\varepsilon\leq 1}|v_n^\varepsilon|_{\beta+1}^{\beta+1}>M)\leq -R. \tag{5.47}$$

Let $p = \frac{2}{\varepsilon}$ in (5.42), we have the following inequalities

$$\begin{aligned}
&\varepsilon log P(\sup_{0\le t\le 1} |u_n^\varepsilon(t) - v_n^\varepsilon(t)|^2 > \delta, (|u_n^\varepsilon|_V^H(1))^2 \le M, \sup_{0\le t\le 1} ||u_n^\varepsilon(t)||^2 \le M, \\
&\qquad \sup_{0\le t\le 1} ||v_n^\varepsilon(t)||^2 \le M, \sup_{0\le t\le 1} |v_n^\varepsilon(t)|_{\beta+1}^{\beta+1} \le M) \\
&\le \varepsilon log P(\sup_{0\le t\le 1\wedge \tau_{\varepsilon,M}^n} |u_n^\varepsilon(t) - v_n^\varepsilon(t)|^2 > \delta) \\
&\le log(\mathbb{E} \sup_{0\le t\le 1\wedge \tau_{\varepsilon,M}^n} |u_n^\varepsilon(t) - v_n^\varepsilon(t)|^{2p})^{\frac{2}{p}} - log\delta^2 \\
&\le 2cM^2 + 2\varepsilon K + log(2c^2M^2\varepsilon^2 + 2c^2M^6\varepsilon^2 + 8C^2\varepsilon^2M^2 + 32\varepsilon^2M^2) \\
&+ C_{\varepsilon,M} - log\delta^2 \\
&\to -\infty, \quad as\ \varepsilon \to 0. \qquad (5.48)
\end{aligned}$$

For any M, (5.48) implies that there exists a positive number ε_0, such that for any $\varepsilon < \varepsilon_0$,

$$\begin{aligned}
&P(\sup_{0\le t\le 1} |u_n^\varepsilon(t) - v_n^\varepsilon(t)|^2 > \delta, (|u_n^\varepsilon|_V^H(1))^2 \le M, \sup_{0\le t\le 1} ||u_n^\varepsilon(t)||^2 \le M, \\
&\sup_{0\le t\le 1} ||v_n^\varepsilon(t)||^2 \le M, \sup_{0\le t\le 1} |v_n^\varepsilon(t)|_{\beta+1}^{\beta+1} \le M) \le e^{-\frac{R}{\varepsilon}}. \qquad (5.49)
\end{aligned}$$

Combining (5.44)–(5.47) and (5.49) together, we can get a positive number ε_0, such that for $\varepsilon \le \varepsilon_0$,

$$P(\sup_{0\le t\le 1} |u_n^\varepsilon(t) - v_n^\varepsilon(t)|^2 > \delta) \le 3e^{-\frac{R}{\varepsilon}}.$$

Since R is arbitrary, this completes the proof of Lemma 2.16.

Proof of (5.6) By Lemma 2.14 and 2.15, we get that for any $R > 0$, there exists an N_0 which satisfies

$$P(\sup_{0\le t\le 1} |u^\varepsilon(t) - u_{N_0}^\varepsilon(t)|^2 > \delta) \le e^{-\frac{R}{\varepsilon}}, \quad for\ any\ \varepsilon \in (0,1], \qquad (5.50)$$

$$P(\sup_{0\le t\le 1} |v^\varepsilon(t) - v_{N_0}^\varepsilon(t)|^2 > \delta) \le e^{-\frac{R}{\varepsilon}}, \quad for\ any\ \varepsilon \in (0,1]. \qquad (5.51)$$

By virtue of Lemma 2.16, there exists ε_0 such that for any $\varepsilon \in (0, \varepsilon_0]$,

$$P(\sup_{0\le t\le 1} |u_{N_0}^\varepsilon(t) - v_{N_0}^\varepsilon(t)|^2 > \delta) \le e^{-\frac{R}{\varepsilon}}. \qquad (5.52)$$

Hence, for any $\varepsilon \in (0, \varepsilon_0]$,

$$P(\sup_{0\le t\le 1} |u^\varepsilon(t) - v^\varepsilon(t)|^2 > \delta) \le 3e^{-\frac{R}{\varepsilon}}. \qquad (5.53)$$

Since R is arbitrary, we deduce

$$\lim_{\varepsilon\to 0} \varepsilon log P(\sup_{0\le t\le 1} |u^\varepsilon(t) - v^\varepsilon(t)|^2 > \delta) = -\infty.$$

This completes the proof of Theorem 2.4.

References

[1] M. T. Barlow and M. Yor, Semimartingale inequalities via the Garsia–Rodemich–Rumsey lemma, and applications to local time, J. Funct. Anal., 49(1982), 198-229.

[2] H. Bessaih, Inviscid limit of stochastic damped 2D Navier–Stokes equations, Nonlinearity, 27(2014), 1-15.

[3] H. Bessaih, Martingale solutions for stochastic Euler equations, Stochastic Anal. Appl., 17(1999), 713-725.

[4] Z. Brzeźniak and E. Motyl, Existence of a martingale solution of the stochastic Navier–Stokes equations in unbounded 2D and 3D domains, J. Differential Equations, 254(2013), 1627-1685.

[5] Z. Brzeźniak and S. Peszat, Stochastic two-dimensional Euler equations, Ann. Probab., 29(2001), 1796-1832.

[6] X. Cai and Q. Jiu, Weak and strong solutions for the incompressible Navier–Stokes equations with damping, J. Math. Anal. Appl., 343(2008), 799-809.

[7] M. Capiński and N. Cutland, Stochastic Euler equations on the torus, Ann. Appl. Probab., 9(1998), 688-705.

[8] G. Da Prato and J. Zabczyk, Ergodicity for infinite dimensional systems, London Math. Soc. Lecture Notes Ser. vol. 229, Cambridge University press, 1996.

[9] G. Da Prato and J. Zabczyk, Stochastic equations in infinite dimensions, Encyclopedia of Mathematics and its applications, Cambridge University press, 1992.

[10] B. Davis, On the L^p-norms of stochastic integrals and other martingale, Duke Math. J., 43(1976), 697-704.

[11] A. Dembo and A. Zeitouni, Large Deviations Techniques and Applications, Springer–Verlag, Berlin, Heidelberg, 1998.

[12] Z. Dong and J. L. Zhai, Martingale solutions and Markov selection of stochastic 3D Navier–Stokes equations with jump, J. Differential Equations, 250(2011), 2737-2778.

[13] F. Flandoli and D. Gatarek, Martingale and stationary solutions for stochastic Navier–Stokes equations, Probab. Theory Related Fields, 102(1995), 367-391.

[14] F. Flandoli and M. Romito, Markov selections for 3D stochastic Navier–Stokes equations, Probab. Theory Related Fields, 140(2008), 407-458.

[15] B. Goldys, M. Rockner and X. C. Zhang, Martingale solutions and Markov selections for stochastic partial differential equations, Stochastic Process. Appl., 119(2009), 1725-1764.

[16] S. M. Griffies, Fundamentals of Ocean Climate Models, Princeton University Press, Princeton, NJ, 2004.

[17] K. Holly and M. Wiciak, Compactness method applied to an abstract nonlinear parabolic equation, Selected Problems of Mathematics, Cracow University of Technology, 1995, 95-160.

[18] L. Hsiao, Quasilinear Hyperbolic Systems and Dissipative Mechanisms, World Scientific, 1997.

[19] Y. Jia, X. Zhang and B. Q. Dong, The asymptotic behavior of solutions to three-dimensional Navier–Stokes equations with nonlinear damping, Nonlinear Anal. Real World Appl., 12(2011), 1736-1747.

[20] W. Liu and M. Röckner, Local and global well-posedness of SPDE with generalized coercivity conditions, J. Differential Equations, 254(2013), 725-755.

[21] M. Métivier, Stochastic partial differential equations in infinite dimensional spaces, Scuola Normale Superiore. Pisa., 1988.

[22] M. Métivier, Semimartingales, Gruyter, 1982.

[23] R. Mikulevicius and B. L. Rozovskii, Global L_2-solutions of stochastic Navier–Stokes equations, Ann. Probab., 33(2005), 137-176.

[24] H. B. de Oliveira, Existence of weak solutions for the generalized Navier–Stokes equations with damping, NoDEA Nonlinear Differential Equations Appl., 20(2013), 797-824.

[25] M. Röckner, B. Schmuland and X. Zhang, Yamada–Watanabe theorem for stochastic evolution equations in infinite dimensions, Condensed Matter Physics, 11(2008), 247-259.

[26] M. Röckner and T. S. Zhang, Stochastic 3D tamed Navier–Stokes equations: existence, uniqueness and small time large deviation principles, J. Differential Equations, 252(2012), 716-744.

[27] M. Röckner, T. S. Zhang and X. C. Zhang, Large deviations for stochastic tamed 3D Navier–Stokes equations, Appl. Math. Optim., 61(2010), 267-285.

[28] M. Röckner and X. Zhang, Stochastic tamed 3D Navier–Stokes equation: existence, uniqueness and ergodicity, Probab. Theory Related Fields, 145(2009), 211-267.

[29] X. Song and Y. Hou, Attractors for the three-dimensional incompressible Navier–Stokes equations with damping, Discrete Contin. Dyn. Syst., 31(2011), 239-252.

[30] X. Song and Y. Hou, Uniform attractors for three-dimensional Navier–Stokes equations with nonlinear damping, J. Math. Anal. Appl., 31(2015), 337-351.

[31] R. Temam, Navier–Stokes Equations, Theory and Numerical Analysis, North-Holland Publishing Company, Amsterdam, New York, Oxford, 1979.

[32] H. Touchette, The large deviation approach to statistical mechanics, Phys. Rep., 478(2009), 1-69.

[33] W. Wang and G. Zhou, Remarks on the Regularity Criterion of the Navier–Stokes Equations with Nonlinear Damping, Math. Probl. Eng., 2015(2015), Article ID 310934, 5 pages.

[34] T. G. Xu and T. S. Zhang, On the small time asymptotics of the two-dimensional stochastic Navier–Stokes equations, Ann. Inst. Henri poincaré Probab. Stat., 45(2009), 1002-1019.

[35] Y. Zhou, Regularity and uniqueness for the 3D incompressible Navier–Stokes equations with damping, Appl. Math. Lett., 25(2015), 1822-1825.

Chapter 3

Model Reduction in Stochastic Environments

E. Forgoston[1,*], L. Billings[2] and I. B. Schwartz[3]

[1]*Department of Mathematical Sciences, Montclair State University, Montclair, NJ 07043, USA*
[*]*eric.forgoston@montclair.edu*

[2]*Department of Mathematical Sciences, Montclair State University, Montclair, NJ 07043, USA*

[3]*Nonlinear Systems Dynamics Section, Plasma Physics Division, Code 6792, U.S. Naval Research Laboratory, Washington, DC 20375, USA*

We present a general theory of stochastic model reduction which is based on a normal form coordinate transform method of A. J. Roberts. This nonlinear, stochastic projection allows for the deterministic and stochastic dynamics to interact correctly on the lower-dimensional manifold so that the dynamics predicted by the reduced, stochastic system agrees well with the dynamics predicted by the original, high-dimensional stochastic system. The method may be applied to any system with well-separated time scales. In this article, we consider a physical problem that involves a singularly perturbed Duffing oscillator as well as a biological problem that involves the prediction of infectious disease outbreaks.

1. Introduction

It is well-known that noise can have a significant effect on deterministic dynamical systems. As an example, given an initial state starting in a basin of attraction, noise can cause the initial state to cross the basin boundary and move into another, distinct basin of attraction [1–5]. Many researchers have investigated how noise affects physical and biological phenomena at a wide variety of levels including switching between the magnetisation states in magnets [6], and voltage and current states in Josephson junctions [7], sub-cellular processes, tissue dynamics, large-scale population dynamics [8], and genetic switching [9].

Stochasticity manifests itself as either external or internal noise. In this article, we shall consider only external noise, which comes from a source outside the system being considered (e.g., population growth under the influence of climatic effects, or a random signal fed into a transmission line), and often is modeled by replacing an external parameter with a random process. Mathematically, the effect of external

[*]Corresponding author.

noise is often described using a Langevin equation or the associated Fokker–Planck equation (though the dynamics of external noise may sometimes be described by a master equation [10]).

The reduction in high-dimensional systems is an important and fundamental problem in nonlinear dynamical systems. Moreover, normal form coordinate transforms provide a way to simplify multiscale nonlinear dynamics by separating the long-term dynamics of interest from the insignificant dynamics [11]. In this article, we present a general theory of stochastic model reduction that is based on a normal form coordinate transform method of A. J. Roberts. This nonlinear, stochastic projection allows for the deterministic and stochastic dynamics to interact correctly on the lower-dimensional manifold so that exactly transformed dynamics reproduce fully the original dynamics. The method may be applied to any system with well-separated time scales. Here, we consider a physical problem that is associated with the use of unmanned sensors operating in the ocean as well as a biological problem that involves the prediction of infectious disease outbreaks.

Section 2 provides an overview of the theory including the use of center manifold theory for deterministic problems (Sec. 2.1) and the normal form coordinate transform for stochastic problems (Sec. 2.2). Sections 3 and 4 provide two examples — the first involves a singularly perturbed stochastic Duffing oscillator, while the second involves a stochastic Susceptible-Exposed-Infectious-Recovered (SEIR) epidemic model. In the first example, we show that one can use deterministic theory as the noise effects occur at such high order so that the stochastic correction is negligible. The reduced system is used to understand a variety of system behaviour including the optimal escape path and escape rate. In the second example, we show how one must use the stochastic theory to obtain long-time predictions of disease outbreak. Conclusions are found in Sec. 5.

2. General theory

We consider the following general $(m+n)$-dimensional system of Stratonovich stochastic differential equations

$$\dot{\mathbf{x}} = \mathbf{A}\mathbf{x} + \mathbf{F}(\mathbf{x}, \mathbf{y}, \boldsymbol{\Phi}), \tag{2.1a}$$

$$\dot{\mathbf{y}} = \mathbf{B}\mathbf{y} + \mathbf{G}(\mathbf{x}, \mathbf{y}, \boldsymbol{\Psi}), \tag{2.1b}$$

where $\mathbf{x}(t) \in \mathbb{R}^m$, $\mathbf{y}(t) \in \mathbb{R}^n$, $\boldsymbol{\Phi}(t)$ and $\boldsymbol{\Psi}(t)$ describe stochastic forces with adjustable intensity, $\mathbf{A}$ and $\mathbf{B}$ are constant matrices, and $\mathbf{F}$ and $\mathbf{G}$ are stochastic, nonlinear functions. Stochastic singular perturbation systems have been explored by Berglund and Gentz [12, 13] and have the form

$$\dot{\mathbf{x}} = \mathbf{A}\mathbf{x} + \mathbf{F}(\mathbf{x}, \mathbf{y}, \boldsymbol{\Phi}), \tag{2.2a}$$

$$\epsilon\dot{\mathbf{y}} = \mathbf{B}\mathbf{y} + \mathbf{G}(\mathbf{x}, \mathbf{y}, \boldsymbol{\Psi}), \tag{2.2b}$$

where ϵ is a small parameter.

2.1. *Deterministic center manifold*

To begin, we remove the stochastic terms from Eqs. (2.1a)–(2.1b) so that $\mathbf{F} = \mathbf{F}(\mathbf{x}, \mathbf{y})$ and $\mathbf{G} = \mathbf{G}(\mathbf{x}, \mathbf{y})$. A general nonlinear system may be transformed so that the system's linear part has a block diagonal form consisting of three matrix blocks. The first matrix block will possess eigenvalues with positive real part; the second matrix block will possess eigenvalues with negative real part; and the third matrix block will possess eigenvalues with zero real part. These three matrix blocks are respectively associated with the unstable eigenspace, the stable eigenspace, and the center eigenspace. If there are no eigenvalues with positive real part, then the orbits will rapidly decay to the center eigenspace.

It is often the case that a system of equations can not be written in a block diagonal form with one matrix block possessing eigenvalues with negative real part and the other matrix block possessing eigenvalues with zero real part. Even though it is possible to construct a center manifold from a system not in separated block form [14], it is much easier to apply the center manifold theory to a system with separated stable and center directions. Therefore, we generally transform the original system of equations to a new system of equations that will have the eigenvalue structure that is needed to apply center manifold theory. The theory allows one to find an invariant center manifold that passes through a fixed point and to which one can restrict the new transformed system.

To make the ideas more concrete, consider the singularly perturbed system given by Eqs. (2.2a)–(2.2b), and let $t = \epsilon\tau$. Denoting $\dot{}$ as d/dt and $'$ as $d/d\tau$, then the deterministic form of Eqs. (2.2a)–(2.2b) is transformed to the following system of equations:

$$\mathbf{x}' = \epsilon\left(\mathbf{A}\mathbf{x} + \mathbf{F}(\mathbf{x}, \mathbf{y})\right), \tag{2.3a}$$

$$\mathbf{y}' = \mathbf{B}\mathbf{y} + \mathbf{G}(\mathbf{x}, \mathbf{y}), \tag{2.3b}$$

$$\epsilon' = 0. \tag{2.3c}$$

We recast the problem by treating ϵ as a state variable, and we let $\bar{\mathbf{A}} = \epsilon\mathbf{A}$ and $\bar{\mathbf{F}} = \epsilon\mathbf{F}$. Equations (2.3a)–(2.3c) can thus be rewritten as

$$\mathbf{x}' = \bar{\mathbf{A}}\mathbf{x} + \bar{\mathbf{F}}(\mathbf{x}, \mathbf{y}, \epsilon), \tag{2.4a}$$

$$\mathbf{y}' = \mathbf{B}\mathbf{y} + \mathbf{G}(\mathbf{x}, \mathbf{y}), \tag{2.4b}$$

$$\epsilon' = 0. \tag{2.4c}$$

If $\bar{\mathbf{A}}$ and $\mathbf{B}$ are constant matrices such that all of the eigenvalues of $\bar{\mathbf{A}}$ have zero real parts, while all of the eigenvalues of $\mathbf{B}$ have negative real parts, then the system will rapidly collapse onto a lower-dimensional manifold given by center manifold theory [15].

If the center manifold is given by

$$\mathbf{y} = \mathbf{h}(\mathbf{x}, \epsilon), \tag{2.5}$$

then substitution of Eq. (2.5) into Eq. (2.4b) leads to the following center manifold condition:

$$\mathbf{h}_{\mathbf{x}} \left(\bar{\mathbf{A}}\mathbf{x} + \bar{\mathbf{F}}(\mathbf{x}, \mathbf{h}(\mathbf{x}, \epsilon), \epsilon) \right) = \mathbf{B}\mathbf{h}(\mathbf{x}, \epsilon) + \mathbf{G}(\mathbf{x}, \mathbf{h}(\mathbf{x}, \epsilon)), \tag{2.6}$$

where $\mathbf{h}_{\mathbf{x}}$ denotes the partial derivative of $\mathbf{h}$ with respect to $\mathbf{x}$. Although it is generally not possible to solve Eq. (2.6) for $\mathbf{h}$, one can approximate the center manifold by expanding $\mathbf{h}$ in the following way:

$$\mathbf{h}(\mathbf{x}, \epsilon) = \mathbf{h}_0(\mathbf{x}) + \epsilon\mathbf{h}_1(\mathbf{x}) + \epsilon^2\mathbf{h}_2(\mathbf{x}) + \mathcal{O}(\epsilon^3). \tag{2.7}$$

Typically, this approximation of $\mathbf{h}(\mathbf{x}, \epsilon)$ is found by substituting Eq. (2.7) into the center manifold condition (Eq. (2.6)) and matching coefficients.

2.2. *Stochastic center manifold and the normal form coordinate transform*

In a manner similar to that shown in Sec. 2.1, the singularly perturbed stochastic system given by Eqs. (2.2a)–(2.2b) can be transformed [12] to the form given by Eqs. (2.4a)–(2.4c), where now $\bar{\mathbf{F}} = \bar{\mathbf{F}}(\mathbf{x}, \mathbf{y}, \epsilon, \mathbf{\Phi})$ and $\mathbf{G} = \mathbf{G}(\mathbf{x}, \mathbf{y}, \mathbf{\Psi})$. If $\bar{\mathbf{A}}$ and $\mathbf{B}$ satisfy the same spectral conditions as for the deterministic system, and if the stochastic time dependence found in $\bar{\mathbf{F}}$ and $\mathbf{G}$ is due to independent white noise processes, then there exists a stochastic center manifold for the original stochastic system [16].

One method for computing the stochastic center manifold for systems with both fast and slow dynamics uses the construction of a normal form coordinate transform that not only reduces the dimension of the dynamics, but also separates all of the fast processes from all of the slow processes [11]. While this type of normal form coordinate transform may be used to find deterministic center manifolds, the application of this transform to stochastic systems is particularly interesting since white noise has fluctuations on all scales.

There are many publications [17–20] which deal with the simplification of a stochastic dynamical system using a stochastic normal form transformation. In these articles, the noise term is multiplied by a small parameter, and therefore, the resulting stochastic normal form is a perturbation of the deterministic normal form. Furthermore, one can find in Coullet et al., [18] and Namachchivaya et al., [20] normal form transformations that involve anticipative noise processes. However, these integrals of the noise process into the future were not dealt with rigorously.

Rigorous, theoretical analysis to support normal form coordinate transforms (and center manifold reduction) was developed by Arnold and Imkeller [21] and Arnold [22]. In this work, the technical problem of the anticipative noise integrals also was dealt with rigorously. Later, another stochastic normal form transformation was developed by Roberts [11]. This new method is such that "anticipation can ... always [be] removed from the slow modes with the result that no anticipation is required after the fast transients decay" (Roberts [11], pp. 13). An advantage of removing anticipation is the simplification of the normal form. Nonetheless, this simpler normal form retains its accuracy with the original stochastic system. Furthermore, when modeling the macroscopic behaviour of microscopic, stochastic systems, it is desirable to avoid anticipation in the normal form [11]. It is important to note that the normal form is valid for all time since it is just a coordinate transform. Furthermore, the dynamics also are valid for all time as long as the truncation error is small enough for the problem of interest.

In the examples presented in Sec. 3 and Sec. 4, we shall use the method of Roberts [11] to simplify our stochastic dynamical system to one that emulates the long-term dynamics of the original, multiple-time-scale system. The method involves five principles, which we recapitulate here for the purpose of clarity. The principles are as follows:

(1) Avoid unbounded, secular terms in both the transformation and the evolution equations to ensure a uniform asymptotic approximation.
(2) Decouple all of the slow processes from the fast processes to ensure a valid long-term model.
(3) Insist that the stochastic slow manifold is precisely the transformed fast processes coordinate being equal to zero.
(4) To simplify matters, eliminate as many as possible of the terms in the evolution equations.
(5) Try to remove all fast processes from the slow processes by avoiding as much as possible the fast time memory integrals in the evolution equations.

In practice, the original stochastic system of equations (which satisfy the necessary spectral requirements) in $(\mathbf{x}, \mathbf{y})$ coordinates is transformed to a new $(\mathbf{X}, \mathbf{Y})$ coordinate system using a stochastic coordinate transform as follows:

$$\mathbf{x} = \mathbf{X} + \boldsymbol{\xi}(\mathbf{X}, \mathbf{Y}, t), \tag{2.8a}$$

$$\mathbf{y} = \mathbf{Y} + \boldsymbol{\eta}(\mathbf{X}, \mathbf{Y}, t), \tag{2.8b}$$

where the specific form of Eqs. (2.8a) and (2.8b) is chosen to simplify the original system according to the five principles listed previously. The terms $\boldsymbol{\xi}(\mathbf{X}, \mathbf{Y}, t)$ and $\boldsymbol{\eta}(\mathbf{X}, \mathbf{Y}, t)$ are found using an iterative procedure that will be explicitly demonstrated using the first example (Sec. 3) of a singularly perturbed, damped, stochastic Duffing oscillator model. Theoretical details of the normal form coordinate transform process can be found in Roberts [11].

3. Example 1: singularly perturbed stochastic Duffing oscillator

An important application in many fields is that of sensing in stochastic environments. Improved environmental sensing and prediction can be achieved through the incorporation of continuous monitoring of the region of interest. For example, one could monitor the stochastic ocean using autonomous underwater gliders [23–25]. However, to do this, one must understand both the dynamics and control of the gliders.

Extending the lifetime (energy optimisation problem) of sensing devices (e.g., gliders) in stochastic environments such as the ocean requires an understanding of the effect of the environmental forces on both the devices and the region being monitored. The ocean dynamics are high-dimensional and stochastic. Therefore, as a first step towards using the underlying ocean structure to optimize a sensor's energy usage, we will use the two methods described in Sec. 2 to obtain a reduction in the dimension of the stochastic system. The reduced system can then be used to understand a variety of system behaviour including the optimal escape path and the escape rate.

We consider the following singularly perturbed, damped, Duffing oscillator system with additive noise:

$$\dot{x} = y + \sqrt{2D}\phi(t), \tag{3.1a}$$

$$\epsilon\dot{y} = (x - x^3 - y), \tag{3.1b}$$

where D is the noise intensity and $\phi(t)$ describes a stochastic white force that is characterized by the following correlation functions:

$$\langle \phi_i(t) \rangle = 0, \tag{3.2a}$$

$$\langle \phi_i(t)\phi_j(t') \rangle = \delta(t - t')\delta_{ij}. \tag{3.2b}$$

In this example, the noise is added only to the x equation. Additionally, one could consider two other scenarios. In the first scenario, noise is added to both the x and y equations, while in the second scenario, noise is added only to the y equation. To implement the first scenario, one adds $\sqrt{2D}\phi_1(t)$ to the right-hand side of the x equation and $\sqrt{2D}\phi_2(t)$ to the right-hand side of the y equation (where $\phi_1(t)$ and $\phi_2(t)$ describe stochastic white forces of intensity D that are characterized by the correlation functions given in Eqs. (3.2a)–(3.2b)). To implement the second scenario, noise is added only to the y equation. However, unlike the example (Eqs. (3.1a)–(3.1b)) and the first scenario, in this case the noise term must be scaled by $\sqrt{\epsilon}$ and the potential function must be scaled by ϵ [26]. Although the following results pertain to Eqs. (3.1a)–(3.1b), we have checked that one obtains similar results for the other two scenarios. The system given by Eqs. (3.1a)–(3.1b) is very strongly damped when $\epsilon \ll 1$. For the case of an under-damped system, one should consider the dynamics in the limit of weak damping [27].

3.1. *Deterministic center manifold*

Following the general theory of Sec. 2.1, we consider the deterministic form of Eqs. (3.1a)–(3.1b) by setting $\phi(t) = 0$. The slow manifold is found by setting $\epsilon = 0$

in Eq. (3.1b). Solving for y gives the equation of the slow manifold as $y = x - x^3$ (which corresponds to $h_0(x)$ in Eq. (2.7)). Substitution of this into the deterministic form of Eq. (3.1a) gives the dynamics along the slow manifold as $\dot{x} = x - x^3$.

If, as in Sec. 2.1, we let $t = \epsilon\tau$ and denote $\dot{}$ as d/dt and $'$ as $d/d\tau$, then Eqs. (3.1a)–(3.1b) (with $\phi(t) = 0$) are transformed to the following system:

$$x' = \epsilon y, \tag{3.3a}$$

$$y' = x - x^3 - y, \tag{3.3b}$$

$$\epsilon' = 0. \tag{3.3c}$$

Rearrangement of Eqs. (3.3a)–(3.3c) leads to a system described by constant matrices $\bar{\mathbf{A}}$ and $\mathbf{B}$ that satisfy the spectral requirements of Sec. 2.1. Furthermore, since the x and ϵ variables are associated with the $\bar{\mathbf{A}}$ matrix (eigenvalues with zero real parts), and the y variable is associated with the $\mathbf{B}$ matrix (eigenvalues with negative real parts), we know that the center manifold is given by $y = h(x, \epsilon)$.

The center manifold condition is given by Eq. (2.6), and we approximate the center manifold (Eq. (2.7)) as follows:

$$\begin{aligned} h(x,\epsilon) &= h_0(x) + \epsilon h_1(x) + \epsilon^2 h_2(x) + \mathcal{O}(\epsilon^3) && (3.4a)\\ &= c_0 + c_{01}\epsilon + c_{10}x + c_{02}\epsilon^2 + c_{11}x\epsilon + c_{20}x^2 \\ &\quad + c_{03}\epsilon^3 + c_{12}x\epsilon^2 + c_{21}x^2\epsilon + c_{30}x^3 + \mathcal{O}(\gamma^4), && (3.4b) \end{aligned}$$

where c_0, c_{01}, c_{10}, c_{02}, ... are unknown coefficients, and $\gamma = |(x, \epsilon)|$ so that γ provides a count of the number of x and ϵ factors in any one term. The center manifold condition for this example is given by

$$\frac{\partial h(x,\epsilon)}{\partial x}(\epsilon h(x,\epsilon)) = -h(x,\epsilon) + x - x^3. \tag{3.5}$$

By substituting Eq. (3.4b) into Eq. (3.5) and matching the different orders to find the coefficients, one finds the following center manifold equation (expanded to sixth-order):

$$\begin{aligned} h(x,\epsilon) &= x - \epsilon x + 2\epsilon^2 x - x^3 - 5\epsilon^3 x + 4\epsilon x^3 + 14\epsilon^4 x \\ &\quad - 20\epsilon^2 x^3 - 42\epsilon^5 x + 104\epsilon^3 x^3 - 3\epsilon x^5 + \mathcal{O}(\gamma^7) && (3.6a)\\ &= x - x^3 + \epsilon(-x + 4x^3 - 3x^5) + \epsilon^2(2x - 20x^3) \\ &\quad + \epsilon^3(-5x + 104x^3) + \epsilon^4(14x) + \epsilon^5(-42x) + \mathcal{O}(\gamma^7). && (3.6b) \end{aligned}$$

Note that by letting $\epsilon = 0$, one recovers the zero-order approximation, $h_0(x)$ (the slow manifold). In addition, since ϵ is now a state variable, the first nontrivial correction term to the zero-order approximation is a quadratic term.

3.2. *Stochastic center manifold and the normal form coordinate transform*

To more accurately describe the stochastic effects, we will derive the normal form coordinate transform (and thus the stochastic center manifold) for the singularly perturbed, stochastic Duffing system given by Eqs. (3.1a)–(3.1b). As demonstrated previously, use of the $t = \epsilon\tau$ transformation leads to the following system:

$$x' = \epsilon(y + \sqrt{2D}\phi) = \epsilon(y + \sigma\phi), \tag{3.7a}$$

$$y' = x - x^3 - y, \tag{3.7b}$$

$$\epsilon' = 0, \tag{3.7c}$$

where σ is the standard deviation of the noise intensity $D = \sigma^2/2$.

The construction of the normal form is quite tedious and complicated (although it is possible to derive the normal form using a computer algebra system). However, the result allows one to determine if there are any noise terms that cause a significant difference between the average stochastic center manifold (the stochastic center manifold generally fluctuates about an average location) and the deterministic center manifold.

For this problem, it turns out that the noise terms that could lead to a difference between the deterministic and average stochastic center manifolds occur at very high order in the normal form expansion. Therefore, the correction to the deterministic center manifold is minimal, and we expect that one can use the deterministic center manifold results of Sec. 3.1 to accurately solve the problem of interest. In our case, we are interested in understanding the optimal escape path and the escape rate, and we expect that results found using the deterministic center manifold reduction will agree very well with numerical computations using the original stochastic system (Eqs. (3.1a)–(3.1b)).

We proceed by showing how to use the method of Roberts [11] described in Sec. 2.2 to construct a normal form coordinate transform that separates the slow and fast dynamics of Eqs. (3.7a)–(3.7b). In what follows, we outline the steps involved in the first iteration, while details regarding the higher iterations can be found in Forgoston and Schwartz [5].

3.2.1. *First iteration*

We begin by letting

$$x \approx X, \tag{3.8a}$$

$$X' \approx 0, \tag{3.8b}$$

and by finding a change to the y coordinate (fast process) with the form

$$y = Y + \eta(\tau, X, Y) + \dots, \tag{3.9a}$$

$$Y' = -Y + G(\tau, X, Y) + \dots, \tag{3.9b}$$

where η and G are small corrections to the coordinate transform and the corresponding evolution equation. Substitution of Eqs. (3.8a)–(3.9b) into Eq. (3.7b) gives the following equation:

$$Y' + \frac{\partial \eta}{\partial \tau} + \frac{\partial \eta}{\partial X}\frac{\partial X}{\partial \tau} + \frac{\partial \eta}{\partial Y}\frac{\partial Y}{\partial \tau} = -Y - \eta + X - X^3. \tag{3.10}$$

Replacing $Y' = \partial Y/\partial \tau$ with $-Y + G$ (Eq. (3.9b)), noting that $\partial X/\partial \tau = 0$ (Eq. (3.8b)), and ignoring the term $\partial \eta/\partial Y \cdot G$ since it is a product of small corrections leads to the following:

$$G + \frac{\partial \eta}{\partial \tau} - Y\frac{\partial \eta}{\partial Y} + \eta = X - X^3. \tag{3.11}$$

Equation (3.11) must now be solved for G and η. In order to keep the evolution equation (Eq. (3.9b)) as simple as possible (principle (4) of Sec. 2.2), we let $G = 0$, which means that the coordinate transform (Eq. (3.9a)) is modified by $\eta = X - X^3$. Therefore, the new approximation of the coordinate transform and its dynamics are given by

$$y = Y + X - X^3 + \mathcal{O}(\zeta^2), \tag{3.12a}$$

$$Y' = -Y + \mathcal{O}(\zeta^2), \tag{3.12b}$$

where $\zeta = |(X, Y, \epsilon, \sigma)|$ so that ζ provides a count of the number of X, Y, ϵ, and σ factors in any one term.

3.2.2. *Higher iterations*

The construction of the normal form continues by seeking corrections, ξ and F, to the x coordinate transform and the X evolution using the updated residual of the x equation (Eq. (3.7a)), and by seeking corrections, η and G, to the y coordinate transform and the Y evolution equation using the updated residual of the y equation (Eq. (3.7b)). Details regarding the second iteration can be found in Forgoston and Schwartz [5].

The derivation of ξ and F in the second and fourth iterations along with the derivation of η and G in the third iteration leads to the following updated approximation of the coordinate transforms and their corresponding evolution equations:

$$\begin{aligned} y =& Y + X - X^3 + \epsilon\left(-X + 4X^3 - 3X^5\right) \\ &+ \epsilon\sigma\left(-e^{-\tau} * \phi + 3X^2 e^{-\tau} * \phi\right) + 3\epsilon^2 XY^2 + \mathcal{O}(\zeta^3), \end{aligned} \tag{3.13a}$$

$$Y' = -Y + \epsilon\left(-Y + 3X^2 Y\right) + \mathcal{O}(\zeta^3), \tag{3.13b}$$

$$\begin{aligned} x =& X - \epsilon Y + \epsilon^2\left(Y - 3X^2 Y\right) \\ &+ \epsilon^2\sigma\left(e^{-\tau} * \phi - 3X^2 e^{-\tau} * \phi\right) + \mathcal{O}(\zeta^4), \end{aligned} \tag{3.13c}$$

$$\begin{aligned} X' =& \epsilon\left(X - X^3\right) + \epsilon\sigma\phi + \epsilon^2\left(-X + 4X^3 - 3X^5\right) \\ &+ \epsilon^2\sigma\left(-\phi + 3X^2\phi\right) + \mathcal{O}(\zeta^4), \end{aligned} \tag{3.13d}$$

where

$$e^{-\tau} * \phi = \int_{-\infty}^{\tau} \exp[-(\tau - s)]\phi(s)\, ds. \tag{3.14}$$

Details regarding the derivation of Eqs. (3.13a)–(3.13d) can be found in Forgoston and Schwartz [5].

One can continue this iterative procedure to obtain higher order terms in the expansions of the coordinate transform and normal form. For the stochastic Duffing system under consideration, the fifth and sixth iterations lead to updated approximations of the x and y coordinate transforms (along with their associated evolution equations) that are extremely long and complicated. These approximations can be found in Forgoston and Schwartz [5].

In the higher order transform found after six iterations, one can see the appearance of quadratic noise terms. For example, one can see terms of the form $e^{-\tau} * (e^{-\tau} * \phi)^2$ in the coordinate transforms, and one can see terms of the form $\phi e^{-\tau} * \phi$ in one of the evolution equations. This quadratic noise is important because it leads to the creation of a deterministic drift within the slow dynamics [11, 20]. Furthermore, the stochastic center manifold generally undergoes fluctuations about a mean or average location. This average stochastic center manifold is usually different from the deterministic center manifold, and it is the quadratic noise process that generates this difference.

3.2.3. *Comparison with deterministic center manifold and effect of quadratic noise*

Letting $Y = 0$ and $\sigma = 0$ in the higher order transform found after six iterations leads to the following deterministic center manifold equation:

$$x = X, \tag{3.15a}$$

$$\begin{aligned} y =& X - X^3 + \epsilon(-X + 4x^3 - 3X^5) + \epsilon^2(2X - 20X^3 + 42X^5) \\ &+ \epsilon^3(-X + 16X^3 - 66X^5 + 96X^7 - 45X^9) + \mathcal{O}(\epsilon^3). \end{aligned} \tag{3.15b}$$

Comparison of Eqs. (3.15a) and (3.15b) with Eq. (3.6b) shows agreement through the $\mathcal{O}(\epsilon^2)$ terms. There appears to be a discrepancy at order $\mathcal{O}(\epsilon^3)$. However, we have checked that this apparent discrepancy is resolved by expanding the stochastic normal form coordinate transform to even higher order. For example, the seventh iteration will yield a $-4\epsilon^3 X$ term in the y coordinate transform. When added to the existing $-\epsilon^3 X$ term, there is an agreement with the $-5\epsilon^3 x$ term in Eq. (3.6b).

Letting only $Y = 0$ in the higher order set of equations leads to the stochastic center manifold equation. If one takes the expectation of this stochastic center manifold equation and uses the following identities [11]:

$$E[e^{\pm\tau} * \phi] = e^{\pm\tau} * E[\phi], \tag{3.16}$$

$$E[(e^{\pm\tau} * \phi)^2] = \frac{1}{2}, \tag{3.17}$$

then one obtains the following:

$$\begin{aligned} E[y] =& X - X^3 + \epsilon(-X + 4X^3 - 3X^5) + \epsilon^2(2X - 20X^3 + 42X^5) \\ &+ \epsilon^3(-X + 16X^3 - 66X^5 + 96X^7 - 45X^9) \\ &+ \epsilon^4\sigma^2(-3X/2 + 9X^3 - 27X^5/2), \end{aligned} \tag{3.18}$$

where the $\mathcal{O}(\epsilon^4\sigma^2)$ terms are associated with the quadratic noise terms.

The average stochastic center manifold equation given by Eq. (3.18) can now be used to solve the problem of interest. As mentioned previously, we are interested in computing escape rates and optimal escape paths. In particular we have analytically found an expression for the escape rate using the theory of large fluctuations [28, 29] using both the average stochastic center manifold and a naïve approach wherein the noise is added after the fact to the deterministic center manifold. Because the noise effects occur at such high order ($\mathcal{O}(\epsilon^4\sigma^2)$), the correction to the naïve approach using the deterministic center manifold is minimal. Details can be found in Forgoston and Schwartz [5].

We have performed numerical computations of escape time and escape path to compare with the analytical results. When $\sigma = 0$ (no noise), the original, singularly perturbed problem given by Eqs. (3.1a)–(3.1b), has three equilibrium points given by $(-1, 0)$, $(0, 0)$, and $(1, 0)$. At the initial time, $t = 0$, a particle is randomly placed near the stable, attracting point $(1, 0)$ within a circle of radius 0.1 centered at $(1, 0)$. Equations (3.1a)–(3.1b) are numerically integrated using a stochastic integrator with a constant time step size, δt, that depends on the value of ϵ, and the time needed for the particle to escape from the basin of attraction is determined. This escape time is based on either the time it takes the particle to cross the $x < -0.2$ barrier, which means the particle has escaped across the unstable saddle, and has entered the second basin of attraction with stable, attracting point $(-1, 0)$, or when the assigned maximum time has been reached. This computation was performed for 10 000 particles, and the mean escape time was determined. Figure 4 in Forgoston and Schwartz [5] demonstrates excellent agreement between the analytical and numerical escape times.

In addition, for each of the 10 000 particles that were initially placed in one of the attracting basins and which later escaped from this basin, across the saddle, and into the other basin of attraction, we retain $t = 200$ worth of the particle's path prior to escape. By creating a histogram representing the probability density, p_h, of this escape prehistory [2], one can see which regions of the phase space are associated with a high or low probability of particle escape.

Figure 1 shows a histogram of escape path prehistory for $\epsilon = 0.1$ and $\sigma = 0.3$ (so that $D = \sigma^2/2 = 0.045$). The color-bar values of Fig. 1 have been normalized by 10^5. The threshold 0 value in the figure is actually about 9 000. Therefore, any histogram box containing less than 9 000 events shows up as white on the histogram. Overlaid on top of the histogram are the graphs of the third-order, fourth-order,

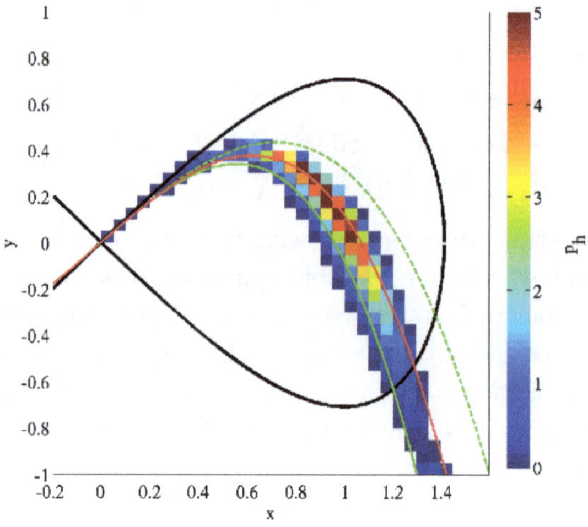

Fig. 1. Escape path prehistory histogram for $\epsilon = 0.1$ and $\sigma = 0.3$ overlaid with the graphs of the third-order (solid, green line), fourth-order (dashed, green line), and fifth-order (solid, red line) center manifold equations given by Eq. (3.6a). The color-bar values have been normalized by 10^5, and the threshold 0 value is about 9 000. Reproduced from Forgoston and Schwartz [5].

and fifth-order center manifold equations. Each of these equations may be found by including terms of the appropriate order from Eq. (3.6a).

One can see from Fig. 1 that the third-order and fourth-order center manifolds essentially bound the entire region of escape path prehistory, while the fifth-order manifold lies along the region of highest probability of escape. Although it is not shown, it should be noted that the optimal escape path (found using large fluctuation theory) associated with the third-order center manifold is a heteroclinic orbit from $x = \sqrt{1 - \epsilon + 2\epsilon^2}$ to $x = 0$ that lies directly on top of a section of the third-order center manifold (solid, green line in Fig. 1). Similarly, the optimal escape paths associated with higher order center manifolds lie directly on top of a section of the corresponding center manifold.

Additionally, one could overlay the histogram of escape path prehistory with the average stochastic center manifold given by Eq. (3.18). However, since the stochastic correction appears at order $\mathcal{O}(\epsilon^4\sigma^2)$, there is no noticeable difference from the manifolds shown in Fig. 1. Therefore, plots of the average stochastic manifold are not shown.

Although corrections due to noise occur at high order in the current example, other problems with quadratic nonlinearities have noise corrections occurring at much lower order [30], as seen in the following example. In such cases, one must use the normal form coordinate transform method rather than relying on the deterministic center manifold reduction.

4. Example 2: stochastic SEIR epidemiological model

The interaction between deterministic and stochastic effects in population dynamics has played, and continues to play, an important role in the modeling of infectious diseases. The mechanistic modeling side of population dynamics is well-known and established [31, 32]. These models typically are assumed to be useful for infinitely large, homogeneous populations, and arise from the mean field analysis of probabilistic models. On the other hand, when one considers finite populations, random interactions give rise to internal noise effects, which may introduce new dynamics. Stochastic effects are quite prominent in finite populations, which can range from ecological dynamics [33] to childhood epidemics in cities [34, 35]. For homogeneous populations with seasonal forcing, noise also comes into play in the prediction of large outbreaks [36–38]. Specifically, external random perturbations change the probabilistic prediction of epidemic outbreaks as well as its control [39].

One aspect of epidemic models which is often of interest involves the inclusion of a time delay. The delay term makes the analysis significantly more complicated. However, it is possible to approximate the delay by creating a cascade consisting of a large number of compartments [40]. For example, one could simulate the delay associated with a disease exposure time with several hundred "exposed" compartments. This model example is just one of several types of very high-dimensional models that are currently of interest. As a result of the high-dimensionality, there is much computation involved, and the analysis is quite difficult. In particular, real-time computation is not currently possible. However, there are usually many time scales that are well-separated (due typically to a large range in order of magnitude of the parameters) when considering such high-dimensional problems. In the presence of well-separated time scales, a model reduction method is needed to examine the dynamics on a lower-dimensional space. It is known that deterministic model reduction methods may not work well in the stochastic realm, which includes epidemic models [5]. However, one can use stochastic model reduction methods including the normal form coordinate transform method of Roberts [11] to correctly project the noise onto the lower-dimensional manifold. Because the noise affects the timing of outbreaks, it is essential to produce a low-dimensional system which captures the correct timing of the outbreaks as well as the amplitude and phase of any recurrent behaviour.

As a first study of a high-dimensional system, we consider the Susceptible-Exposed-Infected-Recovered (SEIR) epidemiological model with stochastic forcing. As previously mentioned, we could easily consider a SEIR-type model where the exposed class was modeled using hundreds of compartments. Since the analysis is similar, we consider the simpler standard SEIR model to demonstrate the power of the method.

We begin by describing the stochastic version of the SEIR model found in Schwartz and Smith [41]. We assume that a given population may be divided into the following four classes which evolve in time:

(1) Susceptible class, $s(t)$, consists of those individuals who may contract the disease.
(2) Exposed class, $e(t)$, consists of those individuals who have been infected by the disease but are not yet infectious.
(3) Infectious class, $i(t)$, consists of those individuals who are capable of transmitting the disease to susceptible individuals.
(4) Recovered class, $r(t)$, consists of those individuals who are immune to the disease.

Furthermore, we assume that the total population size, denoted as N, is constant and can be normalized to $S(t) + E(t) + I(t) + R(t) = 1$, where $S(t) = s(t)/N$, $E(t) = e(t)/N$, $I(t) = i(t)/N$, and $R(t) = r(t)/N$. Therefore, the population class variables S, E, I, and R represent fractions of the total population. The governing equations for the stochastic SEIR model are

$$\dot{S}(t) = \mu - \beta I(t)S(t) - \mu S(t) + \sigma_1\phi_1(t), \quad (4.1a)$$

$$\dot{E}(t) = \beta I(t)S(t) - (\alpha + \mu)E(t) + \sigma_2\phi_2(t), \quad (4.1b)$$

$$\dot{I}(t) = \alpha E(t) - (\gamma + \mu)I(t) + \sigma_3\phi_3(t), \quad (4.1c)$$

$$\dot{R}(t) = \gamma I(t) - \mu R(t) + \sigma_4\phi_4(t), \quad (4.1d)$$

where σ_i is the standard deviation of the noise intensity $D_i = \sigma_i^2/2$. Each of the noise terms, ϕ_i, describes a stochastic, Gaussian white force that is characterized by the correlation functions given by Eqs. (3.2a)–(3.2b).

Additionally, μ represents a constant birth and death rate, β is the contact rate, α is the rate of infection, so that $1/\alpha$ is the mean latency period, and γ is the rate of recovery, so that $1/\gamma$ is the mean infectious period. Although the contact rate β could be given by a time-dependent function (e.g., due to seasonal fluctuations), for simplicity, we assume β to be constant. Throughout this article, we use the following parameter values: $\mu = 0.02(\text{year})^{-1}$, $\beta = 1575.0(\text{year})^{-1}$, $\alpha = 1/0.0279(\text{year})^{-1}$, and $\gamma = 1/0.01(\text{year})^{-1}$. Disease parameters correspond to typical measles values [41, 42]. Note that any other biologically meaningful parameters may be used as long as the basic reproductive rate $R_0 = \alpha\beta/[(\alpha + \mu)(\gamma+\mu)] > 1$. The interpretation of R_0 is the number of secondary cases produced by a single infectious individual in a population of susceptibles in one infectious period.

As a first approximation of stochastic effects, we have considered additive noise. This type of noise may result from migration into and away from the population being considered [43]. Since it is difficult to estimate fluctuating migration rates [44], it is appropriate to treat migration as an arbitrary external noise source. Also, fluctuations in the birth rate manifest itself as additive noise. Furthermore, as we are not interested in extinction events in this article, it is not necessary to use multiplicative noise. In general, for the problem considered here, it is possible that a rare event in the tail of the noise distribution may cause one or more of the S, E, and I components of the solution to become negative. Here, we will always assume that the noise is sufficiently small so that a solution remains positive for a long enough time to gather sufficient statistics. Even though it is difficult to accurately estimate the appropriate noise level from real data, our choices of noise intensity lie within the huge confidence intervals computed in Bjørnstad et al. [44].

Although $S + E + I + R = 1$ in the deterministic system, one should note that the dynamics of the stochastic SEIR system will not necessarily have all of the components sum to unity. However, since the noise has zero mean, the total population will remain close to unity on average. Therefore, we assume that the dynamics are sufficiently described by Eqs. (4.1a)–(4.1c). It should be noted that even if $E(t) + I(t) = 0$ for some t, the noise allows for the re-emergence of the epidemic.

4.1. *Deterministic center manifold analysis*

As seen in Sec. 2.1 one can reduce the dimension of a system of equations using deterministic center manifold theory. In order to make use of the center manifold theory, we transform the deterministic version of Eqs. (4.1a)–(4.1c) to a new system of equations that has the necessary spectral structure. The theory will allow us to find an invariant center manifold passing through the fixed point to which we can restrict the transformed system.

The transformed evolution equations are given by

$$\begin{aligned}\frac{dU}{d\tau} =& -\alpha_0 U + \frac{\mu^2(\gamma_0 V - \alpha_0 U)}{\alpha_0+\gamma_0} - \\ & \frac{(\gamma_0+\mu^2)(\alpha_0+\mu^2)\left[(\alpha_0+\gamma_0)U+\gamma_0 W\right]}{\alpha_0(\alpha_0+\gamma_0)} - \\ & \frac{\mu\beta}{\alpha_0+\gamma_0}\left(\gamma_0 W + (\alpha_0+\gamma_0)U + \frac{\mu^2\alpha_0\gamma_0}{(\gamma_0+\mu^2)(\alpha_0+\mu^2)}\right)(U+V), \qquad (4.2a)\\ \frac{dV}{d\tau} =& \alpha_0 U - \frac{\mu^2(\gamma_0 V - \alpha_0 U)}{\alpha_0+\gamma_0} - \\ & \frac{(\gamma_0+\mu^2)(\alpha_0+\mu^2)\left[(\alpha_0+\gamma_0)U+\gamma_0 W\right]}{\gamma_0(\alpha_0+\gamma_0)} - \\ & \frac{\mu\beta\alpha_0}{\gamma_0(\alpha_0+\gamma_0)}\left(\gamma_0 W + (\alpha_0+\gamma_0)U + \frac{\mu^2\alpha_0\gamma_0}{(\gamma_0+\mu^2)(\alpha_0+\mu^2)}\right) * \\ & (U+V), \qquad (4.2b)\end{aligned}$$

$$\begin{aligned}\frac{dW}{d\tau} =& -\alpha_0 U - \left(\gamma_0 + \mu^2\right)(U+W) + \\ & \frac{\left(\gamma_0+\mu^2\right)\left(\alpha_0+\mu^2\right)\left[\left(\alpha_0+\gamma_0\right)U + \gamma_0 W\right]}{\alpha_0\gamma_0} - \\ & \mu^2 V + \frac{\mu\beta}{\gamma_0}\left(\gamma_0 W + \left(\alpha_0+\gamma_0\right)U + \frac{\mu^2\alpha_0\gamma_0}{\left(\gamma_0+\mu^2\right)\left(\alpha_0+\mu^2\right)}\right) * \\ & (U+V), \end{aligned} \tag{4.2c}$$

$$\frac{d\mu}{d\tau} = 0, \tag{4.2d}$$

where

$$U = \frac{-\gamma_0}{\alpha_0+\gamma_0}\bar{E}, \tag{4.3a}$$

$$V = \bar{S} + \frac{\gamma_0}{\alpha_0+\gamma_0}\bar{E}, \tag{4.3b}$$

$$W = \bar{I} + \bar{E}, \tag{4.3c}$$

$\bar{S}(t) = S(t) - S_0$, $\bar{E}(t) = E(t) - E_0$, and $\bar{I}(t) = I(t) - I_0$,

$$(S_0, E_0, I_0) = \left(\frac{(\gamma+\mu)(\alpha+\mu)}{\beta\alpha}, \frac{\mu}{\alpha+\mu} - \frac{\mu(\gamma+\mu)}{\alpha\beta}, \frac{\mu\alpha}{(\gamma+\mu)(\alpha+\mu)} - \frac{\mu}{\beta}\right) \tag{4.4}$$

is a fixed point corresponding to the endemic state, and $\alpha = \alpha_0/\mu$ and $\gamma = \gamma_0/\mu$, where α_0 and γ_0 are $\mathcal{O}(1)$. Details regarding the transformation can be found in Forgoston et al. [30].

The Jacobian of Eqs. (4.2a)–(4.2d) to zero-order in μ and evaluated at the origin is

$$\left[\begin{array}{c|ccc} -(\alpha_0+\gamma_0) & 0 & -\frac{\gamma_0^2}{(\alpha_0+\gamma_0)} & 0 \\ \hline 0 & 0 & -\frac{\alpha_0\gamma_0}{(\alpha_0+\gamma_0)} & 0 \\ 0 & 0 & 0 & 0 \\ 0 & 0 & 0 & 0 \end{array}\right], \tag{4.5}$$

which shows that Eqs. (4.2a)–(4.2d) may be rewritten in the form

$$\frac{d\mathbf{x}}{d\tau} = \mathbf{A}\mathbf{x} + \mathbf{f}(\mathbf{x}, \mathbf{y}, \mu), \tag{4.6}$$

$$\frac{d\mathbf{y}}{d\tau} = \mathbf{B}\mathbf{y} + \mathbf{g}(\mathbf{x}, \mathbf{y}, \mu), \tag{4.7}$$

$$\frac{d\mu}{d\tau} = 0, \tag{4.8}$$

where $\mathbf{x} = (U)$, $\mathbf{y} = (V, W)$, $\mathbf{A}$ is a constant matrix with eigenvalues that have negative real parts, $\mathbf{B}$ is a constant matrix with eigenvalues that have zero real parts, and $\mathbf{f}$ and $\mathbf{g}$ are nonlinear functions in $\mathbf{x}$, $\mathbf{y}$ and μ. In particular,

$$\mathbf{A} = \left[-(\alpha_0 + \gamma_0)\right], \qquad \mathbf{B} = \begin{bmatrix} 0 & -\frac{\alpha_0\gamma_0}{(\alpha_0+\gamma_0)} \\ 0 & 0 \end{bmatrix}. \tag{4.9}$$

Therefore, the system will rapidly collapse onto a lower-dimensional manifold given by center manifold theory [15]. Furthermore, we know that the center manifold is given by

$$U = h(V, W, \mu), \tag{4.10}$$

where h is an unknown function.

Substitution of Eq. (4.10) into Eq. (4.2a) leads to the following center manifold condition:

$$\begin{aligned} &\frac{\partial h(V,W,\mu)}{\partial V}\frac{dV}{d\tau} + \frac{\partial h(V,W,\mu)}{\partial W}\frac{dW}{d\tau} = -\alpha_0 h(V,W,\mu) + \\ &\frac{\mu^2\left[\gamma_0 V - \alpha_0 h(V,W,\mu)\right]}{\alpha_0+\gamma_0} - \\ &\frac{\left(\gamma_0+\mu^2\right)\left(\alpha_0+\mu^2\right)\left[\left(\alpha_0+\gamma_0\right)h(V,W,\mu)+\gamma_0 W\right]}{\alpha_0\left(\alpha_0+\gamma_0\right)} - \\ &\frac{\mu\beta}{\alpha_0+\gamma_0}\left(\gamma_0 W + \left(\alpha_0+\gamma_0\right)h(V,W,\mu) + \frac{\mu^2\alpha_0\gamma_0}{\left(\gamma_0+\mu^2\right)\left(\alpha_0+\mu^2\right)}\right) * \\ &\left(h(V,W,\mu)+V\right). \end{aligned} \tag{4.11}$$

In general, it is not possible to solve the center manifold condition for the unknown function, $h(V, W, \mu)$. Therefore, we perform the following Taylor series expansion of $h(V, W, \mu)$ in V, W, and μ:

$$\begin{aligned} h(V,W,\mu) = {} & h_0 + h_2 V + h_3 W + h_\mu \mu + h_{22}V^2 + h_{23}VW + h_{33}W^2 + \\ & h_{\mu 2}\mu V + h_{\mu 3}\mu W + h_{\mu\mu}\mu^2 + \ldots, \end{aligned} \tag{4.12}$$

where h_0, h_2, h_3, h_μ, $\ldots$ are unknown coefficients that are found by substituting the Taylor series expansion into the center manifold condition and equating terms of the same order. By carrying out this procedure using a second-order Taylor series expansion of h, the center manifold equation is

$$U = -\frac{\gamma_0^2}{(\alpha_0+\gamma_0)^2} W + \mathcal{O}(\epsilon^3), \tag{4.13}$$

where $\epsilon = |(V, W, \mu)|$ so that ϵ provides a count of the number of V, W, and μ factors in any one term. Substitution of Eq. (4.13) into Eqs. (4.2b) and (4.2c) leads to the following reduced system of evolution equations which describe the dynamics on the center manifold:

$$\frac{dV}{d\tau} = -\frac{\mu^2{\gamma_0}^2\alpha_0 W}{(\alpha_0+\gamma_0)^3} - \frac{\mu^4\alpha_0 W}{(\alpha_0+\gamma_0)^2} - \frac{\gamma_0\mu^2 V}{\alpha_0+\gamma_0} - \frac{(\gamma_0+\mu^2)\,\alpha_0 W}{\alpha_0+\gamma_0} - \frac{\beta{\alpha_0}^2\mu}{(\alpha_0+\gamma_0)^2}\left(W + \frac{\mu^2(\alpha_0+\gamma_0)}{(\gamma_0+\mu^2)(\alpha_0+\mu^2)}\right)\left(V - \frac{{\gamma_0}^2 W}{(\alpha_0+\gamma_0)^2}\right), \tag{4.14a}$$

$$\frac{dW}{d\tau} = \frac{\mu^2{\gamma_0}^2 W}{(\alpha_0+\gamma_0)^2} + \frac{\mu^4 W}{\alpha_0+\gamma_0} - \mu^2 V + \frac{\beta\mu\alpha_0}{\alpha_0+\gamma_0}\left(W + \frac{\mu^2(\alpha_0+\gamma_0)}{(\gamma_0+\mu^2)(\alpha_0+\mu^2)}\right)\left(V - \frac{{\gamma_0}^2 W}{(\alpha_0+\gamma_0)^2}\right). \tag{4.14b}$$

4.2. *Projection of the noise onto the stochastic center manifold*

The stochastic SEIR system given by Eqs. (4.1a)–(4.1c) may be transformed in a manner similar to what was done in the preceding section for the deterministic form of the system. The transformation will have an effect on the noise terms. The transformed stochastic terms are still additive, Gaussian noise processes. However, the transformation has acted on the original stochastic terms ϕ_1, ϕ_2, and ϕ_3 to create new noise processes which have a variance different from that of the original noise processes. This transformed system of equations is an exact transformation of the system of equations given by Eqs. (4.1a)–(4.1c) and can be found in Forgoston et al. [30].

4.2.1. *Reduction of the stochastic SEIR model*

It is tempting to consider the reduced stochastic model found by substitution of Eq. (4.13) into the tranformed equations, so that one has the following stochastic evolution equations (that hopefully describe the dynamics on the stochastic center manifold):

$$\frac{dV}{d\tau} = -\frac{\mu^2{\gamma_0}^2\alpha_0 W}{(\alpha_0+\gamma_0)^3} - \frac{\mu^4\alpha_0 W}{(\alpha_0+\gamma_0)^2} - \frac{\gamma_0\mu^2 V}{\alpha_0+\gamma_0} - \frac{(\gamma_0+\mu^2)\,W\alpha_0}{\alpha_0+\gamma_0} - \frac{\beta{\alpha_0}^2\mu}{(\alpha_0+\gamma_0)^2}\left(W + \frac{\mu^2(\alpha_0+\gamma_0)}{(\gamma_0+\mu^2)(\alpha_0+\mu^2)}\right)\left(V - \frac{{\gamma_0}^2 W}{(\alpha_0+\gamma_0)^2}\right) + \sigma_5\phi_5, \tag{4.15a}$$

$$\frac{dW}{d\tau} = \frac{\mu^2{\gamma_0}^2 W}{(\alpha_0+\gamma_0)^2} + \frac{\mu^4 W}{\alpha_0+\gamma_0} - \mu^2 V + \frac{\beta\mu\alpha_0}{\alpha_0+\gamma_0}\left(W + \frac{\mu^2(\alpha_0+\gamma_0)}{(\gamma_0+\mu^2)(\alpha_0+\mu^2)}\right)\left(V - \frac{{\gamma_0}^2 W}{(\alpha_0+\gamma_0)^2}\right) + \sigma_6\phi_6, \tag{4.15b}$$

where

$$\sigma_5\phi_5 = \mu\sigma_1\phi_1 + \frac{\mu\gamma_0}{\alpha_0+\gamma_0}\sigma_2\phi_2, \tag{4.16a}$$

$$\sigma_6\phi_6 = \mu\sigma_2\phi_2 + \mu\sigma_3\phi_3 \tag{4.16b}$$

are transformed stochastic terms.

One should note that Eqs. (4.15a)–(4.15b) also can be found by naïvely adding the stochastic terms to the reduced system of evolution equations for the deterministic problem (Eqs. (4.14a)–(4.14b)). This type of stochastic center manifold reduction has been done for the case of discrete noise [42]. Additionally, in many other fields (e.g., oceanography, solid mechanics, fluid mechanics), researchers have performed stochastic model reduction using a Karhunen–Loève expansion (principal component analysis, proper orthogonal decomposition) [45, 46]. However, this linear projection does not properly capture the nonlinear effects. Furthermore, one must subjectively choose the number of modes needed for the expansion. Therefore, even though the solution to the reduced model found using this technique may have the correct statistics, individual solution realisations will not agree with the original, complete solution.

It can be seen in Fig. 2 that Eqs. (4.15a)–(4.15b) do not contain the correct projection of the noise onto the center manifold. Therefore, when solving the reduced system, one does not obtain the correct solution. Such errors in stochastic epidemic modeling impact the prediction of disease outbreak when modeling the spread of a disease in a population.

Figures 2(a)–(b) compares the time series of the fraction of the population that is infected with a disease, I, computed using the complete, stochastic system of transformed equations of the SEIR model with the time series of I computed using the reduced system of equations of the SEIR model that is based on the deterministic center manifold with a replacement of the noise terms (Eqs. (4.15a) and (4.15b)). Figure 2(a) shows the initial transients, while Fig. 2(b) shows the time series after the transients have decayed. One can see that the solution computed using the reduced system quickly becomes out of phase with the solution of the complete system. Use of this reduced system would lead to an incorrect prediction of the initial disease outbreak. Additionally, the predicted amplitude of the initial outbreak would be incorrect. The poor agreement, both in phase and amplitude, between the two solutions continues for long periods of time as seen in Fig. 2(b). We also have computed the cross-correlation of the two time series shown in Fig. 2(a)–(b) to be approximately 0.34. Since the cross-correlation measures the similarity between the two time series, this low value quantitatively suggests poor agreement between the two solutions.

4.2.2. *Correct projection of the noise onto the stochastic center manifold*

To project the noise correctly onto the center manifold, we derive a normal form coordinate transform using the principles of Sec. 2.2 for the complete, stochastic system of transformed equations of the SEIR model.

The stochastic system of equations (which satisfies the necessary spectral requirements) in $(U, V, W)^T$ coordinates is transformed to a new $(Y, X_1, X_2)^T$

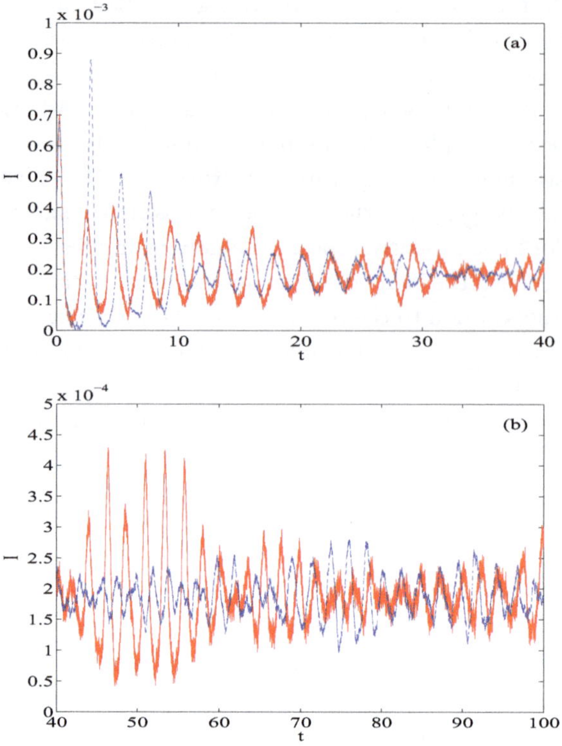

Fig. 2. Time series of the fraction of the population that is infected with a disease, I, computed using the complete, stochastic system of transformed equations of the SEIR model (red, solid line), and computed using the reduced system of equations of the SEIR model that is based on the deterministic center manifold with a replacement of the noise terms (Eqs. (4.15a)–(4.15b)) (blue, dashed line). The standard deviation of the noise intensity used in the simulation is $\sigma_i = 0.0005$, $i = 4, 5, 6$. The time series is shown for (a) $t = 0$ to $t = 40$, and for (b) $t = 40$ to $t = 100$. Reproduced from Forgoston et al. [30].

coordinate system using a near-identity stochastic coordinate transform given as

$$U = Y + \xi(Y, X_1, X_2, \tau), \tag{4.17a}$$

$$V = X_1 + \eta(Y, X_1, X_2, \tau), \tag{4.17b}$$

$$W = X_2 + \rho(Y, X_1, X_2, \tau), \tag{4.17c}$$

where the specific form of $\xi(Y, X_1, X_2, \tau)$, $\eta(Y, X_1, X_2, \tau)$, and $\rho(Y, X_1, X_2, \tau)$ is found using an iterative procedure.

Several iterations lead to coordinate transforms for U, V, and W along with evolution equations describing the Y-dynamics, X_1-dynamics, and X_2-dynamics in the new coordinate system. The Y-dynamics have exponential decay to the $Y = 0$ slow manifold. Substitution of $Y = 0$ leads to complicated expressions for the coordinate transforms — they can be found in Forgoston et al. [30].

We are interested in the long-term slow processes. Since the memory integrals are fast-time processes, we neglect them along with the higher-order multiplicative terms to obtain the following expressions for U, V, and W:

$$U = -\frac{\gamma_0^2 X_2}{(\alpha_0+\gamma_0)^2} - \frac{\mu\beta X_1}{(\alpha_0+\gamma_0)}\left(\frac{\mu^2}{(\alpha_0+\gamma_0)} + \frac{\gamma_0 X_2}{(\alpha_0+\gamma_0)} + \mu^2 X_2\right) + \frac{\mu^2\gamma_0}{(\alpha_0+\gamma_0)}\left(X_1 - \frac{2X_2}{\alpha_0}\right), \tag{4.18a}$$

$$V = X_1, \tag{4.18b}$$

$$W = X_2. \tag{4.18c}$$

Note that Eq. (4.18a) is the deterministic center manifold equation, and at first-order, matches the center manifold equation that was found previously (Eq. (4.13)).

Substitution of $Y = 0$ and neglecting all multiplicative noise terms and memory integrals using the argument from above (so that we consider only first-order noise terms) leads to the following reduced system of evolution equations on the center manifold:

$$\frac{dX_1}{d\tau} = F(X_1(\tau), X_2(\tau)), \tag{4.19a}$$

$$\frac{dX_2}{d\tau} = G(X_1(\tau), X_2(\tau)). \tag{4.19b}$$

The specific form of F and G in Eqs. (4.19a) and (4.19b) are complicated, and can be found in Forgoston et al. [30].

Figures 3(a)–(b) compare the time series of the fraction of the population that is infected with a disease, I, computed using the complete, stochastic system of transformed equations of the SEIR model with the time series of I computed using the reduced system of equations of the SEIR model that is found using the stochastic normal form coordinate transform (Eqs. (4.19a)–(4.19b)). Figure 3(a) shows the initial transients, while Fig. 3(b) shows the time series after the transients have decayed. One can see that there is excellent agreement between the two solutions. The initial outbreak is successfully captured by the reduced system. Furthermore, Fig. 3(b) shows that the reduced system accurately predicts recurrent outbreaks for a time scale that is orders of magnitude longer than the relaxation time. This is not surprising since the solution decays exponentially throughout the transient and then remains close to the lower-dimensional center manifold. Since we are not looking at periodic orbits, there are no secular terms in the asymptotic expansion, and the result is valid for all time. Additionally, any noise drift on the center manifold results in bounded solutions due to sufficient dissipation transverse to the manifold. The cross-correlation of the two time series shown in Fig. 3 is approximately 0.98, which quantitatively suggests there is excellent agreement between the two solutions.

Unlike example 1, where the average stochastic center manifold and deterministic center manifold are virtually identical, the SEIR model considered in example 2 has terms at low order in the normal form transform which causes a significant difference between the average stochastic center manifold and the deterministic manifold. Therefore, as we have demonstrated, when working with the SEIR model, one must use the normal form coordinate transform method to correctly project the noise onto the center manifold.

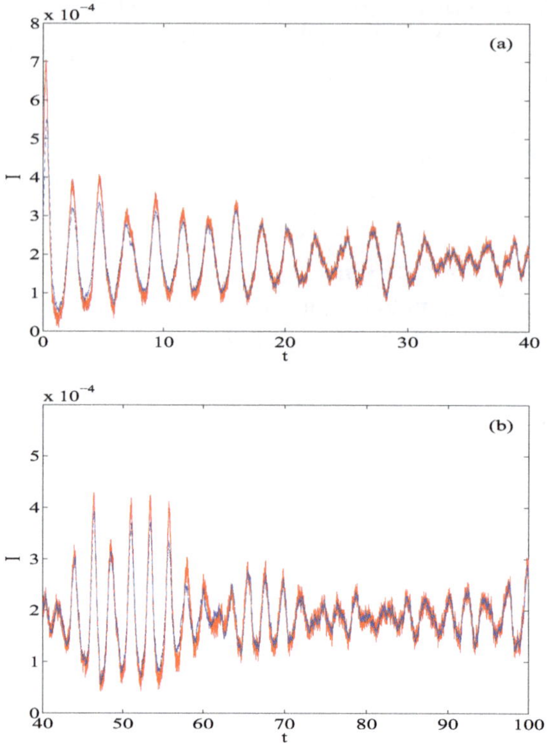

Fig. 3. Time series of the fraction of the population that is infected with a disease, I, computed using the complete, stochastic system of transformed equations of the SEIR model (red, solid line), and computed using the reduced system of equations of the SEIR model that is found using the stochastic normal form coordinate transform (Eqs. (4.19a)–(4.19b)) (blue, dashed line). The standard deviation of the noise intensity used in the simulation is $\sigma_i = 0.0005$, $i = 4, 5, 6$. The time series is shown for (a) $t = 0$ to $t = 40$, and for (b) $t = 40$ to $t = 100$. Reproduced from Forgoston et al. [30].

5. Conclusions

This review highlights results in stochastic model reduction that use the normal form coordinate transform method of A. J. Roberts. It is increasingly apparent that small external effects can play a significant role in the dynamics of models used for real world applications. Therefore, researchers are revisiting existing methods and developing new ones to quantify noise-induced phenomena in stochastic models. Time series prediction is a classic problem in deterministic dynamical systems. As we develop methods to approach its counterpart in stochastic models, we recognize the limitations of high dimensional analysis. Reduced models provide tractable systems to analyze and test, both analytically and computationally. In this work, we demonstrate the importance of properly projecting the noise source through the reduction. The method of Roberts is an important contribution as it not only enables one to reduce the dimensionality of the dynamics, but also provides a way to separate all slow processes from all fast processes. We have demonstrated in these two examples how deriving the deterministic manifold may not be sufficient, and in this case, how one can employ the normal form coordinate transform method to properly capture the stochastic manifold. The result in the SEIR example is a closed form system that can be analyzed and tested for extended time series prediction. Stochastic model reduction is a general approach that can be used to analyze the dynamics of generic stochastic models with separated time scales.

Acknowledgments

EF and LB are supported by the National Science Foundation award DMS #1418956. We thank Tony (A. J. Roberts) for his comments and insight over the years regarding the work presented in this article.

References

[1] M. I. Dykman, Large fluctuations and fluctuational transitions in systems driven by coloured gaussian noise: A high-frequency noise, Phys. Rev. A **42**, pp. 2020–2029 (1990).

[2] M. I. Dykman, P. V. E. McClintock, V. N. Smelyanski, N. D. Stein and N. G. Stocks, Optimal paths and the prehistory problem for large fluctuations in noise-driven systems, Phys. Rev. Lett. **68**, pp. 2718–2721 (1992).

[3] M. Millonas (ed.), Fluctuations and Order: The New Synthesis. Springer–Verlag (1996).

[4] D. G. Luchinsky, P. V. E. McClintock and M. I. Dykman, Analogue studies of nonlinear systems, Rep. Prog. Phys. **61**, pp. 889–997 (1998).

[5] E. Forgoston and I. B. Schwartz, Escape rates in a stochastic environment with multiple scales, SIAM J. Appl. Dyn. Syst. **8**, 3, pp. 1190–1217 (2009).

[6] R. V. Kohn, M. G. Reznikoff and E. Vanden-Eijnden, Magnetic elements at finite temperature and large deviation theory, J. Nonlinear Sci. **15**, pp. 223–253 (2005).

[7] T. A. Fulton and L. N. Dunkelberger, Lifetime of the zero-voltage state in Josephson tunnel junctions, Phys. Rev. B **9**, p. 4760 (1974).

[8] L. S. Tsimring, Noise in biology, Reports on Progress in Physics **77**, 2, p. 026601 (2014).

[9] M. Assaf, E. Roberts and Z. Luthey-Schulten, Determining the stability of genetic switches: Explicitly accounting for mRNA noise, Phys. Rev. Lett. **106**, p. 248102 (2011).

[10] E. Roberts, S. Be'er, C. Bohrer, R. Sharma and M. Assaf, Dynamics of simple gene-network motifs subject to extrinsic fluctuations, Phys. Rev. E **92**, p. 062717 (2015).

[11] A. J. Roberts, Normal form transforms separate slow and fast modes in stochastic dynamical systems, Physica A **387**, 1, pp. 12–38 (2008).

[12] N. Berglund and B. Gentz, Geometric singular perturbation theory for stochastic differential equations, J. Differ. Equations **191**, pp. 1–54 (2003).

[13] N. Berglund and B. Gentz, Noise-induced phenomena in slow–fast dynamical systems: a sample-paths approach. Springer Science & Business Media (2006).

[14] C. Chicone and Y. Latushkin, Center manifolds for infinite dimensional nonautonomous differential equations, J. Differ. Equations **141**, pp. 356–399 (1997).
[15] J. Carr, Applications of Centre Manifold Theory. Springer–Verlag (1981).
[16] P. Boxler, A stochastic version of center manifold theory, Probab. Theory Rel. **83**, pp. 509–545 (1989).
[17] E. Knobloch and K. A. Wiesenfeld, Bifurcations in fluctuating systems: The center-manifold approach, J. Stat. Phys. **33**, 3, pp. 611–637 (1983).
[18] P. H. Coullet, C. Elphick and E. Tirapegui, Normal form of a Hopf bifurcation with noise, Phys. Lett. A **111**, pp. 277–282 (1985).
[19] N. S. Namachchivaya, Stochastic bifurcation, Appl. Math. Comput. **38**, pp. 101–159 (1990).
[20] N. S. Namachchivaya and Y. K. Lin, Method of stochastic normal forms, Int. J. Nonlinear Mech. **26**, pp. 931–943 (1991).
[21] L. Arnold and P. Imkeller, Normal forms for stochastic differential equations, Probab. Theory Rel. **110**, pp. 559–588 (1998).
[22] L. Arnold, Random Dynamical Systems. Springer–Verlag (1998).
[23] D. C. Webb, P. J. Simonetti and C. P. Jones, Slocum: An underwater glider propelled by environmental energy, IEEE J. Oceanic Eng. **26**, pp. 447–452 (2001).
[24] J. Sherman, R. E. Davis, W. B. Owens and J. Valdes, The autonomous underwater glider 'Spray', IEEE J. Oceanic Eng. **26**, pp. 437–446 (2001).
[25] C. C. Eriksen, T. J. Osse, R. D. Light, T. Wen, T. W. Lehman, P. L. Sabin, J. W. Ballard and A. M. Chiodi, Seaglider: A long-range autonomous underwater vehicle for oceanographic research, IEEE J. Oceanic Eng. **26**, pp. 424–436 (2001).
[26] Z. Schuss and A. Spivak, Where is the exit point? Chem. Phys. **235**, pp. 227–242 (1998).
[27] M. I. Dykman, I. B. Schwartz and M. Shapiro, Scaling in activated escape of underdamped systems, Phys. Rev. E **72**, p. 021102 (2005).
[28] M. I. Freidlin and A. D. Wentzell, Random Perturbations of Dynamical Systems. Springer–Verlag (1984).
[29] E. Forgoston and R. O. Moore, A primer on noise-induced transitions in applied dynamical systems, SIAM Review (accepted).
[30] E. Forgoston, L. Billings and I. B. Schwartz, Accurate noise projection for reduced stochastic epidemic models, Chaos **19**, p. 043110 (2009).
[31] R. M. Anderson and R. M. May, Infectious Diseases of Humans. Oxford University Press (1991).
[32] N. T. J. Bailey, The Mathematical Theory of Infectious Diseases. Charles Griffin, London (1975).
[33] G. Marion, E. Renshaw and G. Gibson, Stochastic modelling of environmental variation for biological populations, Theor. Popul. Biol. **57**, 3, pp. 197–217 (2000).
[34] H. T. H. Nguyen and P. Rohani, Noise, nonlinearity and seasonality: The epidemics of whooping cough revisited, J. Roy. Soc. Interface **5**, 21, pp. 403–413 (2008).
[35] P. Rohani, M. J. Keeling and B. T. Grenfell, The interplay between determinism and stochasticity in childhood diseases, Am. Nat. **159**, 5, pp. 469–481 (2002).
[36] D. A. Rand and H. B. Wilson, Chaotic stochasticity — A ubiquitous source of unpredictability in epidemics, P. Roy. Soc. B - Biol. Sci. **246**, 1316, pp. 179–184 (1991).
[37] L. Billings, E. M. Bollt and I. B. Schwartz, Phase-space transport of stochastic chaos in population dynamics of virus spread, Phys. Rev. Lett. **88**, p. 234101 (2002).
[38] L. Stone, R. Olinky and A. Huppert, Seasonal dynamics of recurrent epidemics, Nature **446**, pp. 533–536 (2007).

[39] I. B. Schwartz, L. Billings and E. M. Bollt, Dynamical epidemic suppression using stochastic prediction and control, Phys. Rev. E **70**, p. 046220 (2004).
[40] W. T. Mocek, R. Rudnicki and E. O. Voit, Approximation of delays in biochemical systems, Math. BioSci. **198**, 2, pp. 190–216 (2005).
[41] I. Schwartz and H. Smith, Infinite subharmonic bifurcations in an SEIR epidemic model, J. Math. Biol. **18**, pp. 233–253 (1983).
[42] L. Billings and I. B. Schwartz, Exciting chaos with noise: unexpected dynamics in epidemic outbreaks, J. Math. Biol. **44**, pp. 31–48 (2002).
[43] F. Brauer, P. van den Driessche and J. Wu (eds.), Mathematical Epidemiology. Springer–Verlag (2008).
[44] O. N. Bjørnstad, B. F. Finkenstädt and B. T. Grenfell, Dynamics of measles epidemics: Estimating scaling, Ecol. Monogr. **72**, 2, pp. 169–184 (2002).
[45] A. Doostan, R. G. Ghanem and J. Red-Horse, Stochastic model reduction for chaos representations, Comput. Methods Appl. Mech. Engrg. **196**, pp. 3951–3966 (2007).
[46] D. Venturi, X. Wan and G. E. Karniadakis, Stochastic low-dimensional modelling of a random laminar wake past a circular cylinder, J. Fluid Mech. **606**, pp. 339–367 (2008).

Chapter 4

An Averaging Principle for Multi-valued Stochastic Differential Equations Driven by G-Brownian Motion

Yong Xu[1,*], Min Han[1,†] and Bin Pei[2,‡]

[1]*Department of Applied Mathematics, Northwestern Polytechnical University, Xi'an, 710072, P.R. China*

[2]*School of Mathematical Sciences, Fudan University, Shanghai, 200433, P.R. China*

[*]*hsux3@nwpu.edu.cn;* [†]*hanmin7727@163.com;* [‡]*binpei@hotmail.com*

1. Introduction

Motivated by uncertainty problems and risk measures, stochastic differential equations (SDEs) driven by G-Brownian motion have attracted researchers' great interest ([14], [6], [18] and [17]). For example, consider a Black–Scholes like a market with uncertain volatilities, the basic securities consist of 2 assets, a riskless one, the bond, and a risky securities, the stock. Their prices are governed by

$$dS_t^0 = rS_t^0 dt,$$

for the bond with a constant interest rate $r > 0$. The price S_t^1 satisfied Black–Scholes–Merton's model driven by G-Brownian motion:

$$dS_t^1 = b_t S_t^1 dt + \beta_t S_t^1 d\langle B\rangle_t + \sigma_t S_t^1 dB_t, S_0^1 = x$$

for the stock. b_t, β_t, σ_t are assumed to be deterministic functions of t. $\{\langle B\rangle_t\}_{t\geq 0}$ is the quadratic variation process of G-Brownian motion $\{B_t\}_{t\geq 0}$. In this market, the stock price evolution involves not only risk modeled by the noise part but also ambiguity about the risk due to the unknown deviation of the process B (G-Brownian motion) from its mean. Thus, the G-Brownian motion is useful to measure the mean-uncertainty of risky positions. For the theory of G-Brownian motion, Peng ([14]) firstly studied the notion of sublinear expectation space and the fundamental theory of G-expectation and G-conditional expectation. Peng ([16]) also introduced the so-called G-Gaussian distribution and the G-Brownian motion. Gao ([6, 7]) investigated the existence, uniqueness and large deviations for Lipschitz SDEs driven by G-Brownian motion. Ren ([20]) investigated the exponential stability of solutions for SDEs with G-Brownian motion.

*Corresponding author.

For the theory of multi-valued stochastic differential equations (MSDEs), Cépa ([2]) first studied the existence and uniqueness theorem for MSDEs driven by Brownian motion. Ren, Xu and Zhang ([22]) investigated the existence, uniqueness and large deviations for MSDEs driven by continuous semimartingales. Ren, Wu and Zhang ([22]) considered the exponential ergodicity of non-Lipschitz MSDEs. Ren and Xu ([23]) studied a transfer principle for multivalued SDEs. Very recently, an averaging principle for MSDEs driven by Brownian motion was obtained by Xu and Liu ([30]). Guo and Pei ([8]) considered the averaging principle for MSDEs driven by Poisson point processes.

In recent years, the averaging principle has become a powerful tool for simplifying nonlinear dynamical systems. It has received significant attention from many practical areas, such as mechanics, automatic control and many other areas ([4], [5]). Khasminskii ([10]) first considered the averaging principle for Itô SDEs. Xu and his co-workers derived averaging principles for SDEs driven by Lévy noise ([25, 26]) and fractional Brownian motion (fBm) ([12, 13, 27–29]). A natural question is: Is there an averaging principle for MSDEs driven by G-Brownian motion? Can we establish an averaging principle to pave a way for reduction of computational complexity? To the authors' knowledge, the averaging principle for MSDEs driven by G-Brownian motion have not been considered. Therefore, in this work, we consider an averaging principle for MSDEs driven by G-Brownian motion in the sense of p-th moment and convergence in capacity.

This work is organized as follows. Section 2 introduces some notations and preliminaries which will be useful in what follows. Section 3 is devoted to establish an averaging principle for MSDEs driven by G-Brownian motion.

2. Preliminaries

The main purpose of this section is to recall some preliminary results in G-framework which are needed in the sequel ([3], [11], [14], [15] and [16]).

2.1. *Sublinear expectation*

Definition 4.1. Let Ω be a given set and let $\mathcal{H}$ be a vector lattice of real valued functions defined on Ω, namely $c \in \mathcal{H}$ for each constant c and $|X| \in \mathcal{H}$, if $X \in \mathcal{H}$. $\mathcal{H}$ is considered as the space of random variables. A sublinear expectation $\hat{\mathbb{E}}$ on $\mathcal{H}$ is a function $\hat{\mathbb{E}} : \mathcal{H} \to \mathbb{R}$ satisfying the following properties: for all $X, Y \in \mathcal{H}$, we have

(i) monotonicity: if $X \geq Y$, then $\hat{\mathbb{E}}[X] \geq \hat{\mathbb{E}}[Y]$;
(ii) constant preserving: $\hat{\mathbb{E}}[c] = c$;
(iii) sub-additivity: $\hat{\mathbb{E}}[X + Y] \leq \hat{\mathbb{E}}[X] + \hat{\mathbb{E}}[Y]$;
(iv) positive homogeneity: $\hat{\mathbb{E}}[\lambda X] = \lambda \hat{\mathbb{E}}[X]$ for each $\lambda \geq 0$.

The triple $(\Omega, \mathcal{H}, \hat{\mathbb{E}})$ is called a sublinear expectation space. $X \in \mathcal{H}$ is called a random variable in $(\Omega, \mathcal{H}, \hat{\mathbb{E}})$.

Definition 4.2. (G-Normal Distribution). A random variable X on a sublinear expectation space $(\Omega, \mathcal{H}, \hat{\mathbb{E}})$ is called (centralized) G-normal distributed if for any $a, b \geq 0$

$$aX + b\bar{X} \overset{d}{=} \sqrt{a^2 + b^2}X,$$

where $\bar{X}$ is an independent copy of X. The letter G denotes the function

$$G(a) := \frac{1}{2}(\overline{\sigma}^2 a^+ - \underline{\sigma}^2 a^-), a \in \mathbb{R},$$

with $\underline{\sigma}^2 := -\hat{\mathbb{E}}[-X^2] \leq \hat{\mathbb{E}}[X^2] =: \overline{\sigma}^2$.

2.2. *G-Brownian motion and G-expectation*

Definition 4.3. (G-Brownian motion) A process $(B_t \in \mathcal{H})_{t \geq 0}$ on a sublinear expectation space $(\Omega, \mathcal{H}, \hat{\mathbb{E}})$ is called a G-Brownian motion if the following properties are satisfied:

(a) $B_0 = 0$,

(b) For each $t, s \geq 0$, the increment $B_{t+s} - B_t \overset{d}{=} \sqrt{s}X$ are independent from $(B_{t_1}, B_{t_2}, \cdots, B_{t_n})$ for each $n \in \mathbb{N}, 0 \leq t_1 \leq t_2 \leq \cdots t_n \leq t$, where X is G-normal distributed.

To proceed, let $\Omega = C_0([0, \infty), \mathbb{R}^d)$, the space of all $\mathbb{R}^d$-valued continuous functions on $[0, \infty)$ with $\omega_0 = 0$, be equipped with the distance

$$\rho(\omega^1, \omega^2) = \sum_{i=1}^{\infty} 2^{-i}[(\max_{t \in [0,i]} |\omega_t^1 - \omega_t^2|) \wedge 1],$$

and for every $\omega \in \Omega$, let $B_t(\omega) = \omega_t$ be the canonical process.

For each $T > 0$, denote

$$Lip\,(\Omega_T) := \{\varphi(B_{t_1}, \ldots, B_{t_n}) : n \geq 1, t_1 \ldots t_n \in [0, T], \\ \varphi \in C_{b.Lip}(\mathbb{R}^{d \times n})\},$$

where $C_{b,Lip}(\mathbb{R}^{d \times n})$ is the space of all bounded Lipschitz functions defined on $\mathbb{R}^n$ and $Lip\,(\Omega) = \bigcup\limits_T Lip\,(\Omega_T)$.

Denote by $\mathbb{S}_d$ the collection of all $d \times d$ symmetric matrices. For each given monotonic and sublinear function $G : \mathbb{S}_d \to \mathbb{R}$, we can construct a G-expectation $\hat{\mathbb{E}}$.

In the following, we want to construct a sublinear expectation on $(\Omega, Lip(\Omega))$, such that the canonical process $(B_t)_{t \geq 0}$ is a G-Brownian motion. For this we first construct a sequence of d-dimensional random vectors $(\xi_i)_{i=1}^{\infty}$ on a sublinear expectation space $(\tilde{\Omega}, \tilde{\mathcal{H}}, \tilde{\mathbb{E}})$ such that ξ_i is G-normal distributed and ξ_{i+1} is independent from $(\xi_1, \cdots \xi_i)$ for each $i = 1, 2, \cdots$.

We now introduce a sublinear expectation $\hat{\mathbb{E}}$ defined on $Lip(\Omega)$ via the following procedure: for each $X \in Lip(\Omega)$ with

$$X = \varphi(B_{t_1} - B_{t_0}, B_{t_2} - B_{t_1}, \cdots, B_{t_n} - B_{t_{n-1}}),$$

for some $\varphi \in C_{b,Lip}(\mathbb{R}^{d\times n})$ and $0 = t_0 < t_1 < \cdots < t_n < \infty$, we set

$$\begin{aligned}&\hat{\mathbb{E}}[\varphi(B_{t_1} - B_{t_0}, B_{t_2} - B_{t_1}, \cdots, B_{t_n} - B_{t_{n-1}})]\\ &:= \tilde{\mathbb{E}}[\varphi(\sqrt{t_1 - t_0}\xi_1, \cdots, \sqrt{t_n - t_{n-1}}\xi_n)].\end{aligned}$$

The related conditional expectation of

$$X = \varphi(B_{t_1} - B_{t_0}, B_{t_2} - B_{t_1}, \cdots, B_{t_n} - B_{t_{n-1}}),$$

under Ω_{t_j} is defined by

$$\begin{aligned}\hat{\mathbb{E}}[X|\Omega_{t_j}] &= \hat{\mathbb{E}}[\varphi(B_{t_1}, B_{t_2} - B_{t_1}, \cdots, B_{t_n} - B_{t_{n-1}})|\Omega_{t_j}]\\ &:= \varphi(B_{t_1}, \cdots, B_{t_{j-1}}),\end{aligned}$$

where

$$\varphi(x_1, \cdots, x_j) = \tilde{\mathbb{E}}[\varphi(x_1, \cdots, x_j, \sqrt{t_{j+1} - t_j}\xi_{j+1}, \cdots, \sqrt{t_n - t_{n-1}}\xi_n)].$$

It is easy to check that $\hat{\mathbb{E}}[\cdot]$ consistently defines a sublinear expectation on $Lip(\Omega)$ and $(B_t)_{t\geq 0}$ is a G-Brownian motion. Since $Lip(\Omega_T) \subset Lip(\Omega)$, $\hat{\mathbb{E}}$ is also a sublinear expection on $Lip(\Omega_T)$. We denote by $L_G^p(\Omega_T)$ the completion of $Lip(\Omega_T)$ under the norm $(\hat{\mathbb{E}}[|\cdot|^p])^{\frac{1}{p}}$, for $p \geq 1$. Therefore, it can be extended continuously to the completion $L_G^p(\Omega_T)$.

Denis et al. [3] proved the following representation theorem of G-expectation on $L_G^1(\Omega_T)$.

Lemma 4.1. *There exists a weakly compact set $\mathcal{P} \subset \mathcal{M}_1(\Omega_T)$, the set of all probability measures on $(\Omega_T, \mathcal{B}(\Omega_T))$, such that*

$$\hat{\mathbb{E}}[\xi] = \sup_{P\in\mathcal{P}} \mathbb{E}_P[\xi]$$

for all $\xi \in L_G^1(\Omega_T)$, $\mathcal{P}$ is called a set that represents $\hat{\mathbb{E}}$.

Let $\mathcal{P}$ be a weakly compact set that represents $\hat{\mathbb{E}}$. For this $\mathcal{P}$, we define the capacity

$$\overline{\mathbb{C}}(A) := \sup_{P\in\mathcal{P}} P(A), A \in \mathcal{B}(\Omega_T).$$

Lemma 4.2. *([16]) Let $X \in L_G^p$. Then, for each $\alpha > 0$, it holds that*

$$\overline{\mathbb{C}}\{|X| > \alpha\} \leq \frac{\hat{\mathbb{E}}[|X|^p]}{\alpha^p}.$$

2.3. *G-Itô calculus*

Peng ([16]) also introduced the related stochastic calculus of Itôs type with respect to G-Brownian motion. Let $T \in \mathbb{R}^+$ be fixed.

Definition 4.4. For fixed $p \geq 1, T \in \mathbb{R}^+$, define the space $M_G^{p,0}([0,T])$ of simple process by

$$M_G^{p,0}([0,T]) := \{\eta_t(\omega) := \sum_{j=0}^{N-1} \zeta_{t_j}(\omega) I_{[t_j,t_{j+1})}; \zeta_{t_j}(\omega) \in L_G^p(\Omega_{t_j}),$$
$$\forall N \geq 1, 0 = t_0 < t_1 < \cdots < t_N = T, j = 0, 1, \cdots, N-1\}.$$

Denote by $M_G^p([0,T])$ the completion of $M_G^{p,0}([0,T])$ under the norm

$$\|\eta\|_{M_G^p([0,T])} = \left| \int_0^T \hat{\mathbb{E}}[|\eta_s|^p] ds \right|^{\frac{1}{p}}.$$

Definition 4.5. For $\eta_t(\omega) = \sum_{j=1}^{N-1} \zeta_{t_j}(\omega) I_{[t_j,t_{j+1})} \in M_G^{p,0}([0,T])$, define

$$I(\eta) = \int_0^T \eta_t dB_t := \sum_{j=0}^{N-1} \zeta_j \left(B_{t_{j+1}}^N - B_{t_j}^N \right).$$

The map $I : M_G^{2,0}([0,T]) \to L_G^2(\Omega_T)$ is linear and continuous. Hence, it can be extended continuously to $I : M_G^2([0,T]) \to L_G^2(\Omega_T)$.

For each $\eta \in M_G^p([0,T])$, the stochastic integral is defined by

$$I(\eta) = \int_0^T \eta_t dB_t, \eta_t \in M_G^p([0,T]).$$

Unlike the classical theory, the quadratic variation process of G-Brownian motion B is not always a deterministic process and it can be formulated in $L_G^2(\Omega_t)$ by

$$\langle B \rangle_t = \lim_{N\to\infty} \sum_{i=0}^{N-1} (B_{t_{i+1}^N} - B_{t_i^N})^2 = B_t^2 - 2\int_0^t B_s dB_s,$$

where $t_i^N = \frac{iT}{N}$ for each integer $N \geq 1$.

Definition 4.6. Define a mapping $M_G^{1,0}([0,T]) \to L_G^1(\Omega_T)$:

$$Q(\eta) = \int_0^T \eta_s \langle B \rangle_s := \sum_{i=0}^{N-1} (\langle B \rangle_{t_{i+1}^N} - \langle B \rangle_{t_i^N})^2.$$

Then Q can be uniquely extended to $M_G^1([0,T]) \to L_G^1(\Omega_T)$. We also denote this mapping by

$$Q(\eta) = \int_0^T \eta_s \langle B \rangle_s.$$

In view of the dual formulation of G-expectation as well as the properties of the quadratic variation process $\langle B \rangle$ in G-framework, Gao, Song and Bai ([1, 6, 24]) obtained the following BDG type inequalities.

Lemma 4.3. *Let $p \geq 1, \eta \in M_G^p([0,T])$ and $0 \leq s \leq t \leq T$. Then*

$$\hat{\mathbb{E}}\left[\sup_{0\leq t\leq T}\left|\int_0^t \eta(s)d\langle B\rangle_s\right|^p\right] \leq T^{p-1}\hat{\mathbb{E}}\left[\int_0^T |\eta(s)|^p ds\right].$$

Lemma 4.4. *Let $p \geq 1, \eta \in M_G^p([0,T])$ and $0 \leq s \leq t \leq T$. Then*

$$\hat{\mathbb{E}}\left[\sup_{t\in[0,T]}\left|\int_0^t \eta(r)dB(r)\right|^p\right] \leq C_{p,T}\hat{\mathbb{E}}\left[\left|\int_0^T |\eta(r)|^2 dr\right|^{\frac{p}{2}}\right].$$

Convention. The letter C below with or without subscripts will denote positive constants whose value may change in different occasions. We will write the dependence of constants on parameters explicitly if it is essential.

3. An averaging principle for multi-valued SDEs driven by G-Brownian motion

3.1. *Notations and assumptions*

In this paper, we will consider the stochastic averaging for solutions to the following multi-valued SDEs driven by G-Brownian motion on $\mathbb{R}$:

$$\begin{aligned} dX(t) + \partial\varphi(X(t)) \ni f(t, X(t))dt + g(t, X(t))d\langle B\rangle_t \\ +\sigma(t, X(t))dB(t), t \in [0,T], \end{aligned} \tag{3.1}$$

where the initial condition $X(0) = \xi \in \overline{\mathrm{Dom}(\varphi)}$ with $\hat{\mathbb{E}}[|\xi|^2] < \infty$, and $\partial\varphi$ is the subdifferential operator associated to φ, which is a lower and semicontinuous (l.s.c) function on $\mathbb{R}$. $\langle B\rangle_t$ is the quadratic variation process of G-Brownian motion $B(t)$.

Let us consider the following assumptions:

(H1) The function $\varphi : \mathbb{R}^d \to (-\infty, +\infty)$ is l.s.c. such that

$$\mathrm{Int}(\mathrm{Dom}(\varphi)) \neq \emptyset,$$

where $\mathrm{Dom}(\varphi) \equiv \{x \in \mathbb{R}^d : \mathrm{Dom}(\varphi) < +\infty\}$ and supposes that $0 \in \mathrm{Int}(\mathrm{Dom}(\varphi))$ and $\varphi(x) \geq \varphi(0) = 0$, for all $x \in \mathbb{R}^d$.

We recall that the subdifferential of the function φ is defined by

$$\partial\varphi(x) = \{y \in \mathbb{R}^d : \langle y, z - x\rangle + \varphi(x) \leq \varphi(z), \forall z \in \mathbb{R}^d\}$$

and by $(x, x^*) \in Gr(\partial\varphi)$, which means that $x \in \mathrm{Dom}(\partial\varphi)$ and $x^* \in \mathrm{Dom}(\partial\varphi)$ where

$$\mathrm{Dom}(\partial\varphi) := \{x \in \mathbb{R}^d : \partial\varphi(x) \neq \emptyset\}.$$

Remark 4.1. Under assumption (H1), the subdifferential operator φ becomes a maximal monotone operator, i.e., maximal in the class of operators which satisfy the condition

$$\langle y^* - z^*, y - z\rangle \geq 0, (y, y^*), (z, z^*) \in Gr(\partial\varphi).$$

Conversely (only in the case $d = 1$), we recall that, if A is a given maximal monotone operator on $\mathbb{R}$, then there exists a proper l.s.c. function . such that $A = \partial\varphi$.

Now we give the previous definition of the solution to (4.4).

Definition 4.7. The solution of Eq. (4.4) is a pair of processes (X, K) satisfying that

(i) $X(0) = \xi$ and $X(t) \in \overline{\mathrm{Dom}(\varphi)}$ for every $s \in [t, T], q.s.$;
(ii) for any $t < T < +\infty$, K is continuous and of finite variation on $[t, T]$ with $K(0) = 0$, and for all $s, s' \in [t, T]$,

$$\lim_{s \to s'} \hat{\mathbb{E}}[|K(s) - K'(s)|]^2 = 0;$$

(iii)

$$X(t) + K(t) = X(s) + \int_s^t f(r, X(r))dr + \int_s^t \sigma(r, X(r))dB(r) + \int_s^t g(r, X(r))d\langle B\rangle_r,$$

(iv) $\int_s^t \langle u - X(r), dK(r)\rangle - \int_s^t \varphi(X(r))dr \leq (t-s)\varphi(u), a.s.$.

Lemma 4.5. *In what follows, we shall write $dK(s) \in \varphi(X(s))ds$, q.s., instead of inequality (iv) (see also the below result). Now, taking the processes X, X', K, K' such that $dK(s) \in \varphi(X(s))ds$ and $dK'(s) \in \varphi(X'(s))ds$, we have*

$$\int_0^t \langle X(s) - X'(s), dK(s) - dK'(s)\rangle \geq 0.$$

Lemma 4.6. *Let $\varphi \in C^2([0,T] \times \mathbb{R})$ be a real function with $\partial_s\varphi, \partial_x\varphi, \partial_{xx}\varphi \in C_{b,Lip(\mathbb{R})}$. Let b, σ and f be bounded processes in $M_G^2([0,T])$ and $K \in M_G^2([0,T])$ is continuous, which satisfies for each $s \leq t \leq T$,*

$$\lim_{s \to s'} \hat{\mathbb{E}}[|K(s) - K'(s)|^2] = 0;$$

Then,

$$\begin{aligned}\Phi(X(t), t) - \Phi(X(s), s) = &\int_s^t \left[\frac{\partial \Phi}{\partial t}(X(r)) + \frac{\partial \Phi}{\partial x}(X(r))f(r)\right] dr \\ &+ \int_s^t \frac{\partial \Phi}{\partial t}(X(r))g(r)d\langle B\rangle_r \\ &+ \int_s^t \frac{\partial \Phi}{\partial x}(X(r))\sigma(r)dB(r) \\ &- \int_s^t \frac{\partial \Phi}{\partial x}(X(r))dK(r) \\ &+ \int_s^t \frac{\partial^2 \Phi}{\partial x^2}(X(r))\sigma^2(r)d\langle B\rangle_r.\end{aligned}$$

3.2. *The averaging principle*

We now consider the following integral form of MSDEs (4.4) in $\mathbb{R}$:

$$X(t) = X(0) + \int_0^t f(r, X(r))dr + \int_0^t \sigma(r, X(r))dB(r) + \int_0^t g(r, X(r))d\langle B\rangle_r - K(t), \tag{3.2}$$

where $X(0)$ is a given 1-dimensional random variable. And let us consider the following assumption:

(H2) The functions $f : [0,T]\times\mathbb{R} \to \mathbb{R}$, $g : [0,T]\times\mathbb{R} \to \mathbb{R}$ and $\sigma : [0,T]\times\mathbb{R} \to \mathbb{R}\times\mathbb{R}$ are continuous. There exists a constant C_1 such that for any $x,\ y \in \mathbb{R}$,

$$|f(t,x) - f(t,y)| + |\sigma(t,x) - \sigma(t,y)| + |g(t,x) - g(t,y)| \le C_1|x-y|,$$
$$|f(t,0)| + |g(t,0)| + |\sigma(t,0)| \le C_1.$$

On the basis of Theorem 3.1 in [19], if the coefficients of (3.2) satisfy conditions (H1)–(H2), there exists the unique solution $X(t), t \in [0,T]$ to (3.2).

Now we discuss a standard MSDE driven by G-Brownian motion using an averaging principle in $\mathbb{R}$. The MSDE is defined as:

$$X_\varepsilon(t) = X(0) + \varepsilon \int_0^t f(s, X_\varepsilon(s))ds + \sqrt{\varepsilon}\int_0^t \sigma(s, X_\varepsilon(s))dB(s) + \varepsilon\int_0^t g(s, X_\varepsilon(s))d\langle B\rangle_s - \varepsilon K(t), \tag{3.3}$$

where $X(0) = X_0$ is a given 1-dimensional random variable as the initial condition, $t \in [0,T]$ and the coefficients have the same conditions as in (3.2), and $\varepsilon \in (0, \varepsilon_0]$ is a positive small parameter with ε_0 a fixed number.

Assume that (H1)–(H2) are satisfied. Besides, the functions $\bar{f} : [0,T] \times \mathbb{R} \to \mathbb{R}$, $\bar{\sigma}, \bar{g} : [0,T] \times \mathbb{R} \to \mathbb{R} \times \mathbb{R}$ are continuous. Presuming they meet the following additional inequalities:

(C1)

$$\frac{1}{T_1}\int_0^{T_1} |f(s,x) - \bar{f}(x)|^2 ds \le \varphi_1(T_1)(1+|x|^2),$$

(C2)

$$\frac{1}{T_1}\int_0^{T_1} |g(s,x) - \bar{g}(x)|^2 ds \le \varphi_2(T_1)(1+|x|^2),$$

(C3)

$$\frac{1}{T_1}\int_0^{T_1} |\sigma(s,x) - \bar{\sigma}(x)|^2 ds \le \varphi_3(T_1)(1+|x|^2),$$

where $T_1 \in [0,T]$, $\varphi_i(T_1)$ are positive bounded functions with $\lim_{T_1\to\infty} \varphi_i(T_1) = 0, i = 1,2,3$.

Then, we can obtain the MSDE with the averaging principle:

$$\begin{aligned}Z_\varepsilon(t) = X(0) + \varepsilon\int_0^t \bar{f}(Z_\varepsilon(s))ds + \sqrt{\varepsilon}\int_0^t \bar{\sigma}(Z_\varepsilon(s))dB(s)\\ +\varepsilon\int_0^t \bar{g}(Z_\varepsilon(s))d\langle B\rangle_s - \varepsilon\bar{K}(t).\end{aligned} \tag{3.4}$$

This MSDE is called the averaged MSDE of the original standard MSDE (3.2). Under the similar conditions such as $X(t)$ in (3.3), this equation will have a unique solution $Z_\varepsilon(t)$.

Lemma 4.7. *Suppose that the original MSDE* (3.3) *satisfies the assumptions* (H1)–(H2). *Then, we have*

$$\hat{\mathbb{E}}\bigg[\sup_{t\in[0,T]}|X(t)|^{2p}\bigg] \le C.$$

Proof: By Itô's formula, we have

$$\begin{aligned}|X(t)|^{2p} = |X_0|^{2p} + 2p\int_0^t |X(s)|^{2p-2}\langle X(s), f(s,X(s))\rangle ds\\ +2p\int_0^t |X(s)|^{2p-2}\langle X(s), g(s,X(s))\rangle d\langle B\rangle_s\\ +2p\int_0^t |X(s)|^{2p-2}\langle X(s), \sigma(s,X(s))\rangle dB(s)\\ +2p(p-1)\int_0^t |X(s)|^{2p-4}|X^T(s)\sigma(s,X(s))|^2 d\langle B\rangle_s\\ +p\int_0^t |X(s)|^{2p-2}|\sigma(s,X(s))|^2 d\langle B\rangle_s\\ -2p\int_0^t |X(s)|^{2p-2}\langle X(s), dK(s)\rangle.\end{aligned}$$

Firstly, from Lemma 4.5, we have

$$\int_0^t |X(s)|^{2p-2}\langle X(s), dK(s)\rangle \ge 0.$$

Then, we have

$$\begin{aligned}\hat{\mathbb{E}}\bigg[\sup_{t\in[0,T]}|X(t)|^{2p}\bigg] \le \hat{\mathbb{E}}[|X_0|^{2p}] + C\hat{\mathbb{E}}\bigg[\sup_{t\in[0,T]}\bigg|\int_0^t |X(s)|^{2p-2}\langle X(s), f(s,X(s))\rangle ds\bigg|\bigg]\\ +C\hat{\mathbb{E}}\bigg[\sup_{t\in[0,T]}\bigg|\int_0^t |X(s)|^{2p-2}\langle X(s), g(s,X(s))\rangle d\langle B\rangle_s\bigg|\bigg]\\ +C\hat{\mathbb{E}}\bigg[\sup_{t\in[0,T]}\bigg|\int_0^t |X(s)|^{2p-2}\langle X(s), \sigma(s,X(s))\rangle dB(s)\bigg|\bigg]\\ +C\hat{\mathbb{E}}\bigg[\sup_{t\in[0,T]}\bigg|\int_0^t |X(s)|^{2p-4}|X^T(s)\sigma(s,X(s))|^2 d\langle B\rangle_s\bigg|\bigg]\\ +C\hat{\mathbb{E}}\bigg[\sup_{t\in[0,T]}\bigg|\int_0^t |X(s)|^{2p-2}|\sigma(s,X(s))|^2 d\langle B\rangle_s\bigg|\bigg]\end{aligned}$$

$$
\begin{aligned}
&\leq \hat{\mathbb{E}}[|X_0|^{2p}] + C\int_0^T \hat{\mathbb{E}}[|X(s)|^{2p-1}|f(s,X(s))|]ds \\
&+C\int_0^T \hat{\mathbb{E}}[|X(s)|^{2p-1}|g(s,X(s))|]ds \\
&+C\int_0^T \hat{\mathbb{E}}[|X(s)|^{2p-2}|\sigma(s,X(s))|^2]ds \\
&+C\hat{\mathbb{E}}\Bigg[\sup_{t\in[0,T]}\Bigg|\int_0^t |X(s)|^{2p-2}\langle X(s),\sigma(s,X(s))\rangle dB(s)\Bigg|\Bigg].
\end{aligned}
$$

Next, by Lemma 1.3, we have

$$
\begin{aligned}
&\hat{\mathbb{E}}\Bigg[\sup_{t\in[0,T]}\Bigg|\int_0^t |X(s)|^{2p-2}\langle X(s),\sigma(s,X(s))\rangle dB(s)\Bigg|\Bigg] \\
&\leq C\hat{\mathbb{E}}\Bigg[\int_0^T |X(s)|^{4p-2}|\sigma(s,X(s))|^2 d\langle B\rangle_s\Bigg]^{\frac{1}{2}} \\
&\leq \frac{1}{2p}\hat{\mathbb{E}}\Bigg[\sup_{t\in[0,T]}|X(t)|^{2p}\Bigg] + C\int_0^T \hat{\mathbb{E}}[|X(s)|^{2p-2}\cdot|\sigma(s,X(s))|^2]ds.
\end{aligned}
$$

Finally, we have

$$
\begin{aligned}
\hat{\mathbb{E}}\Bigg[\sup_{t\in[0,T]}|X(t)|^{2p}\Bigg] &\leq C(1+\hat{\mathbb{E}}[|X_0|^{2p}]) + C\int_0^T \hat{\mathbb{E}}[\sup_{s\in[0,t]}|X(s)|^{2p}]ds \\
&\quad + \frac{1}{2p}\hat{\mathbb{E}}\Bigg[\sup_{t\in[0,T]}|X(t)|^{2p}\Bigg].
\end{aligned}
$$

Therefore, by Gronwall's inequality, we have

$$
\hat{\mathbb{E}}\Bigg[\sup_{t\in[0,T]}|X(t)|^{2p}\Bigg] \leq C.
$$

□

Now we claim a strong limit theorem to show relationship between solution processes $X_\varepsilon(t)$ and $Z_\varepsilon(t)$. It shows that the solution of averaged (3.4) converges to that of the original (3.3) in the sense of p-th moments and convergence in capacity.

Theorem 4.1. *Suppose that the original MSDE* (3.3) *and the averaged MSDE* (3.4) *both satisfy the assumptions* (H1)–(H2) *and* (C1)–(C3). *For a given arbitrarily small number* $\delta_1 > 0$, *there exist* $L > 0$, $\alpha \in (0,1)$, $\varepsilon_1 \in (0,\varepsilon_0]$, *such that for any* $\varepsilon \in (0,\varepsilon_1]$, *each* $t \in [0, L\varepsilon^{\frac{1}{2}-\alpha}(1-\sqrt{\varepsilon})]$,

$$
\hat{\mathbb{E}}\Bigg[\sup_{t\in[0,L\varepsilon^{\frac{1}{2}-\alpha}(1-\sqrt{\varepsilon})]}|X_\varepsilon(t)-Z_\varepsilon(t)|^{2p}\Bigg] \leq \delta_1.
$$

Proof: Consider the difference $X_\varepsilon(t) - Z_\varepsilon(t)$. By (3.3) and (3.4), we have

$$
\begin{aligned}
X_\varepsilon(t) - Z_\varepsilon(t) &= \varepsilon\int_0^t [f(s,X_\varepsilon(s)) - \bar{f}(Z_\varepsilon(s))]ds \\
&\quad + \sqrt{\varepsilon}\int_0^t [\sigma(s,X_\varepsilon(s)) - \bar{\sigma}(Z_\varepsilon(s))]dB(s)
\end{aligned}
$$

$$
\begin{aligned}
&+\varepsilon\int_0^t [g(s,X_\varepsilon(s))-\bar{g}(Z_\varepsilon(s))]d\langle B\rangle_s\\
&-\varepsilon[K(t)-\bar{K}(t)].
\end{aligned}
$$

By Itô's formula, let $\Delta(s)=X_\varepsilon(s)-Z_\varepsilon(s)$, we have

$$
\begin{aligned}
|X_\varepsilon(t)-Z_\varepsilon(t)|^{2p} &= 2p\varepsilon\int_0^t |\Delta(s)|^{2p-2}\langle\Delta(s), f(s,X_\varepsilon(s))-\bar{f}(Z_\varepsilon(s))\rangle ds\\
&\quad +2p\varepsilon\int_0^t |\Delta(s)|^{2p-2}\langle\Delta(s), g(s,X_\varepsilon(s))-\bar{g}(Z_\varepsilon(s))\rangle d\langle B\rangle_s\\
&\quad +2p\sqrt{\varepsilon}\int_0^t |\Delta(s)|^{2p-2}\langle\Delta(s), \sigma(s,X_\varepsilon(s))-\bar{\sigma}(Z_\varepsilon(s))\rangle dB(s)\\
&\quad +2p(p-1)\varepsilon\int_0^t |\Delta(s)|^{2p-4}|\Delta^T(s)(\sigma(s,X_\varepsilon(s))\\
&\quad -\bar{\sigma}(Z_\varepsilon(s)))|^2 d\langle B\rangle_s\\
&\quad +p\varepsilon\int_0^t |\Delta(s)|^{2p-2}|\sigma(s,X_\varepsilon(s))-\bar{\sigma}(Z_\varepsilon(s))|^2 d\langle B\rangle_s\\
&\quad -2p\varepsilon\int_0^t |\Delta(s)|^{2p-2}\langle X_\varepsilon(s)-Z_\varepsilon(s), dK(s)-d\bar{K}(s)\rangle\\
&:= \sum_{i=1}^{6} I_i.
\end{aligned}
$$

Firstly, for I_1, by the elementary inequality, we can obtain

$$
\begin{aligned}
&\hat{\mathbb{E}}[\sup_{0\le t\le u} I_1]\\
&\le 2p\varepsilon\hat{\mathbb{E}}\left[\sup_{0\le t\le u}\left|\int_0^t |\Delta(s)|^{2p-2}\langle\Delta(s), f(s,X_\varepsilon(s))-\bar{f}(Z_\varepsilon(s))\rangle ds\right|\right]\\
&\le 2p\varepsilon\hat{\mathbb{E}}\left[\sup_{0\le t\le u}\left|\int_0^t |\Delta(s)|^{2p-2}\langle\Delta(s), f(s,X_\varepsilon(s))-f(s,Z_\varepsilon(s))\rangle ds\right|\right]\\
&\quad +2p\varepsilon\hat{\mathbb{E}}\left[\sup_{0\le t\le u}\left|\int_0^t |\Delta(s)|^{2p-2}\langle\Delta(s), f(s,Z_\varepsilon(s))-\bar{f}(Z_\varepsilon(s))\rangle ds\right|\right]\\
&:= I_{11}+I_{12}.
\end{aligned}
$$

By (H2), for I_{11}, one has

$$
\begin{aligned}
I_{11} &= 2p\varepsilon\hat{\mathbb{E}}\left[\sup_{0\le t\le u}\left|\int_0^t |\Delta(s)|^{2p-2}\langle\Delta(s), f(s,X_\varepsilon(s))-f(s,Z_\varepsilon(s))\rangle ds\right|\right]\\
&\le 2p\varepsilon L\int_0^u \hat{\mathbb{E}}\left[\sup_{0\le r\le s}|X_\varepsilon(r)-Z_\varepsilon(r)|^{2p}\right]ds. \qquad (3.5)
\end{aligned}
$$

Next, about I_{12}, by Young's inequality, and (C1), we get

$$
\begin{aligned}
I_{12} &\le 2p\varepsilon\hat{\mathbb{E}}\left[\sup_{0\le t\le u}\left|\int_0^t |\Delta(s)|^{2p-2}\langle\Delta(s), f(s,Z_\varepsilon(s))-\bar{f}(Z_\varepsilon(s))\rangle ds\right|\right]\\
&\le p\varepsilon\hat{\mathbb{E}}\left[\sup_{0\le t\le u}\int_0^t |X_\varepsilon(s)-Z_\varepsilon(s)|^{2p} ds\right]
\end{aligned}
$$

$$
\begin{aligned}
&+p\varepsilon\hat{\mathbb{E}}\left[\sup_{0\le t\le u}\int_0^t|\Delta(s)|^{2p-2}|f(s,Z_\varepsilon(s))-\bar{f}(Z_\varepsilon(s))|^2ds\right]\\
&\le p\varepsilon\hat{\mathbb{E}}\left[\sup_{0\le t\le u}\int_0^t|X_\varepsilon(s)-Z_\varepsilon(s)|^{2p}ds\right]\\
&\quad+\varepsilon\hat{\mathbb{E}}\left[\sup_{0\le t\le u}\int_0^t[(p-1)|\Delta(s)|^{2p}+1]|f(s,Z_\varepsilon(s))-\bar{f}(Z_\varepsilon(s))|^2ds\right]\\
&\le p\varepsilon\hat{\mathbb{E}}\left[\sup_{0\le t\le u}\int_0^t|X_\varepsilon(s)-Z_\varepsilon(s)|^{2p}ds\right]\\
&\quad+(C_{x,p}(p-1)+1)\varepsilon\hat{\mathbb{E}}\left[\sup_{0\le t\le u}t\frac{1}{t}\int_0^t|f(s,Z_\varepsilon(s))-\bar{f}(Z_\varepsilon(s))|^2ds\right]\\
&\le p\varepsilon\int_0^u\hat{\mathbb{E}}\left[\sup_{0\le r\le s}|X_\varepsilon(r)-Z_\varepsilon(r)|^{2p}\right]ds\\
&\quad+(C_{x,p}(p-1)+1)\varepsilon u[\sup_{0\le t\le u}\varphi_1(t)]\left(1+\hat{\mathbb{E}}\left[\sup_{0\le s\le u}|Z_\varepsilon(s)|^2\right]\right).
\end{aligned}
$$

By the properties of solution for MSDE, we know that if $\hat{\mathbb{E}}[|X_0|^2]<\infty$, then for each $T\ge 0$,

$$\hat{\mathbb{E}}\left[\sup_{0\le s\le u}|Z_\varepsilon(s)|^2\right]<\infty.$$

This property combines with the boundedness of $\varphi_1(t)$. We can further estimate that there exists a constant C_{12} such that

$$I_{12}\le p\varepsilon\int_0^u\hat{\mathbb{E}}\left[\sup_{0\le r\le s}|X_\varepsilon(r)-Z_\varepsilon(r)|^{2p}\right]ds+\varepsilon uC_{12}. \tag{3.6}$$

So putting (3.5) and (3.6) together, we have

$$I_1\le(2L+1)p\varepsilon\int_0^u\hat{\mathbb{E}}\left[\sup_{0\le r\le s}|X_\varepsilon(r)-Z_\varepsilon(r)|^{2p}\right]ds+\varepsilon uC_{12}. \tag{3.7}$$

Then for I_2, we get

$$
\begin{aligned}
\hat{\mathbb{E}}[\sup_{0\le t\le u}I_2]&\le 2p\varepsilon\hat{\mathbb{E}}\left[\sup_{0\le t\le u}\left|\int_0^t|\Delta(s)|^{2p-2}\langle\Delta(s),g(s,X_\varepsilon(s))-\bar{g}(Z_\varepsilon(s))\rangle d\langle B\rangle_s\right|\right]\\
&\le 2p\varepsilon\hat{\mathbb{E}}\left[\int_0^u|\Delta(s)|^{2p-2}\langle\Delta(s),g(s,X_\varepsilon(s))-\bar{g}(Z_\varepsilon(s))\rangle ds\right]\\
&\le 2p\varepsilon\hat{\mathbb{E}}\left[\int_0^u|\Delta(s)|^{2p-2}\langle\Delta(s),g(s,X_\varepsilon(s))-g(s,Z_\varepsilon(s))\rangle ds\right]\\
&\quad+2p\varepsilon\hat{\mathbb{E}}\left[\int_0^u|\Delta(s)|^{2p-2}\langle\Delta(s),g(s,Z_\varepsilon(s))-\bar{g}(Z_\varepsilon(s))\rangle ds\right]\\
&\le 2pL\varepsilon\hat{\mathbb{E}}\left[\int_0^u|X_\varepsilon(s)-Z_\varepsilon(s)|^{2p}ds\right]\\
&\quad+p\varepsilon\hat{\mathbb{E}}\left[\int_0^u(|\Delta(s)|^{2p}+|\Delta(s)|^{2p-2}|g(s,X_\varepsilon(s))-\bar{g}(Z_\varepsilon(s))|^2)ds\right]\\
&\le(2L+1)p\varepsilon\hat{\mathbb{E}}\left[\int_0^u|X_\varepsilon(s)-Z_\varepsilon(s)|^{2p}ds\right]
\end{aligned}
$$

$$
\begin{aligned}
&+p\varepsilon\hat{\mathbb{E}}\Big[\int_0^u |\Delta(s)|^{2p-2}|g(s,Z_\varepsilon(s))-\bar{g}(Z_\varepsilon(s))|^2 ds\Big]\\
&:= I_{21}+I_{22}. \qquad (3.8)
\end{aligned}
$$

Thanks to condition (H2), one can get

$$
\begin{aligned}
I_{21} &\le (2L+1)p\varepsilon\hat{\mathbb{E}}\Big[\int_0^u |X_\varepsilon(s)-Z_\varepsilon(s)|^{2p} ds\Big]\\
&\le \varepsilon C_{21}\int_0^u \hat{\mathbb{E}}\Big[\sup_{0\le r\le s}|X_\varepsilon(r)-Z_\varepsilon(r)|^{2p}\Big]ds.
\end{aligned}
$$

Due to condition (C2), we have

$$
\begin{aligned}
I_{22} &= p\varepsilon\hat{\mathbb{E}}\Big[\int_0^u |\Delta(s)|^{2p-2}|g(s,Z_\varepsilon(s))-\bar{g}(Z_\varepsilon(s))|^2 ds\Big]\\
&\le \varepsilon\hat{\mathbb{E}}\Big[\int_0^u [(p-1)|\Delta(s)|^{2p}+1]|g(s,Z_\varepsilon(s))-\bar{g}(Z_\varepsilon(s))|^2 ds\Big]\\
&\le (C_{x,p}(p-1)+1)\varepsilon\hat{\mathbb{E}}\Big[u\frac{1}{u}\int_0^u |g(s,Z_\varepsilon(s))-\bar{g}(Z_\varepsilon(s))|^2 ds\Big]\\
&\le (C_{x,p}(p-1)+1)\varepsilon u[\sup_{0\le t\le u}\varphi_3(t)]\Big(1+\hat{\mathbb{E}}\Big[\sup_{0\le s\le u}|Z_\varepsilon(s)|^2\Big]\Big)\\
&\le \varepsilon u C_{22}.
\end{aligned}
$$

Then

$$
\hat{\mathbb{E}}[\sup_{0\le t\le u} I_2] \le \varepsilon C_{21}\int_0^u \hat{\mathbb{E}}\Big[\sup_{0\le r\le s}|X_\varepsilon(r)-Z_\varepsilon(r)|^{2p}\Big]ds + \varepsilon u C_{22}. \qquad (3.9)
$$

To proceed, using the Young's inequality, we have

$$
\begin{aligned}
I_3 &\le 2p\sqrt{\varepsilon}\hat{\mathbb{E}}\left[\sup_{0\le t\le u}\left|\int_0^t |\Delta(s)|^{2p-2}\langle\Delta(s),\sigma(s,X_\varepsilon(s))-\bar{\sigma}(Z_\varepsilon(s))\rangle dB(s)\right|\right]\\
&\le 2p\sqrt{\varepsilon}\hat{\mathbb{E}}\left[\int_0^u |\Delta(s)|^{4p-4}|\langle X_\varepsilon(s)-Z_\varepsilon(s),\sigma(s,X_\varepsilon(s))-\bar{\sigma}(Z_\varepsilon(s))\rangle|^2 ds\right]^{\frac{1}{2}}\\
&\le 2p\sqrt{\varepsilon}\hat{\mathbb{E}}\left[\sup_{0\le s\le u}|\Delta(s)|^{2p}\int_0^u |\Delta(s)|^{2p-2}|\sigma(s,X_\varepsilon(s))-\bar{\sigma}(Z_\varepsilon(s))|^2 ds\right]^{\frac{1}{2}}\\
&\le \sqrt{\varepsilon}\hat{\mathbb{E}}\Big[\sup_{0\le t\le u}|X_\varepsilon(t)-Z_\varepsilon(t)|^{2p}\Big]\\
&\quad +C_p\sqrt{\varepsilon}\hat{\mathbb{E}}\left[\int_0^u |\Delta(s)|^{2p-2}|\sigma(s,X_\varepsilon(s))-\bar{\sigma}(Z_\varepsilon(s))|^2 ds\right]\\
&:= I_{31}+I_{32}.
\end{aligned}
$$

According to (3.8) and (3.9), we get

$$
I_{32} \le C_{31}\sqrt{\varepsilon}\int_0^u \hat{\mathbb{E}}\Big[\sup_{0\le r\le s}|X_\varepsilon(r)-Z_\varepsilon(r)|^{2p}\Big]ds + C_{32}u\sqrt{\varepsilon}.
$$

Thus, we have

$$
\hat{\mathbb{E}}[\sup_{0\le t\le u} I_3] \le \sqrt{\varepsilon}\hat{\mathbb{E}}\Big[\sup_{0\le s\le u}|X_\varepsilon(s)-Z_\varepsilon(s)|^{2p}\Big] + C_{32}u\sqrt{\varepsilon}
$$

$$+C_{31}\sqrt{\varepsilon}\int_0^u \hat{\mathbb{E}}\Big[\sup_{0\le r\le s}|X_\varepsilon(r)-Z_\varepsilon(r)|^{2p}\Big]ds. \tag{3.10}$$

For the term I_4, we have

$$\begin{aligned}\hat{\mathbb{E}}[\sup_{0\le t\le u} I_4] &\le 2p(p-1)\varepsilon\hat{\mathbb{E}}\Big[\sup_{0\le t\le u}\Big|\int_0^t |\Delta(s)|^{2p-4}|\Delta^T(s)(\sigma(s,X_\varepsilon(s))\\ &\quad -\bar{\sigma}(Z_\varepsilon(s)))|^2 d\langle B\rangle_s\Big|\Big]\\ &\le 2p(p-1)\varepsilon\hat{\mathbb{E}}\Big[\int_0^u |\Delta(s)|^{2p-2}|(\sigma(s,X_\varepsilon(s))-\bar{\sigma}(Z_\varepsilon(s)))|^2 ds\Big].\end{aligned}$$

Similar to the proof of I_{32}, we have

$$\hat{\mathbb{E}}[\sup_{0\le t\le u} I_4] \le C_{41}\varepsilon\int_0^u \hat{\mathbb{E}}\Big[\sup_{0\le r\le s}|X_\varepsilon(r)-Z_\varepsilon(r)|^{2p}\Big]ds + C_{42}u\varepsilon.$$

Next, for the term I_5, it is easy to obtain

$$\begin{aligned}\hat{\mathbb{E}}[\sup_{0\le t\le u} I_5] &\le p\varepsilon\hat{\mathbb{E}}\Big[\sup_{0\le t\le u}\Big|\int_0^t |\Delta(s)|^{2p-2}|\sigma(s,X_\varepsilon(s))-\bar{\sigma}(Z_\varepsilon(s))|^2 d\langle B\rangle_s\Big|\Big]\\ &\le p\varepsilon\hat{\mathbb{E}}\Big[\int_0^u |\Delta(s)|^{2p-2}|\sigma(s,X_\varepsilon(s))-\bar{\sigma}(Z_\varepsilon(s))|^2 ds\Big]\\ &\le C_{51}\varepsilon\int_0^u \hat{\mathbb{E}}\Big[\sup_{0\le r\le s}|X_\varepsilon(r)-Z_\varepsilon(r)|^{2p}\Big]ds + C_{52}u\varepsilon.\end{aligned}$$

Then, according to (4.7) and (4.5), it is immediate to conclude that

$$\int_0^t |\Delta(s)|^{2p-2}\langle X_\varepsilon(s)-Z_\varepsilon(s), dK(s)-d\bar{K}(s)\rangle \geqslant 0. \tag{3.11}$$

Finally, we have

$$\begin{aligned}&\hat{\mathbb{E}}\Big[\sup_{0\le t\le u}|X_\varepsilon(t)-Z_\varepsilon(t)|^{2p}\Big]\\ &\le C_{11}\varepsilon\int_0^u \hat{\mathbb{E}}\Big[\sup_{0\le r\le s}|X_\varepsilon(r)-Z_\varepsilon(r)|^{2p}\Big]ds + \varepsilon u C_{12}\\ &\quad +\varepsilon C_{21}\int_0^u \hat{\mathbb{E}}\Big[\sup_{0\le r\le s}|X_\varepsilon(r)-Z_\varepsilon(r)|^{2p}\Big]ds + \varepsilon u C_{22}\\ &\quad +\sqrt{\varepsilon}\hat{\mathbb{E}}\Big[\sup_{0\le s\le u}|X_\varepsilon(s)-Z_\varepsilon(s)|^{2p}\Big] + C_{32}u\sqrt{\varepsilon}\\ &\quad +C_{31}\sqrt{\varepsilon}\int_0^u \hat{\mathbb{E}}\Big[\sup_{0\le r\le s}|X_\varepsilon(r)-Z_\varepsilon(r)|^{2p}\Big]ds\\ &\quad +C_{41}\varepsilon\int_0^u \hat{\mathbb{E}}\Big[\sup_{0\le r\le s}|X_\varepsilon(r)-Z_\varepsilon(r)|^{2p}\Big]ds + C_{42}u\varepsilon\\ &\quad +C_{51}\varepsilon\int_0^u \hat{\mathbb{E}}\Big[\sup_{0\le r\le s}|X_\varepsilon(r)-Z_\varepsilon(r)|^{2p}\Big]ds + C_{52}u\varepsilon.\end{aligned}$$

So, we have

$$(1-\sqrt{\varepsilon})\hat{\mathbb{E}}\Big[\sup_{0\le t\le u}|X_\varepsilon(t)-Z_\varepsilon(t)|^{2p}\Big] \le \Upsilon\sqrt{\varepsilon}\int_0^u \hat{\mathbb{E}}\Big[\sup_{0\le r\le s}|X_\varepsilon(r)-Z_\varepsilon(r)|^{2p}\Big]ds$$

$$+\sqrt{\varepsilon}u\Theta,$$

where

$$\Upsilon = ((C_{11} + C_{21} + C_{41} + C_{51})\sqrt{\varepsilon} + C_{31}),$$

and

$$\Theta = ((C_{12} + C_{22} + C_{42} + C_{52})\sqrt{\varepsilon} + C_{32}).$$

Then by Grönwall's inequality

$$\hat{\mathbb{E}}\left[\sup_{0\le t\le u} |X_\varepsilon(t) - Z_\varepsilon(t)|^{2p}\right] \le \frac{\sqrt{\varepsilon}u\Theta}{1-\sqrt{\varepsilon}} \exp\left(\frac{u\sqrt{\varepsilon}\Upsilon}{1-\sqrt{\varepsilon}}\right).$$

Choose $\alpha \in (0,1)$ and $L > 0$ such that for every $t \in [0, L\varepsilon^{\frac{1}{2}-\alpha}(1-\sqrt{\varepsilon})] \subseteq [0,T]$, we have

$$\hat{\mathbb{E}}\left[\sup_{0\le t\le u} |X_\varepsilon(t) - Z_\varepsilon(t)|^{2p}\right] \le QL\varepsilon^{1-\alpha},$$

where $Q = \Theta \exp(L\varepsilon^{1-\alpha}\Upsilon)$ is a constant.

Constantly, given any number $\delta_1 > 0$, we can choose $\varepsilon_1 \in (0, \varepsilon_0]$ such that for each $\varepsilon \in (0, \varepsilon_1]$, and for every $t \in [0, L\varepsilon^{\frac{1}{2}-\alpha}(1-\sqrt{\varepsilon})] \subseteq [0,T]$, the inequality

$$\hat{\mathbb{E}}\left[\sup_{0\le t\le u} |X_\varepsilon(t) - Z_\varepsilon(t)|^{2p}\right] \le \delta_1$$

holds. □

Remark 4.2. When we take $K(t) = 0$, (3.3) will reduce to the standard SDEs driven by G-Brownian motion and the averaging principle for this equation has been established ([9]).

We also have the following result on uniform convergence in capacity.

Lemma 4.8. *Suppose that all assumptions* (H1)–(H2) *and* (C1)–(C2) *are satisfied. Then for any number $\delta_2 > 0$, each $t \in [0, L\varepsilon^{\frac{1}{2}-\alpha}(1-\sqrt{\varepsilon})]$, we have*

$$\lim_{\varepsilon\to 0} \overline{\mathbb{C}}\left[\sup_{t\in[0,L\varepsilon^{\frac{1}{2}-\alpha}(1-\sqrt{\varepsilon})]} |X_\varepsilon(t) - Z_\varepsilon(t)| > \delta_2\right] = 0,$$

where L and α are the same to Theorem 4.1.

Proof: By Lemma 4.2 and Theorem 4.1, for any given number $\delta_2 > 0$, one can find

$$\begin{aligned}\overline{\mathbb{C}}\left[\sup_{t\in[0,L\varepsilon^{\frac{1}{2}-\alpha}(1-\sqrt{\varepsilon})]} |X_\varepsilon(t) - Z_\varepsilon(t)| > \delta_2\right] &\le \frac{1}{\delta_2^p}\hat{\mathbb{E}}\left[\sup_{0\le t\le u} |X_\varepsilon(t) - Z_\varepsilon(t)|^p\right] \\ &\le \frac{Q}{\delta_2^p}\varepsilon^{1-\alpha}.\end{aligned}$$

Let $\varepsilon \to 0$ and the required result follows. □

Remark 4.3. If the system satisfies the non-Lipschitz condition ([9]), we can also obtain the similar convergence result.

Acknowledgments

This work was supported by the NSF of China (11572247), the Fundamental Research Funds for the Central Universities. The authors would like to thank the referees for their careful reading of the manuscript and for the clarification comments which lead to an improvement of the presentation of the paper.

References

[1] Bai X. and Lin Y., (2014), On the existence and uniqunenss of solutions to the stochastic differential equations driven by G-Brownian motion with integral Lipschitz coefficients, Acta Math. Appl.Sin. Engl. Ser. 30, pp. 589-610.
[2] Cépa E., (1998), Problàme de Skorohod multivoque, Ann. Probab. 26(2), pp. 500-532.
[3] Denis L., Hu M. and Peng S., (2011), Function spaces and capacity related to a sublinear expectation: application to G-Brownian motion pathes, Potential Anal. **34**, pp. 139-161.
[4] Freidlin M. and Wentzell A., Random Perturbations of Dynamical Systems, New York: Springer, (1998).
[5] Duan J. and Wang W., Effective dynamics of stochastic partial differential equations, Elsevier, (2014).
[6] Gao F., (2009), Pathwise properties and homeomorphic flows for stochastic differential equations driven by G-Brownian motion, Stochastic Processes and Their Applications 119, pp. 3356-3382.
[7] Gao F. and Jiang H., (2010). Large deviations for stochastic differential equations driven by G-Brownian motion, Stochastic Processes and Their Applications 120, pp. 2212-2240.
[8] Guo R. and Pei B., (2018), Stochastic averaging principles for multi-valued stochastic differential equations driven by Poisson point processes, Stoch. Anal. Appl. DOI: 10.1080/07362994.2018.1461567.
[9] Han M. and Liu Y., (2017), Stochastic averaging for non-Lipschitz stochastic differential equations with G-Brownian Motion, Chinese Journal of Applied Probability and Statistics 33(3), pp. 297-309.
[10] Khasminskii R., (1968), On the averaging principle for stochastic differential Itô equations, Kybernetika 4, pp. 260-279.
[11] Lin Y., (2013), Stochastic differential equations driven by G-Brownian motion with reflecting boundary conditions, Electronic. J. Probbab **18**, pp. 1-23.
[12] Pei B., Xu Y. and Yin G., (2017), Stochastic averaging for a class of two-time-scale systems of stochastic partial differential equations, Nonlinear. Anal. 160, pp. 159-176.
[13] Pei B., Xu Y. and Yin G., (2018), Averaging principles for SPDEs driven by fractional Brownian motions with random delays modulated by two-time-scale Markov switching processes, Stoch. Dynam. 18(3), pp. 1850023.
[14] Peng S., (2007), G-expectation, G-Brownian motion and related stochastic calculus of Itô type, Stochastic Analysis and Applications **2**, pp. 541-567.
[15] Peng S., (2008a), Multi-dimensional G-Brownian motion and related stochastic calculus under G-expectation, Stochastic Processes and Their Applications 118, pp. 2223-2253.
[16] Peng S., (2008b), Nonlinear expectations and stochastic calculus under uncertainty, arxiv:1002.4546v1.

[17] Ren Y., Bi Q. and Sakthivel R., (2013), Stochastic functional differential equations with infinite delay driven by G-Brownian motion, Math. Methods Appl. Sci 36, pp. 1746-1759.
[18] Ren Y. and Hu L., (2011), A note on the stochastic differential equations driven by G-Brownian motion, Stat. Probab. Lett 81, pp. 580-585.
[19] Ren Y., Wang J. and Hu L., (2017), Multi-valued stochastic differential equations driven by G-Brownian motion and related stochastic control problems, International Journal of Control, 90(5), pp. 1132-1154.
[20] Ren Y., Jia X. and Hu L., (2015), Exponential stability of solutions to impulsive stochastic differential equations with G-Brownian motion, Discrete. Cont. Dyn-B 7, pp. 2157-2169.
[21] Ren J., Xu S. and Zhang H., (2014), On existence, uniqueness and convergence of multi-valued stochastic differential equations driven by continuous semimartingales, Sci. China. Math. 57(3), pp. 589-607.
[22] Ren J., Wu J. and Zhang H., (2010), Exponential ergodicity of non-Lipschitz multivalued stochastic differential equations, Bull. Sci. Math. 134, pp. 391-404.
[23] Ren J. and Xu S., (2009), A transfer principle for multivalued stochastic defferential equations, J. Funct. Anal. 256(9), pp. 2780-2814.
[24] Song Y., (2011), Some properties on G-evaluation and its applications to G-martingale decomposition, Science China Mathematics 54(2), pp. 287-300.
[25] Xu Y., Xu W. and Duan J., (2011), An averaging principle for stochastic dynamical systems with Lévy noise, Physica D. 240, pp. 1395-1401.
[26] Xu Y., Pei B. and Li Y., (2015), Approximation properties for solutions to non-Lipschitz stochastic differential equations with Lévy noise, Math. Method. Appl. Sci. 38, pp. 2120-2131.
[27] Xu Y., et al., (2014), Stochastic averaging principle for dynamical systems with fractional Brownian motion, Discrete. Cont. Dyn-B. 19(4), pp. 1197-1212.
[28] Xu Y., Pei B. and Guo R., (2015), Stochastic averaging for slow–fast dynamical systems with fractional Brownian motion, Discrete. Cont. Dyn-B. 20(7), pp. 2257-2267.
[29] Xu Y., Pei B. and Wu J., (2017), Stochastic averaging principle for differential equations with non-Lipschitz coefficients driven by fractional Brownian motion, Stoch. Dynam. 17(2), pp. 1750013.
[30] Xu J. and Liu J., (2014), An averaging principle for multivalued stochastic defferential equations, Stoch. Anal. Appl. 32, pp. 962-974.

Chapter 5

Optimal Control for the Nonlocal Backward Heat Equation

Xiaoli Wang*, Jinchun He†, Haoyuan Xu‡ and Meihua Yang

School of Mathematics and Statistics, Huazhong University of Science and Technology, Wuhan 430074, China

Hubei Key Laboratory of Engineering Modeling and Scientific Computing, Huazhong University of Science and Technology, Wuhan, 430074, China

**xiaoliw@hust.edu.cn; †taoismnature@hust.edu.cn; ‡hyxu@hust.edu.cn*

In this paper, we consider a nonlocal backward heat equation in a bounded domain $\Omega \subset \mathbb{R}^n$ which has a Lipschitz boundary. The well-posedness is well known for the nonlocal forward heat equation, but the corresponding backward equation is ill-posed, at least in ordinary spaces. In this paper, we obtain the well-posedness of the nonlocal backward heat equation in a dense subspace of $L^2(\Omega)$. Based on the well-posedness result, we obtain a singular optimal system (SOS) for the nonlocal backward heat system.

Keywords: Nonlocal backward heat equation; well-posedness; optimal control.

1. Introduction

In this paper, we consider the following nonlocal backward heat equation which defined on a bounded domain Ω with a Lipschitz boundary in the Euclidean space $\mathbb{R}^n$

$$\begin{cases} z_t - A_\alpha z = v & in \;\; \Omega \times (0,T) := Q, \\ z \equiv 0 & in \;\; \Omega^c \times (0,T), \\ z(x,0) = z_0 & in \;\; \Omega \end{cases} \tag{1.1}$$

where v is an inhomogeneous term and

$$A_\alpha z(x) := C(n,\alpha)\ P.V. \int_{\mathbb{R}^n} \frac{z(x)-z(y)}{|x-y|^{n+2\alpha}} dy, x \in \Omega, \alpha \in (0,1)$$

with a constant $C(n,\alpha)$ given by

$$C(n,\alpha) = \frac{\alpha 2^{2\alpha}\Gamma(\frac{n+2\alpha}{2})}{\pi^{\frac{n}{2}}\Gamma(1-\alpha)}.$$

And the specific properties of A_α will be introduced in Section 2.

*Corresponding author.

Inverse problem, where one seeks for causes for observed or desired effects, is usually "ill-posed" which means their solution may not exist and even if it exists, it is not unique or is unstable with respect to data perturbations. The theory of ill-posed problems has various applications: interpretation of readings of many physical instruments and of geophysical, geological, and astronomical observations, optimization of control, management and planning, synthesis of automatic systems, etc. Moreover, a variety of areas of the theory of ill-posed problems can be included in traditional areas of mathematics such as functional analysis, function theory, linear algebra and differential equations.

The backward heat equation is a classical inverse problem in connection with heat equation, and it models the problem of determining the temperature in the past from observation of the present distribution. The determination of the unknown initial temperature from the observable scattered final data is needed in many practical applications. The backward heat equation appears in the modeling of heat propagation in thermophysics and the mechanics of continuous media widely. The backward heat equation is typically ill-posed in the sense of Hadamard [3]. However, sometimes the well-posedness can be obtained in a smaller space. In fact, W. L. Miranker in [11] showed that in a subspace of $L^2(\mathbb{R})$, the one-dimensional backward heat equation leads to a well-posed problem in the sense of Hadamard. In [12] and [13], J. L. Lions got the well-posed result of the backward heat equation in an n-dimensional bounded region Ω when the inhomogeneous term v and the initial value z_0 were taken in a dense subspace of $L^2(\Omega)$.

In the last years, more and more attention has been given to the analysis of the nonlocal operators and many interesting results have been proved, e.g., see [4] for the well-posedness of a nonlocal Fokker–Planck equation and see [1] for the dynamics of a nonlocal system. Indeed, it has been shown that the nonlocal operator is particularly appropriate for the study of a huge spectrum of phenomena in practical applications, such as market fluctuations ([10]), dislocation dynamics in crystals ([2]), population dynamics ([16]), anomalous transport and diffusion ([9]). Hence we study the nonlocal backward heat equation (1.1). Inspired by [12] and [13], we show that the problem (1.1) is well-posed in a dense subspace W of $L^2(\Omega)$ as stated in the following Theorem 5.1.

Before introducing Theorem 5.1, it is necessary to give the notations throughout the paper first. We denote the scalar product in $L^2(\Omega)$ by $(\cdot,\cdot)_\Omega$, and the norm in $L^2(\Omega)$ and $L^2(Q)$ shall be denoted by $\|\cdot\|_\Omega$ and $\|\cdot\|_Q$. In addition, $\|\cdot\|_1$ and $\|\cdot\|_{X_0}$ denote the norm in spaces W and X_0, where the spaces W, X_0 will be introduced in detail in section 2.

Theorem 5.1. *For $\forall z_0 \in W$, $\forall v \in L^2(0,T;W)$, the system (1.1) has a unique solution $z \in C([0,T];L^2(\Omega))$, which $z' \in L^2(0,T;L^2(\Omega))$. Moreover, there exists $C > 0$ such that $\int_0^T \|z(s)\|^2_{X_0} ds \leq C$.*

One of the important applications of backward problem is the control theory. Optimal control problem is one interesting part of the control theory and has been studied much. J. L. Lions introduced the optimal control for wave equation in [7] and the optimal control of singular distributed systems in [6]. Inspired by the results in [6] and [5], which deal with the backward heat equation, we consider the optimal control problem for the nonlocal backward heat equation system (1.1).

Let admissible set $\mho_{ad}$ be a closed convex subset of $L^2(\Omega)$. $\forall v \in \mho_{ad} \cap W$, by Theorem 5.1, there exists $z = z(v) \in L^2(Q)$ satisfying (1.1) (we pick $z_0 = 0$). For given $z_d \in L^2(Q)$, we can define the cost function

$$J(v, z(v)) = \frac{1}{2}\|z(v) - z_d\|_Q^2 + \frac{N}{2}\|v\|_Q^2, \quad \text{for } v \in \mho_{ad} \cap W,$$

therefore it is easy to see that J is densely defined in $\mho_{ad}$.

Assume $\inf\limits_{v \in \mho_{ad}} J(v, z(v))$ is achieved by the optimal couple $\{u, y(u)\}$ (it is easy to see that it is unique). Then we have the following theorem.

Theorem 5.2. *Let $\{u, y\}$ be the optimal couple of problem $\inf\limits_{v \in \mho_{ad}} J(v, z(v))$. Then there exists $p \in L^2(0, T; W')$ satisfying the following SOS:*

$$\begin{cases} y_t - A_\alpha y = u & in \;\; \Omega \times (0, T) := Q \\ y \equiv 0 & in \;\; \Omega^c \times (0, T) \\ y(x, 0) = 0 & in \;\; \Omega, \end{cases}$$

$$\begin{cases} p_t + A_\alpha p = z_d - y & in \;\; \Omega \times (0, T) := Q \\ p \equiv 0 & in \;\; \Omega^c \times (0, T) \\ p(x, T) = 0 & in \;\; \Omega \end{cases}$$

and

$$(p, v)_Q + N(u, v - u)_Q \geq (y - z_d, y)_Q, \forall v \in \mho_{ad} \cap L^2(0, T; W).$$

The rest of this paper is organized as follows. In section 2, we give some preliminaries, mainly introduce the nonlocal operator A_α and the space X_0 and W in detail. In section 3, we prove Theorem 5.1 by Galekin method. In section 4, we prove Theorem 5.2.

2. Preliminary

$\forall z \in H^\alpha(\mathbb{R}^n)$, we define the operator A_α:

$$A_\alpha z(x) := C(n, \alpha)\ P.V. \int_{\mathbb{R}^n} \frac{z(x) - z(y)}{|x - y|^{n+2\alpha}} dy, x \in \mathbb{R}^n, \alpha \in (0, 1).$$

For $\alpha \in (0,1)$, we denote

$$X_0 := \Big\{ z | z : \mathbb{R}^n \to \mathbb{R} \ is\ measurable, z|_\Omega \in L^2(\Omega), \frac{z(x)-z(y)}{|x-y|^{\frac{n}{2}+\alpha}} \in L^2(G), z = 0 \ in\ \Omega^c \Big\},$$

where $G = \mathbb{R}^{2n} \backslash (\Omega^c \times \Omega^c)$.

$\forall z \in X_0 \subset H^\alpha(\mathbb{R}^n)$ (this is a result in [14], Lemma 5), therefore $A_\alpha z$ is well defined. Obviously, it is a self-adjoint positive operator. By [14],

$$\|z\|_{X_0} = \Big(\int_G \frac{|z(x)-z(y)|^2}{|x-y|^{n+2\alpha}} dxdy \Big)^{\frac{1}{2}}$$

is a norm on X_0, and X_0 is a Hilbert space with the inner product

$$(A_\alpha z, z)_\Omega = \|z\|^2_{X_0}.$$

See more about the space X_0 in [14].

Next, we introduce the space W in Theorem 5.1. Let $D(A_\alpha)$ denote the domain of A_α restricted on the functions with support in Ω, that is $D(A_\alpha) = X_0$. Define

$$W = \Big\{ z \in \bigcap_{\nu=0}^{\infty} D(A_\alpha^\nu) : \sum_{\nu=0}^{\infty} \frac{1}{\nu!} \|A_\alpha^\nu z\|^2_\Omega < +\infty \Big\}.$$

And W is endowed with the inner product

$$(\varphi, \psi)_W = \sum_{\nu=0}^{\infty} (\nu!)^{-1} (A_\alpha^\nu \varphi, A_\alpha^\nu \psi) \ \ for\ each\ \ \varphi, \psi\ in\ W. \tag{2.1}$$

The norm induced by (2.1) is given by

$$\|\varphi\|_1^2 = \sum_{\nu=0}^{\infty} (\nu!)^{-1} \|A_\alpha^\nu \varphi\|^2_\Omega.$$

Lemma 5.1. *The space W is a nonempty Hilbert space, which is, in fact, dense in $L^2(\Omega)$.*

Proof. The proof can be easily found in L. A. Medeiros's paper [8] once the operator A_α is a self-adjoint positive operator. So we omit the proof. □

Remark 5.1. If $u \in W$, then there exists a constant C such that $\|u\|_\Omega \le C\|u\|_1$.

Proof. If $u \in W$, by Lemma 5.1, let $u = \sum_i a_i w_i$, $\{w_i\}, i = 1, 2, \cdots$ is a basis of $L^2(\Omega)$. We have

$$\|u\|_1^2 = \sum_{k=0}^{\infty} \frac{1}{k!} \int_\Omega (\sum_{i=1}^{\infty} a_i \lambda_i^k w_i)^2 dx = \int_\Omega \sum_{i=1}^{\infty} \sum_{k=0}^{\infty} a_i^2 \frac{\lambda_i^{2k}}{k!} = \int_\Omega \sum_{i=1}^{\infty} a_i^2 e^{\lambda_i^2}.$$

Hence

$$\|u\|_1^2 \ge \int_\Omega \sum_{i=1}^{\infty} a_i^2 e^{\lambda_1^2} \ge \int_\Omega \sum_{i=1}^{\infty} a_i^2 = \|u\|^2_\Omega.$$

□

3. The proof of Theorem 5.1

In this section, we prove Theorem 5.1 by Galekin method.

Proof of Theorem 5.1. We prove the existence of the solution of (1.1) first.

Let $\lambda_1, \cdots, \lambda_\nu, \cdots$ be the eigenvalues of the operator A_α, and the corresponding eigenfunctions are $w_1, \cdots, w_\nu, \cdots$. Since the space spanned by all eigenvalues is dense in $L^2(\Omega)$, the eigenfunctions $\{w_j\}$, $j = 1, 2, \cdots, \nu, \cdots$ is a basis of $L^2(\Omega)$. Note that $\{w_j\} \in W$, $j = 1, 2, \cdots, \nu, \cdots$. Let

$$z_\nu(t) = \sum_{j=1}^{\nu} z_{\nu,j}(t) w_j,$$

where $z_{\nu,j}(t)$ is the solution of the following ordinary differential equation $z'_{\nu,j}(t) - \lambda_\nu z_{\nu,j} = (v(t), w_j)_\Omega$ with $z_{\nu,j}(0) = (z_0, w_j)_\Omega$. By solving the above ordinary differential equation, we can easily obtain:

$$\begin{aligned} z_\nu(t) &= \sum_{j=1}^{\nu} (z_0, w_j)_\Omega e^{\lambda_j t} w_j + \int_0^t \sum_{j=1}^{\nu} (v(s), w_j)_\Omega e^{\lambda_j (t-s)} w_j ds \\ &\triangleq \sum_{j=1}^{\nu} f_j(t) w_j + \sum_{j=1}^{\nu} g_j(t) w_j, \end{aligned}$$

and $z_\nu(t)$ satisfies

$$z'_\nu(t) - A_\alpha z_\nu = \sum_{j=1}^{\nu} (v(t), w_j)_\Omega w_j \tag{3.1}$$

with

$$z_\nu(0) = \sum_{j=1}^{\nu} (z_0, w_j)_\Omega w_j.$$

We will prove the limit of $z_\nu(t)$ as $\nu \to \infty$ is the solution of (1.1).

Step 1. We claim that

$$\begin{aligned} &(i) |f_j(t)|^2 \le \alpha_1 \sum_{k=0}^{\infty} (k!)^{-1} |(A_\alpha^k z_0, w_j)_\Omega|^2 \triangleq S_j, \\ &(ii) |g_j(t)|^2 \le T\alpha_1 \sum_{k=0}^{\infty} (k!)^{-1} \int_0^T |(A_\alpha^k v(s), w_j)_\Omega|^2 ds \triangleq D_j, \end{aligned} \tag{3.2}$$

where $\alpha_1 = e^{T^2}$. Simple calculation for (3.2)(i) is the following:

$$\begin{aligned}
|f_j(t)|^2 &= |(z_0, w_j)_\Omega|^2 e^{2\lambda_j t} \\
&\le |(z_0, w_j)_\Omega|^2 e^{\lambda_j^2 + t^2} \le e^{T^2} |(z_0, w_j)_\Omega|^2 \sum_{k=0}^{\infty} (k!)^{-1} \lambda_j^{2k} \\
&= \alpha_1 \sum_{k=0}^{\infty} (k!)^{-1} |(z_0, \lambda_j^k w_j)_\Omega|^2 = \alpha_1 \sum_{k=0}^{\infty} (k!)^{-1} |(z_0, A_\alpha^k w_j)_\Omega|^2 \\
&= \alpha_1 \sum_{k=0}^{\infty} (k!)^{-1} |(A_\alpha^k z_0, w_j)_\Omega|^2.
\end{aligned}$$

Similarly, we get the estimate of $|g_j(t)|^2$.
Next, based on (3.2), we give the estimate of $\|z_\nu(t)\|_\Omega^2$.

$$\begin{aligned}
\|z_\nu(t)\|_\Omega^2 &= \int_\Omega \sum_{j=1}^{\nu} (f_j(t) + g_j(t))^2 w_j^2(x) dx \\
&\le 2 \int_\Omega \sum_{j=1}^{\nu} (f_j^2(t) + g_j^2(t)) w_j^2(x) dx \\
&\le 2 \int_\Omega \sum_{j=1}^{\nu} \alpha_1 \sum_{k=0}^{\infty} (k!)^{-1} |(A_\alpha^k z_0, w_j)_\Omega|^2 w_j^2(x) dx \\
&+ 2 \int_\Omega \sum_{j=1}^{\nu} \alpha_1 T \sum_{k=0}^{\infty} (k!)^{-1} \int_0^T |(A_\alpha^k v(s), w_j)_\Omega|^2 ds w_j^2(x) dx \\
&= 2\alpha_1 \sum_{k=0}^{\infty} (k!)^{-1} \|A_\alpha^k z_0\|_\Omega^2 + 2\alpha_1 T \int_0^T \sum_{k=0}^{\infty} (k!)^{-1} \|A_\alpha^k v(s)\|_\Omega^2 ds \\
&= 2\alpha_1 \|z_0\|_1^2 + 2\alpha_1 T \int_0^T \|v(s)\|_1^2 ds \\
&\le 2\alpha_1 (1+T)(\|z_0\|_1^2 + \int_0^T \|v(s)\|_1^2 ds).
\end{aligned}$$

That is, we obtain

$$\|z_\nu(t)\|_\Omega^2 \le K_1 K_0, \nu = 1, 2, \cdots, 0 \le t \le T. \tag{3.3}$$

where $K_0 = \|z_0\|_1^2 + \int_0^T \|v(s)\|_1^2 ds$ and $K_1 = 2\alpha_1(1+T)$.

Step 2. We estimate $\displaystyle\int_0^T \|z_\nu(s)\|_{X_0}^2 ds$.

We multiply $z_\nu'(t) - A_\alpha z_\nu = \sum_{j=1}^{\nu} (v(t), w_j)_\Omega w_j$ by z_ν, and integral on $(0, T) \times \Omega$, we get

$$\frac{1}{2}\int_0^T \frac{d}{dt}\|z_\nu\|_\Omega^2 dt - \int_0^T \|z_\nu\|_{X_0}^2 dt = \int_0^T \int_\Omega \sum_{j=1}^{\nu}(v(t), w_j)_\Omega w_j z_\nu dxdt$$

$$\frac{1}{2}\|z_\nu(T)\|_\Omega^2 - \frac{1}{2}\|z_\nu(0)\|_\Omega^2 - \int_0^T \|z_\nu\|_{X_0}^2 dt = \int_0^T \int_\Omega \sum_{j=1}^{\nu}(v(t), w_j)_\Omega w_j z_\nu dxdt$$

$$\int_0^T \|z_\nu\|_{X_0}^2 dt = \frac{1}{2}\|z_\nu(T)\|_\Omega^2 - \frac{1}{2}\|z_\nu(0)\|_\Omega^2 - \int_0^T \int_\Omega \sum_{j=1}^{\nu}(v(t), w_j)_\Omega w_j z_\nu dxdt$$

$$\begin{aligned}\int_0^T \|z_\nu\|_{X_0}^2 dt &\le \frac{1}{2}\|z_\nu(T)\|_\Omega^2 - \frac{1}{2}\|z_\nu(0)\|_\Omega^2 + \frac{1}{2}\int_0^T \int_\Omega (\sum_{j=1}^{\nu}(v(t), w_j)_\Omega w_j)^2 dxdt \\ &\quad + \frac{1}{2}\int_0^T \|z_\nu\|_\Omega^2 dt \\ &\le \frac{1}{2}(K_1 + 1 + K_1 T)K_0,\end{aligned} \tag{3.4}$$

where the denotations K_0, K_1 are the same as (3.3).

Step 3. We claim

$$(i)|f_j'(t)| \le \sum_{k=0}^{\infty}(k!)^{-1}T^k|(A_\alpha^{k+1}z_0, w_j)_\Omega|,$$

$$(ii)|g_j'(t)| \le |(v(t), w_j)_\Omega| + \sum_{k=0}^{\infty}(k!)^{-1}T^k \int_0^T |(A_\alpha^{k+1}v(s), w_j)_\Omega| ds.$$

By simple calculation, we get:

$$f_j'(t) = \lambda_j(z_0, w_j)_\Omega e^{\lambda_j t},$$

$$|f_j'(t| \le |\lambda_j(z_0, w_j)_\Omega e^{\lambda_j T}| = |\lambda_j(z_0, w_j)_\Omega \sum_{k=0}^{\infty}(k!)^{-1}\lambda_j^k T^k|$$

$$= |(z_0, w_j)_\Omega \sum_{k=0}^{\infty}(k!)^{-1}\lambda_j^{k+1}T^k| = |\sum_{k=0}^{\infty}(k!)^{-1}T^k(A_\alpha^{k+1}z_0, w_j)_\Omega|.$$

$$g_j'(t) = \lambda_j \int_0^t (v(s), w_j)_\Omega e^{\lambda_j(t-s)} ds + (v(t), w_j)_\Omega,$$

$$|g_j'(t)| \le |(v(t), w_j)_\Omega| + \lambda_j \int_0^t (v(s), w_j)_\Omega \sum_{k=0}^{\infty}(k!)^{-1}\lambda_j^k T^k ds$$

$$\le |(v(t), w_j)_\Omega| + \sum_{k=0}^{\infty} T^k(k!)^{-1} \int_0^T |(A_\alpha^{k+1}v(s), w_j)_\Omega| ds.$$

$$\begin{aligned}\int_0^T\int_\Omega |z'_\nu|^2 dxdt &= \int_0^T\int_\Omega |\sum_{j=1}^{\nu}(f'_j(t)+g'_j(t))w_j|^2 dxdt\\ &= \int_0^T\int_\Omega |\sum_{j=1}^{\nu} f'_j(t)w_j + \sum_{j=1}^{\nu} g'_j(t)w_j|^2 dxdt\\ &\le \int_0^T\int_\Omega \sum_{j=1}^{\nu}(2f'^2_j(t)+2g'^2_j(t))w_j^2 dxdt\\ &= 2\int_0^T\int_\Omega \sum_{j=1}^{\nu}|f'_j(t)|^2 w_j^2 dxdt + 2\int_0^T\int_\Omega \sum_{j=1}^{\nu}|g'_j(t)|^2 w_j^2 dxdt\\ &\triangleq A+B.\end{aligned}$$

On one hand, in order to estimate A, we choose $\epsilon > 0$ such that $0 < b = 1+\epsilon < 2$, there is $\alpha_0(\epsilon) = \alpha_0 > 0$, which satisfies, for $z_0 \in W$:

$$\sum_{k=0}^{\infty}(k!)^{-b}\|A_\alpha^{k+1}z_0\|_\Omega^2 \le \alpha_0\|z_0\|_1^2. \tag{3.5}$$

We obtain:

$$\begin{aligned}A &= 2\int_0^T\int_\Omega \sum_{j=1}^{\nu}|f'_j(t)|^2 w_j^2 dxdt\\ &\le 2\int_0^T\int_\Omega \sum_{j=1}^{\nu}|\sum_{k=0}^{\infty}(k!)^{-1}T^k(A_\alpha^{k+1}z_0,w_j)_\Omega|^2 w_j^2 dxdt\\ &\le 2\int_0^T\int_\Omega |\sum_{k=0}^{\infty}(k!)^{-1}T^k A_\alpha^{k+1}z_0|^2 w_j^2 dxdt\\ &\le 2\int_0^T (\sum_{k=0}^{\infty}(k!)^{\frac{b-2}{2}}T^k(\int_\Omega |A_\alpha^{k+1}z_0|^2 dx)^{\frac{1}{2}}(k!)^{-\frac{b}{2}})^2 dt\\ &\le 2\int_0^T \sum_{k=0}^{\infty}(k!)^{b-2}T^{2k}\sum_{k=0}^{+\infty}\int_\Omega |A_\alpha^{k+1}z_0|^2 dx(k!)^{-b}dt \le 2T\alpha_0\alpha'_2\|z_0\|_1^2\end{aligned}$$

with $\alpha'_2 = \sum_{k=0}^{\infty}(k!)^{b-2}T^{2k}$. In the last term, we also used (3.5).

On the other hand,

$$
\begin{aligned}
B &= 2\int_0^T \int_\Omega \sum_{j=1}^{\nu} |g_j'(t)|^2 w_j^2 dxdt \\
&\le 2\int_0^T \int_\Omega \sum_{j=1}^{\nu} (|(v(t), w_j)_\Omega| + \sum_{k=0}^{\infty} (k!)^{-1} T^k \int_0^T |(A_\alpha^{k+1} v(s), w_j)_\Omega| ds)^2 w_j^2 dxdt \\
&\le 4\int_0^T \int_\Omega \sum_{j=1}^{\nu} (v(t), w_j)_\Omega^2 w_j^2 dxdt \\
&\quad + 4\int_0^T \int_\Omega \sum_{j=1}^{\nu} (\sum_{k=0}^{\infty} (k!)^{-1} T^k \int_0^T |(A_\alpha^{k+1} v(s), w_j)_\Omega| ds)^2 w_j^2 dxdt \\
&\le 4\int_0^T \|v(s)\|_1^2 ds + 4T^2 \alpha_2' \alpha_0 \int_0^T \|v(s)\|_1^2 ds.
\end{aligned}
$$

Thus, we obtain

$$
\begin{aligned}
\|z_\nu'\|_{L^2(0,T;L^2(\Omega))}^2 &\le 2T\alpha_2'\alpha_0 \|z_0\|_1^2 + 4\int_0^T \|v(s)\|_1^2 ds \\
&\quad + 4T^2 \alpha_2' \alpha_0 \int_0^T \|v(s)\|_1^2 ds \\
&\le [2T\alpha_2'\alpha_0(1+2T) + 4] K_0
\end{aligned} \tag{3.6}
$$

with $K_0 = \|z_0\|_1^2 + \int_0^T \|v(s)\|_1^2 ds$.

Step 4. By the inequalities (3.3) and (3.6), we get $z_\nu \in L^\infty(0,T;L^2(\Omega))$ and $z_\nu' \in L^2(0,T;L^2(\Omega))$ independent of ν. And by (3.4), we also get that z_ν is bounded in $L^2(0,T;X_0)$ independent of ν. By extracting a subsequence (still denoted by z_ν), if necessary, we may assume there exists a $z(t) \in L^2(0,T;X_0)$, such that

$$
z_\nu(t) \rightharpoonup z(t) \quad \text{in } L^2(0,T;X_0),
$$

then $z(t)$ is a weak solution of (1.1). Since the embedding $X_0 \hookrightarrow L^2(\Omega)$ is compact ([15, Lemma 9]), we get

$$
z_\nu(t) \to z(t) \quad \text{in } L^2(0,T;L^2(\Omega)). \tag{3.7}
$$

By (3.3) and (3.6), it is easy to see that

$$
z(t) = \lim_{\nu\to\infty} z_\nu(t) = \sum_{j=1}^{\infty} (z_0, w_j)_\Omega e^{\lambda_j t} w_j + \int_0^t \sum_{j=1}^{\infty} (v(s), w_j)_\Omega e^{\lambda_j (t-s)} w_j ds
$$

By (3.3) and (3.6) again, we can get $z \in L^\infty(0,T;L^2(\Omega))$ and $z_t \in L^2(0,T;L^2(\Omega))$. Then by the semigroup theory, we have $z \in \mathcal{C}(0,T;L^2(\Omega))$.

In a similar way in Step 2, we obtain

$$
\int_0^T \|z\|_{X_0}^2 dt \le \frac{1}{2}(K_1 + 1 + K_1 T) K_0.
$$

Furthermore, we next prove the uniqueness of the solution of (1.1).

Step 5. In order to get the uniqueness, we first prove the uniqueness of the solution for the following system:

$$\begin{cases} z_t + A_\alpha z = -v & in\ \ \Omega \times (0,T) := Q \\ z = 0 & in\ \ \Omega^c \times (0,T) \\ z(x,T) = z_0 & in\ \ \Omega. \end{cases} \tag{3.8}$$

Suppose z and $\tilde{z}$ are both solutions of the system above.
Let $w := z - \tilde{z}$, $e(t) := \int_\Omega w(x,t)^2 dx \quad (0 \le t \le T)$.
Thus

$$\dot{e}(t) = 2\int_\Omega w w_t dx = -2\int_\Omega w A_\alpha w dx = -2\|w\|^2_{X_0}, \tag{3.9}$$

$$\ddot{e}(t) = -2\int_\Omega w_t A_\alpha w dx - 2\int_\Omega w A_\alpha w_t dx = 2\int_\Omega A_\alpha w A_\alpha w dx - 2\int_\Omega w A_\alpha w_t dx \tag{3.10}$$

$$= 4\int_\Omega (A_\alpha w)^2 dx.$$

Now since $w = 0$ on Ω^c,

$$\|w\|^2_{X_0} = \int_\Omega A_\alpha w \cdot w dx \le (\int_\Omega w^2 dx)^{\frac{1}{2}} (\int_\Omega (A_\alpha w)^2 dx)^{\frac{1}{2}}.$$

Thus (3.9) and (3.10) imply

$$(\dot{e}(t))^2 = 4\|w\|^4_{X_0} \le 4\int_\Omega w^2 dx \int_\Omega (A_\alpha w)^2 dx = e(t)\ddot{e}(t).$$

Hence,

$$e(t)\ddot{e}(t) \ge \dot{e}^2(t). \tag{3.11}$$

Now if $e(t) = 0$ for all $0 \le t \le T$, we are done.
Otherwise, there exists an interval $[t_1, t_2] \subset [0,T]$ with

$$\eta(t) > 0 \quad for\ \ t_1 \le t < t_2, \ \ e(t_2) = 0.$$

We write

$$f(t) := \log e(t) \quad (t_1 \le t < t_2), \tag{3.12}$$

then $\ddot{f}(t) = \frac{\ddot{e}(t)}{e(t)} - \frac{\dot{e}^2(t)}{e^2(t)} \ge 0$ *by* (3.11), and so f is convex on the interval (t_1, t_2).
Consequently if $0 < \tau < 1, t_1 < t < t_2$, we have

$$f((1-\tau)t_1 + \tau t) \le (1-\tau)f(t_1) + \tau f(t).$$

Recalling (3.12), we deduce

$$e((1-\tau)t_1+\tau t) \le e^{1-\tau}(t_1)e^{\tau}(t),$$

and so

$$0 \le e((1-\tau)t_1+\tau t_2) \le e^{1-\tau}(t_1)e^{\tau}(t_2) = 0.$$

Thus $e((1-\tau)t_1+\tau t_2) = 0$. That is $e(t) = 0$ for all time $t_1 \le t \le t_2$.
We demonstrate that the uniqueness of system (1.1) is equivalent to the uniqueness of the system (3.8).

Step 6. By variable substitution $t = -t'$, the system (3.8) converts to the following system:

$$\begin{cases} z_{t'} - A_\alpha z = v & in \;\; \Omega \times (0,T) := Q \\ z = 0 & in \;\; \Omega^c \times (0,T) \\ z(x,-T) = z_0 & in \;\; \Omega. \end{cases} \tag{3.13}$$

Continuous variable substitution $t'' = t' + T$, the system (3.13) converts to the following system:

$$\begin{cases} z_{t''} - A_\alpha z = v & in \;\; \Omega \times (0,T) := Q \\ z = 0 & in \;\; \Omega^c \times (0,T) \\ z(x,0) = z_0 & in \;\; \Omega. \end{cases} \tag{3.14}$$

So the uniqueness of system (3.8) is equivalent to the uniqueness of system (3.14) and we complete the proof. □

4. The proof of Theorem 5.2

In this section, based on the well-posed result, we will consider the optimal control problem for nonlocal backward heat equation. In order to obtain an SOS, we divide the following proof into four steps. Firstly, we define a penalized function $J_\epsilon(v,z)$, and assume $\{u_\epsilon, y_\epsilon\}$ is a optimal couple of the penalized problem $\inf J_\epsilon(v,z)$ for each $\epsilon > 0$, we get the uniqueness of the optimal couple. Secondly, we obtain that the optimal couple $\{u_\epsilon, y_\epsilon\}$ of $\inf J_\epsilon(v,z)$ converges to the optimal couple $\{u,y\}$ of $\inf J(v,z(v))$ as $\epsilon \to 0$. Thirdly, since $\{u_\epsilon, y_\epsilon\}$ is the optimal couple of $\inf J_\epsilon(v,z)$, we use variation principle to characterize (u_ϵ, y_ϵ) and obtain the system that p_ϵ (which involves u_ϵ and y_ϵ and will be given in detail in Step 3) satisfies. Finally, we claim p_ϵ is uniformly bounded in $L^2(0,T;W')$, so there exists p such that $p_\epsilon \rightharpoonup p$ as $\epsilon \to 0$, and recall the convergence of u_ϵ and y_ϵ as $\epsilon \to 0$ in Step 2, so we obtain the SOS by letting $\epsilon \to 0$.

Proof. Step 1. We first introduce the penalized problem:

$$Find \ \inf J_\epsilon(v,z), \quad where \ J_\epsilon(v,z) = J(v,z) + \frac{1}{2\epsilon}\|z_t - A_\alpha z - v\|_Q^2, \tag{4.1}$$

where $v \in \mho_{ad}$, $z_t - A_\alpha z$, $z \in L^2(Q)$ and $z(0) = 0$, $z = 0$ in $\Omega^c \times (0,T)$. Let $\{u_\epsilon, y_\epsilon\}$ be the unique solution of the penalized problem:

$$J_\epsilon(u_\epsilon, y_\epsilon) = \inf J_\epsilon(v,z).$$

We denote $J_\epsilon(u_\epsilon, y_\epsilon) = inf J_\epsilon(v,z)$. The uniqueness is easily seen. Indeed, we assume that $\{u_1, y_1\}$ and $\{u_2, y_2\}$ are two different solutions of (4.1), since $u_i \in \mho_{ad}$, $y_{it} - A_\alpha y_i \in L^2(Q)$, i=1, 2, so $\lambda u_1 + (1-\lambda)u_2 \in \mho_{ad}$, $(\lambda y_1 + (1-\lambda)y_2)_t - A_\alpha(\lambda y_1 + (1-\lambda)y_2) \in L^2(Q)$, then we will get $\{\lambda u_1 + (1-\lambda)u_2, \lambda y_1 + (1-\lambda)y_2\}$ is also a solution of (4.1) by the linearity of functional J_ϵ. This will cause a contradiction.

Step 2. By using the properties of J and J_ϵ, we get the convergence of u_ϵ and y_ϵ in $L^2(Q)$.
We get the uniform boundedness of $J_\epsilon(u_\epsilon, y_\epsilon)$ easily

$$J_\epsilon(u_\epsilon, y_\epsilon) \le J_\epsilon(u,y) = J(u,y), \forall \epsilon > 0 \tag{4.2}$$

since $\{u, y\}$ is the optimal couple of the problem $\inf\limits_{v \in \mho_{ad}} J(v, z(v))$. Thus $J(u_\epsilon, y_\epsilon)$ and $\frac{1}{2\epsilon}\|y_{\epsilon t} - A_\alpha y_\epsilon - u_\epsilon\|_Q^2$ is uniformly bounded in $L^2(Q) \times L^2(Q)$. So there exists subsequence (u_ϵ, y_ϵ) (we still denote (u_ϵ, y_ϵ)), such that

$$u_\epsilon \rightharpoonup u_0 \ \ in \ \ L^2(Q) \ \ as \ \ \epsilon \to 0,$$

$$y_\epsilon \rightharpoonup y_0 \ \ in \ \ L^2(Q) \ \ as \ \ \epsilon \to 0,$$

and by the uniform boundedness of $\frac{1}{2\epsilon}\|y_{\epsilon t} - A_\alpha y_\epsilon - u_\epsilon\|_Q^2$, we get

$$y_{\epsilon t} - A_\alpha y_\epsilon - u_\epsilon \to 0 \quad as \ \ \epsilon \to 0.$$

Thus

$$y_{\epsilon t} \rightharpoonup y_{0t} \ \ in \ \ L^2(Q) \ \ as \ \ \epsilon \to 0.$$

Obviously, we can get $\{u_0, y_0\}$ satisfying the following system:

$$\begin{cases} y_{0t} - A_\alpha y_0 = u_0 & in \ \ \Omega \times (0,T) := Q \\ u_0 \in \mho_{ad}, \ y_{0t} - A_\alpha y_0, \ y_0 \in L^2(Q) \\ y_0 \equiv 0 & in \ \ \Omega^c \times (0,T) \\ y_0(x,0) = 0 & in \ \ \Omega. \end{cases}$$

By the lower semicontinuity of J, we have

$$J(u_0, y_0) \le \liminf_{\epsilon \to 0} J(u_\epsilon, y_\epsilon),$$

so we have

$$J(u_0, y_0) \le \liminf_{\epsilon \to 0} J(u_\epsilon, y_\epsilon) \le \liminf_{\epsilon \to 0} J_\epsilon(u_\epsilon, y_\epsilon) \le J(u,y),$$

the last step above we used (4.2), and we obtain

$$J(u_0, y_0) = J(u, y),$$

$$\lim_{\epsilon \to 0} J(u_\epsilon, y_\epsilon) = J(u, y).$$

We have $u_0 = u, y_0 = y$ since $\inf_{v \in \mho_{ad}} J(v, z(v))$ admits a unique optimal couple.

So we get

$$u_\epsilon \to u \; in \; L^2(Q) \;\; as \;\; \epsilon \to 0, \tag{4.3}$$

$$y_\epsilon \to y \; in \; L^2(Q) \;\; as \;\; \epsilon \to 0. \tag{4.4}$$

Step 3. Since $\inf J_\epsilon(v, z) = J_\epsilon(u_\epsilon, y_\epsilon)$, now we use variational principle to characterize (u_ϵ, y_ϵ) and obtain the system that p_ϵ (will be given in detail in the following) satisfies.

On one hand, set $z = y_\epsilon + \tau z'$, we should have

$$\frac{dJ_\epsilon}{d\tau}(v, y_\epsilon + \tau z')\Big|_{\tau=0} = 0.$$

$$\begin{aligned}
J_\epsilon(v, y_\epsilon + \tau z') &= J(v, y_\epsilon + \tau z') + \frac{1}{2\epsilon}\|(y_\epsilon + \tau z')_t - A_\alpha(y_\epsilon + \tau z') - v\|_Q^2 \\
&= \frac{1}{2}\|y_\epsilon + \tau z' - z_d\|_Q^2 + \frac{N}{2}\|v\|_Q^2 \\
&\quad + \frac{1}{2\epsilon}\|(y_{\epsilon t} - A_\alpha y_\epsilon) + \tau(z'_t - A_\alpha z') - v\|_Q^2 \\
&= \frac{1}{2}\int_Q [(y_\epsilon - z_d)^2 + \tau^2 z'^2 + 2\tau z'(y_\epsilon - z_d)]dxdt + \frac{N}{2}\|v\|_Q^2 \\
&\quad + \frac{1}{2\epsilon}\int_Q (y_{\epsilon t} - A_\alpha y_\epsilon - v)^2 + \tau^2(z'_t - A_\alpha z')^2 \\
&\quad + 2\tau(y_{\epsilon t} - A_\alpha y_\epsilon - v)(z'_t - A_\alpha z')dxdt,
\end{aligned}$$

$\frac{dJ_\epsilon}{d\tau}(v, y_\epsilon + \tau z')\Big|_{\tau=0} = (z', y_\epsilon - z_d)_Q + \frac{1}{\epsilon}(y_{\epsilon t} - A_\alpha y_\epsilon - v, z'_t - A_\alpha z')_Q = 0.$

Let $v = u_\epsilon$, since $u_\epsilon \in \mho_{ad}$. We have

$$\frac{1}{\epsilon}(y_{\epsilon t} - A_\alpha y_\epsilon - u_\epsilon, z'_t - A_\alpha z')_Q + (y_\epsilon - z_d, z')_Q = 0.$$

Let $p_\epsilon = -\frac{1}{\epsilon}(y_{\epsilon t} - A_\alpha y_\epsilon - u_\epsilon)$, we obtain

$$-(p_\epsilon, z'_t - A_\alpha z')_Q + (y_\epsilon - z_d, z')_Q = 0. \tag{4.5}$$

On the other hand, set $v = (1-\tau)u_\epsilon + \tau v'$, and $v \in \mho_{ad}$ since $\mho_{ad}$ is a convex subset and $u_\epsilon, v' \in \mho_{ad}$. We have $\frac{d}{d\tau}J_\epsilon((1-\tau)u_\epsilon + \tau v', z)\Big|_{\tau=0} \geq 0$ since u_ϵ is the optimal choice.

$$\begin{aligned}
&J_\epsilon((1-\tau)u_\epsilon + \tau v', z) \\
&= J((1-\tau)u_\epsilon + \tau v', z) + \frac{1}{2\epsilon}\|z_t - A_\alpha z - ((1-\tau)u_\epsilon + \tau v')\|_Q^2 \\
&= \frac{1}{2}\|z - z_d\|_Q^2 + \frac{N}{2}\|(1-\tau)u_\epsilon + \tau v'\|_Q^2 + \frac{1}{2\epsilon}\|z_t - A_\alpha z - u_\epsilon + \tau(u_\epsilon - v')\|_Q^2,
\end{aligned}$$

$$\begin{aligned}
&\frac{d}{d\tau}J_\epsilon((1-\tau)u_\epsilon+\tau v',z)\big|_{\tau=0}\\
=&N\int_Q u_\epsilon(v'-u_\epsilon)dxdt+\frac{1}{\epsilon}\int_Q -(v'-u_\epsilon)(z_t-A_\alpha z-u_\epsilon)dxdt\\
=&(Nu_\epsilon-\frac{1}{\epsilon}(z_t+A_\alpha z-u_\epsilon),v'-u_\epsilon)_Q\\
=&(Nu_\epsilon+p_\epsilon,v'-u_\epsilon)_Q.
\end{aligned}$$

Thus, we obtain

$$(Nu_\epsilon+p_\epsilon,v'-u_\epsilon)_Q\ge 0. \tag{4.6}$$

Moreover,

$$\begin{aligned}
&\int_Q -p_\epsilon(z_t-A_\alpha z)dxdt=\int_Q -p_\epsilon z_t dxdt+\int_Q p_\epsilon A_\alpha z dxdt \qquad (4.7)\\
=&\int_\Omega -p_\epsilon z dx\big|_{t=0}^{t=T}+\int_Q p_{\epsilon t} z dxdt+\int_Q p_\epsilon A_\alpha z dxdt\\
=&\int_Q p_{\epsilon t} z dxdt+\int_Q zA_\alpha p_\epsilon dxdt,
\end{aligned}$$

we need

$$p_\epsilon(x,T)=0 \tag{4.8}$$

and

$$p_\epsilon=0 \quad in \quad \Omega^c \tag{4.9}$$

to obtain the identity above.
It follows from (4.5), (4.7), (4.8) and (4.9) that

$$\int_Q p_{\epsilon t} z dxdt+\int_Q zA_\alpha p_\epsilon dxdt+\int_Q (y_\epsilon-z_d)z dxdt=0,$$

$\forall z$ satisfies $z, z_t-A_\alpha z\in L^2(Q)$.
Hence

$$\begin{cases} p_{\epsilon t}+A_\alpha p_\epsilon=z_d-y_\epsilon & in \quad \Omega\times(0,T):=Q\\ p_\epsilon=0 & in \quad \Omega^c\times[0,T]\\ p_\epsilon(x,T)=0 & in \quad \Omega.\end{cases} \tag{4.10}$$

Taking the scalar product of (4.10)(1) with z, we obtain

$$(p_{\epsilon t}+A_\alpha p_\epsilon,z)_Q=(z_d-y_\epsilon,z)_Q.$$

Integrating by parts, we have

$$(p_\epsilon,z_t-A_\alpha z)_Q=(y_\epsilon-z_d,z)_Q,\forall z \ satisfies \ z,z_t-A_\alpha z\in L^2(Q). \tag{4.11}$$

Set $z = y_\epsilon$ in (4.11), we have

$$(y_\epsilon - z_d, y_\epsilon)_Q = (p_\epsilon, y_{\epsilon t} - A_\alpha y_\epsilon)_Q = (p_\epsilon, u_\epsilon - \epsilon p_\epsilon)_Q.$$

By (4.6), we have

$$(p_\epsilon, v)_Q + N(u_\epsilon, v - u_\epsilon)_Q \geq (p_\epsilon, u_\epsilon)_Q = (y_\epsilon - z_d, y_\epsilon)_Q + \epsilon \|p_\epsilon\|_Q^2 \geq (y_\epsilon - z_d, y_\epsilon)_Q. \tag{4.12}$$

Step 4. We obtain Theorem 5.2 by letting $\epsilon \to 0$ in (4.10) and (4.12) by the weak convergence of p_ϵ (will be proved at once) and the convergence of u_ϵ and y_ϵ. First, we claim that

$$p_\epsilon \textit{ remains in a bounded set of } L^2(0,T;W'). \tag{4.13}$$

In order to obtain (4.13), we consider the following question:

$$\begin{cases} \phi_t - A_\alpha \phi = \psi & in \ \ \Omega \times (0,T) := Q \\ \phi = 0 & in \ \ \Omega^c \times (0,T) \\ \phi(x,0) = 0 & in \ \ \Omega \end{cases}$$

where $\psi \in L^2(0,T;W)$.
The solution is given by

$$\phi(t) = \int_0^t \exp(t-\eta)A_\alpha \cdot \psi(\eta)d\eta = \sum_k \int_0^t \frac{(t-\eta)^k}{k!} A_\alpha^k \psi(\eta)d\eta,$$

hence

$$\|\phi(t)\|_\Omega^2 \leq (\sum_k \int_0^t \|A_\alpha^k \psi(\eta)\|_\Omega^2 \frac{1}{k!} d\eta)(\sum_k \int_0^t \frac{(t-\eta)^{2k}}{k!} d\eta)$$

and

$$\|\phi\|_{L^\infty(0,T;L^2(\Omega))} \leq c\|\psi\|_{L^2(0,T;W)}. \tag{4.14}$$

By (4.11) and (4.14), we have

$$(p_\epsilon, \phi_t - A_\alpha \phi)_Q = (y_\epsilon - z_d, \phi)_Q \leq \|y_\epsilon - z_d\|_Q \cdot \|\phi\|_Q \leq C\|\psi\|_{L^2(0,T;W)}.$$

Furthermore,

$$(p_\epsilon, \phi_t - A_\alpha \phi)_Q = (p_\epsilon, \psi)_Q = \|p_\epsilon\|_{L^2(0,T;W')} \|\psi\|_{L^2(0,T;W)},$$

thus

$$\|p_\epsilon\|_{L^2(0,T;W')} \leq C.$$

Having the uniform boundedness of p_ϵ in $L^2(0,T;W')$, we get

$$p_\epsilon \rightharpoonup p \quad in \quad L^2(0,T;W') \quad as \quad \epsilon \to 0,$$

and combine with (4.3) and (4.4), let $\epsilon \to 0$, we can obtain the system that p satisfies:

$$\begin{cases} p_t + A_\alpha p = z_d - y & in \ \ \Omega \times (0,T) := Q \\ p \equiv 0 & in \ \ \Omega^c \times (0,T) \\ p(x,T) = 0 & in \ \ \Omega \end{cases}$$

and

$$(p, v)_Q + N(u, v-u)_Q \geq (y - z_d, y)_Q, \forall v \in \mho_{ad} \cap L^2(0,T;W).$$

We complete the proof of Theorem 5.2. □

References

[1] L. Bai and M. Yang, *A determining form for a nonlocal system*, Adv. Nonlinear Stud. 17 (2017), 705–713.

[2] S. Dipierro, G. Palatucci and E. Valdinoci, *Dislocation dynamics in crystals: a macroscopic theory in a fractional Laplace setting*, Comm. Math. Phys. 333 (2015), 1061–1105.

[3] J. Hadamard, *Lectures on Cauchy's problem in linear partial differential equations*, Dover Publications, New York, 1953.

[4] J. He, J. Duan and H. Gao, *A nonlocal Fokker–Planck equation for non-Gaussian stochastic dynamical systems*, Appl. Math. Lett. 49 (2015), 1–6.

[5] J. L. Lions, *Some methods in the mathematical analysis of systems and their control*, Science Pr, 1981.

[6] J. L. Lions, *Some remarks on the optimal control of singular distributed systems*, Nonlinear Functional Analysis and Its Applications, Part 2, (1983), 127–153.

[7] J. L. Lions, *Exact controllability, stabilization and perturbations for distributed systems*, SIAM Rev. 30 (1988), 1–68.

[8] L. A. Medeiros, *Remarks on a non-well-posed problem*, Proc. Roy. Soc. Edinburgh Sect. A 102 (1986), 131–140.

[9] M. M. Meerschaert, *Fractional calculus, anomalous diffusion, and probability*, Fractional dynamics (2012), 265–284.

[10] R. C. Merton, *Option pricing when underlying stock returns are discontinuous*, J. Finan. Econ. 3 (1976), 265–284.

[11] W. L. Miranker, *A well posed problem for the backward heat equation*, Proc. Amer. Math. Soc. 12 (1961), 243–247.

[12] P. H. Rivera and C. F. Vasconcellos, *Optimal control for a backward parabolic system*, SIAM J. Control Optim. 25 (1987), 1163–1172.

[13] P. H. Rivera, *On the optimal control of non-well-posed linear evolution systems*, Mat. Apl. Comput. 7 (1988), 55–67.

[14] R. Servadei and E. Valdinoci, *Mountain pass solutions for non-local elliptic operators*, J. Math. Anal. Appl. 389 (2012), 887–898.

[15] R. Servadei and E. Valdinoci, *The Brezis–Nirenberg result for the fractional Laplacian*, Trans. Amer. Math. Soc. 367 (2015), 67–102.

[16] G. M. Viswanathan, V. Afanasyev and S. V. Buldyrev, et al., *Levy flight search patterns of wandering albatrosses*, Nature 381 (1996), 413–415.

[17] J. R. Wang, *Shannon wavelet regularization methods for a backward heat equation*, J. Comput. Appl. Math. 235 (2011), 3079–3085.

Chapter 6

Hölder Estimates for Solutions of Stochastic Nonlocal Diffusion Equations

Guangying Lv[1,*], Hongjun Gao[2,†], Jinlong Wei[3,‡] and Jiang-Lun Wu[4,§]

[1]*Institute of Applied Mathematics, Henan University, Kaifeng, Henan 475001, China*

[2]*Institute of Mathematics, School of Mathematical Science, Nanjing Normal University, Nanjing 210023, China*

[3]*School of Statistics and Mathematics, Zhongnan University of Economics and Law, Wuhan, Hubei 430073, China*

[4]*Department of Mathematics, Swansea University, Swansea SA2 8PP, UK*

[*]*gylvmaths@henu.edu.cn;* [†]*gaohj@njnu.edu.cn*

[‡]*weijinlong.hust@gmail.com;* [§]*j.l.wu@swansea.ac.uk*

In this paper, we aim to obtain local and global L^∞ estimates, and Hölder estimates for solutions of stochastic nonlocal heat equations. By utilising the estimates of kernel functions and stochastic De Giorgi's iteration, we succeed to get the desired results.

Keywords: Nonlocal diffusion; Itô's formula; L^∞ estimates; Hölder estimate.
AMS subject classifications (2010): 35K20; 60H15; 60H40.

1. Introduction

In this paper, we are concerned with regularity of the solutions to the following stochastic nonlocal diffusion equation

$$du_t = [\mathcal{D}^\alpha u + f(x,t,u)]dt + \sum_{i=0}^{\infty} g_i(x,t,u)dW_t^i, \quad t > 0, x \in \mathbb{R}^n \tag{1.1}$$

with initial $u(x,0) = u_0(x)$, where $\alpha \in (0,1)$ and

$$\mathcal{D}^\alpha u(x) := C'_\alpha \int_{\mathbb{R}^n \setminus \{x\}} \frac{u(x) - u(y)}{|x-y|^{n+2\alpha}} Q(x-y)dy,$$

$$C'_\alpha := \int_{\mathbb{R}^n \setminus \{0\}} \frac{(1 - cos(\mathbf{e} \cdot z))dz}{|z|^{n+2\alpha}} > 0,$$

with $\mathbf{e} \in \mathbb{R}^n$ being any unit vector and $Q : x \in \mathbb{R}^n \mapsto Q(x) \in \mathbb{R}$ being an even function, $f, g_i : \mathbb{R}^n \times [0,\infty) \times \mathbb{R} \to \mathbb{R}, i \in \mathbb{N}$, are progressively measurable such that $\sum_{i=0}^{\infty} |g_i(x,t,u)|^2 < \infty$ a.e. for $(x,t,u) \in \mathbb{R}^n \times [0,\infty) \times \mathbb{R}$ (i.e., the sequence $g := \{g_i\}_{i\in\mathbb{N}}$ is l^2-valued), and $\{W_t^i, t \geq 0\}_{i\in\mathbb{N}}$ is a sequence of independent, standard Brownian motions on a filtered probability space $(\Omega, \mathcal{F}, \mathcal{F}_t, \mathbb{P})$. It is easy to see that if $Q(x) \equiv 1$, $\mathcal{D}^\alpha = \Delta^\alpha := -(-\Delta)^\alpha$. There are two fundamental questions regarding the regularity of solutions to stochastic nonlocal diffusion equations. One is concerning the regularity of solutions under weak assumptions on the coefficient $g := (g_i)_{i\in\mathbb{N}}$, that is, what are the least assumptions on g. The other is the regularity of solutions of the equations on bounded domains. The first equation has been well studied by many authors. By using the method of [18], Kim–Kim [13] considered the following stochastic evolution equation

$$\begin{cases} du = (a(\omega,t)\Delta^\alpha u + f(\omega,t,x,u))\,dt + \sum_{i=0}^{\infty} g_i(\omega,t,x,u)dW_t^i, \\ u(0) = u_0, \end{cases} \tag{1.2}$$

where the process $a(\omega,t)$ is predictable and there is a constant $\delta > 0$ such that $\delta < a(\omega,t) < \delta^{-1}$ for all $(\omega,t) \in \Omega \times [0,\infty)$. Denote $f(0) := f(\omega,t,x,u)$, $g_i(0) = g_i(\omega,t,x,u)$, and set

$$\|u\|_{H_p^\gamma} := \|(1-\Delta)^{\frac{\gamma}{2}}u\|_{L^p};\quad \|u\|_{\mathcal{H}_p^\gamma(T)} := \left(\mathbb{E}\left[\int_0^T \|u(\omega,t,\cdot)\|_{H_p^\gamma}^p dt\right]\right)^{1/p};$$

$$\|u\|_{\mathbb{H}_p^\gamma(T,l_2)} := \left(\mathbb{E}\left[\int_0^T \||(1-\Delta)^{\frac{\gamma}{2}}u(\omega,t,\cdot)|_{l_2}\|_{H_p^\gamma}^p dt\right]\right)^{1/p};$$

$$\|u_0\|_{U_p^\gamma} := \left(\mathbb{E}\|u_0\|_{H_p^\gamma}^p\right)^{1/p}.$$

Then, under the assumptions that (i)

$$f(0) \in \mathbb{H}_p^\gamma(T),\ g(0) = (g_i(0))_{i\in\mathbb{N}} \in \mathbb{H}_p^{\gamma+2\alpha-2\alpha/p}(T,l_2);$$

and (ii) for any $\varepsilon > 0$, there exists a constant K_ε such that

$$\begin{aligned} &\|f(\omega,t,\cdot,u) - f(\omega,t,\cdot,v)\|_{H_p^\gamma} + \|g(\omega,t,\cdot,u) - g(\omega,t,\cdot,v)\|_{H_p^{\gamma+\alpha}(l_2)} \\ &\leq \varepsilon\|u-v\|_{H_p^{\gamma+2\alpha}} + K(\varepsilon)\|u-v\|_{H_p^\gamma} \end{aligned}$$

holds for any $u, v \in H_p^{\gamma+2\alpha}$ and for all $(\omega,t) \in \Omega \times [0,\infty)$, the authors obtained the following result.

Proposition 6.1. *Let $p \in [2,\infty)$ and $\gamma \in \mathbb{R}$. Assume that the above assumptions hold and $u_0 \in U_p^{\gamma+\alpha-\alpha/p}$, the equation (1.2) admits a unique solution u in $\mathcal{H}_p^{\gamma+\alpha}(T)$ and it holds that*

$$\|u\|_{\mathcal{H}_p^{\gamma+2\alpha}(T)} \leq c(p,T,\delta)\left(\|f(0)\|_{\mathbb{H}_p^\gamma(T)} + \|g(0)\|_{\mathbb{H}_p^{\gamma+\alpha}(T,l_2)} + \|u_0\|_{U_p^{\gamma+2\alpha-2\alpha/p}}\right),$$

where δ is the constant such that $\delta < a(\omega,t) < \delta^{-1}$.

Recently, Kim–Kim [11] considered the L_p-theory for stochastic partial differential equations driven by Lévy processes, also see [6, 12, 14, 15]. Zhang [23] obtained the L_p-theory of semi-linear SPDEs on general measure spaces. The earliest results about the L_p-theory of SPDEs are the work of Krylov [17, 18]. The Hölder estimate of SPDEs has been studied by many authors. Hsu–Wang [9] obtained the stochastic De Giorgi iteration and regularity of stochastic partial differential equations. Du–Liu [7] got the Schauder estimate for stochastic PDEs. Combining the deterministic theory and convolution properties, Debussche [5] established the regularity result for quasilinear stochastic partial differential equations of parabolic type. Kuksin et al. [16] obtained Hölder estimates for solutions of parabolic SPDEs on bounded domain. More recently, Tian et al. [21] considered the equation (1.1) with f and g independent of u. They obtained the local Hölder estimates of mild solutions.

In this short paper, we consider the Hölder continuity of the solution to (1.1). The method we used here is similar to that in [9]. Set $\|\cdot\|_p := \|\cdot\|_{L^p(\mathbb{R}^n)}$. The assumption on the SPDEs (1.1) is the following condition:

(A) There exist two positive constants q_i $(i = 1, 2)$, a nonnegative function $K \in L^2(\mathbb{R}^n) \cap L^\infty(\mathbb{R}^n)$ and a positive constant Λ such that

$$q_1 \leq Q(x) \leq q_2, \quad \forall x \in \mathbb{R}^n$$
$$|f(x,t,u)| + |g(x,t,u)|_{l^2} \leq K(x) + \Lambda|u|, \quad \forall (x,t,u) \in \mathbb{R}^n \times \mathbb{R}_+ \times \mathbb{R}.$$

Our main result may be formulated as follows:

Theorem 6.1. *Assume that Condition (A) holds. Let $T > 0$ be fixed. Let u be a stochastic strong solution of SPDEs (1.1) with initial data $u(0) = u_0$. Then for every $p > 0$ there is a constant $C = C(n, \Lambda, T, p)$ such that*

$$\mathbb{E}\int_0^T \|u(t)\|_2^p + \mathbb{E}\|u\|^p_{\infty,[T,2T]\times\mathbb{R}^n} \leq C(\|u_0\|_2 + \|K\|_2 + \|K\|_\infty)^p. \tag{1.3}$$

Moreover, there exists a positive constant β such that the solution $u \in C^\beta([T, 2T] \times \mathbb{R}^n)$ almost surely. Furthermore, for every $p > 0$, there is a constant $C = C(n, \Lambda, T, p)$ such that

$$\mathbb{E}\|u\|^p_{C^\beta([T,2T]\times\mathbb{R}^n)} \leq C(\|u_0\|_2 + \|K\|_2 + \|K\|_\infty)^p. \tag{1.4}$$

It is remarked that if we set $\mathcal{D}^*(u)(x,y) := (u(x) - u(y))\frac{x-y}{|x-y|^{\frac{n+2\alpha}{2}+1}}$ and $2|\beta(x,y)| := \frac{C'_\alpha}{|x-y|^{n+2\alpha}}$, then, following the definition of [8], we have

$$(-\Delta)^\alpha u = C'_\alpha \int_{\mathbb{R}} \frac{u(x) - u(y)}{|x-y|^{n+2\alpha}} dy = -\mathcal{D}(\mathcal{D}^* u)(x),$$

where

$$\mathcal{D}(\mathcal{V})(x) := -\int_{\mathbb{R}^n} (\mathcal{V}(x,y) + \mathcal{V}(y,x)) \cdot \beta(x,y) dy, \quad \text{for } x \in \mathbb{R}^n.$$

The above discussion indicates that the operator $(-\Delta)^\alpha$ is a symmetric operator.

Recall that a stochastic process $u = u(x,t,\omega)$ is said to be a stochastic strong solution of (1.1) if it is an almost surely continuous L^2 process belonging to the space $L^2(\Omega \times \mathbb{R}_+, W^{\alpha,2}(\mathbb{R}^n))$ and satisfies the SPDE (1.1) in the following sense $\mathbb{P}$-a.e.,

$$\begin{aligned}\langle u(t),\phi\rangle = \langle u(0),\phi\rangle - \int_0^t \langle \mathcal{D}^\alpha u(s),\phi\rangle ds + \int_0^t \langle f(u(s)),\phi\rangle ds \\ + \int_0^t \langle g_i(u(s)),\phi\rangle dW_s^i\end{aligned}$$

for all $\phi \in C_c(\mathbb{R}^n)$. For simplicity, we will denote $f(x,t,u)$ and $g(x,t,u)$ by $f(u)$ and $g(u)$ respectively. In next section, we will prove our main result.

The rest of this paper is organized as follows. Section 2 is concerned with the proof of the main result. In Section 3, we discuss another type of Hölder continuity which is in the sense of almost sure continuity.

2. Proof of Theorem 6.1

In this section, we show our main result. Following the idea of [9], let us define

$$\begin{aligned}&\|g\|_{p_1,p_2,I} := \|g\|_{L^{p_1}(I,L^{p_2}(\mathbb{R}^n))}, \qquad I_k = [(1-2^{-k})T, 2T] \\ &U_{k,a} := \|u_{k,a}\|^2_{4,2,I_k} = \sqrt{\int_{I_k} \|u_{k,a}\|_2^4 dt},\end{aligned}$$

where $u_{k,a} = (u - a(1-2^{-k}))^+$ and $a \geq 1$.

We first need to prove the following lemmas.

Lemma 6.1. *Assume that the function $K(x)$ in the linear growth condition (A) satisfies $\|K\|_\infty \leq 1$. Then for $n \geq [2\alpha]$, there exists a constant $C = C(n,\Lambda,T)$ such that*

$$U_{k,a} \leq \frac{C^k}{a^{2\alpha/(n+\alpha)}}(U_{k-1,a} + X^*_{k-1,a})U_{k-1,a}^{\alpha/(n+\alpha)},$$

where $[2\alpha]$ denotes the integer part of 2α and

$$X^*_{k-1,a} := \sup_{(1-2^{-k})T \leq s \leq t \leq 2T} \int_s^t \langle g_i(u(\tau)), u_{k,a}(\tau)\rangle dW_\tau^i.$$

Proof. The proof is similar to that of Proposition 2.1 in [9]. Applying Hölder inequality, we get

$$\|u_{k,a}\|_2^2 \leq \|u_{k,a}\|^2_{2(n+\alpha)/n} |\{u_{k,a} > 0\}|^{\alpha/(n+\alpha)}. \tag{2.1}$$

Using Chebyshev's inequality, we have

$$|\{u_{k,a} > 0\}| = |\{u_{k-1,a}(t) > 2^{-k}a\}| \leq \left(\frac{2^k}{a}\right)^2 \|u_{k-1,a}\|_2^2.$$

Squaring (5.2) and integrating with respect to t on I_k, we obtain

$$U_{k,a}^2 \le \left(\frac{2^k}{a}\right)^{4\alpha/(n+\alpha)} \int_{I_k} \|u_{k,a}(t)\|_{2(n+\alpha)/n}^4 \|u_{k-1,a}(t)\|_2^{4\alpha/(n+\alpha)} dt.$$

By using Hölder inequality again, we have

$$U_{k,a} \le \left(\frac{2^k}{a}\right)^{2\alpha/(n+\alpha)} \left(\int_{I_k} \|u_{k,a}(t)\|_{2(n+\alpha)/n}^{4(n+\alpha)/n} dt\right)^{n/[2(n+\alpha)]} \times \left(\int_{I_k} \|u_{k-1,a}(t)\|_2^4 dt\right)^{\alpha/[2(n+\alpha)]}.$$

The third factor of the right hand side can be estimated by $U_{k-1,a}^{\alpha/(n+\alpha)}$. The second factor is exactly $\|u_{k,a}(t)\|_{4(n+\alpha)/n,2(n+\alpha)/n,I_k}$. We then claim that

$$\|u_{k,a}(t)\|_{4(n+\alpha)/n,2(n+\alpha)/n,I_k}^2 \le \sup_{t\in I_k} \|u_{k,a}\|_2^2 + \int_{I_k} \|u_{k,a}(t)\|_{2n/(n-2\alpha)}^2 dt. \tag{2.2}$$

We recall the $L_t^q L_x^p$ interpolation inequality

$$\|u\|_{r_1,r_2,I} \le \|u\|_{p_1,p_2,I}^{\gamma} \|u\|_{q_1,q_2,I}^{1-\gamma}$$

with

$$\frac{1}{r_1} = \frac{\gamma}{p_1} + \frac{1-\gamma}{q_1}, \quad \frac{1}{r_2} = \frac{\gamma}{p_2} + \frac{1-\gamma}{q_2}.$$

Using the above inequality with the following parameters:

$$r_1 = \frac{4(n+\alpha)}{n}, \quad r_2 = \frac{2(n+\alpha)}{n}, \quad p_1 = \infty,$$
$$q_1 = p_2 = 2, \quad q_2 = \frac{2n}{n-2\alpha}, \quad \gamma = \frac{n+2\alpha}{2(n+\alpha)},$$

and the elementary inequality

$$ab \le \frac{a^p}{p} + \frac{b^q}{q} \le a^p + b^q,$$

with $p = 2(n+\alpha)/(n+2\alpha)$ and $q = 2(n+\alpha)/n$ we obtain the inequality (2.2).

Next, applying the Sobolev inequality $W^{\alpha,2}(\mathbb{R}^n) \hookrightarrow L^{\frac{2n}{n-2\alpha}}(\mathbb{R}^n)$, we obtain the following

$$U_{k,a} \le \left(\frac{2^k}{a}\right)^{2\alpha/(n+\alpha)} \left[\sup_{t\in I_k} \|u_{k,a}\|_2^2 + \int_{I_k} \|u_{k,a}(t)\|_{W^{\alpha,2}(\mathbb{R}^n)}^2 dt\right] U_{k-1,a}^{\alpha/(n+\alpha)}. \tag{2.3}$$

We now use Itô formula to bound the terms involving the supermum over I_k and the $W^{\alpha,2}(\mathbb{R}^n)$ norm of u. Notice that the function $h_r(u) = |(u-r)^+|^2$ is piecewise smooth with continuous derivative and its second derivative has a single point of discontinuity at $u = r$. The quadratic variation process of the martingale part of the process $u(t)$ is absolutely continuous with respect to the Lebesgue measure on $\mathbb{R}_+$. Thus Itô formula can be applied to the composition $h_{a_k}(u(t)) = |u_{k,a}|^2$ and we have

$$d\|u_{k,a}\|_2^2 = -2\langle \mathcal{D}^\alpha u_{k,a}, u_{k,a}\rangle dt + 2\langle g_i(u), u_{k,a}(t)\rangle dW_t^i \\ + \left[\int_{\mathbb{R}^n} \{|g(u(t))|^2 + 2u_{k,a}(t)f(u(t))\}1_{u_{k,a}(t)>0}dx\right] dt. \tag{2.4}$$

The validity of the above application of Itô formula can be fully justified, see Remark 2.3 in [9]. For the third term of right hand side of (2.4), we see that if $u_{k,a} > 0$, then the inequalities $1 \le a \le 2^k u_{k-1,a}$ and $0 < u \le u_{k-1,a} + a \le (1+2^k)u_{k-1,a}$ hold. By the assumption (A) and the fact that $u_{k,a} \le u_{k-1,a}$, the third term can be bounded by $C^k\|u_{k-1,a}\|_2^2 dt$ for some constant C. Integrating (2.4) from t_0 to t with $t_0 \in I_{k-1} \setminus I_k$ and $t \in I_k$, by using the symmetry and bounded of Q, we have

$$\|u_{k,a}(t)\|_2^2 + q_1 \int_{t_0}^t \|u_{k,a}(s)\|_{W^{\alpha,2}(\mathbb{R}^n)}^2 ds \\ \le \|u_{k,a}(t_0)\|_2^2 + C^k U_{k-1,a} + \int_{t_0}^t \langle g_i(u(s)), u_{k,a}(s)\rangle dW_s^i.$$

Taking the supremum over I_k, we have for some constant $C = C(n, \Lambda)$

$$\sup_{t\in I_k} \|u_{k,a}(t)\|_2^2 + \int_{t_0}^{2T} \|u_{k,a}(s)\|_{W^{\alpha,2}(\mathbb{R}^n)}^2 ds \\ \le C\|u_{k,a}(t_0)\|_2^2 + C^k U_{k-1,a} + CX_{k-1,a}^*. \tag{2.5}$$

Using the fact that $u_{k,a} \le u_{k-1,a}$, we can find $t_0 \in I_{k-1} \setminus I_k$ by the mean value theorem such that

$$\|u_{k,a}(t_0)\|_2^2 = \frac{1}{I_{k-1} \setminus I_k} \int_{I_{k-1}\setminus I_k} \|u_{k,a}(t)\|_2^2 dt \le 2^k T^{-1} U_{k-1,a}. \tag{2.6}$$

Combining (2.3), (2.5) and (2.6), we get the desired inequality. The proof of Lemma 6.1 is then complete. □

Notice that the definition of $X_{k,a}^*$ in Lemma 6.1 is the same as defined in Proposition 2.1 of [9]. So we have the following result.

Lemma 6.2. [9, Lemma 3.2] *Assume that $\|K\|_\infty \le 1$. There exists a constant $C = C(n, \Lambda)$ such that for all positive γ and β,*

$$\mathbb{P}\{X_{k,a}^* \ge \gamma\beta,\ U_{k,a} \le \beta\} \le Ce^{-\gamma^2/C^k}.$$

By using the iterative inequality of Lemma 6.1 and the comparison result in Lemma 6.2, we can derive the estimation of $\|u^+\|_{\infty,[T,2T]\times\mathbb{R}^n}$.

Lemma 6.3. [9, Proposition 3.3] *Assume that $\|K\|_\infty \le 1$. There exists a constant $M_0 = M_0(n, \Lambda)$ such that for all $a \ge 1$ and $M > M_0$,*

$$\mathbb{P}\{\|u^+\|_{\infty,[T,2T]\times\mathbb{R}^n} > a, M\|u^+\|_{4,2,[0,2T]} \le a\} \le Ce^{-M^\delta}.$$

Here $\delta = \alpha/(n+\alpha)$ when $n \ge [2\alpha]$ and δ can be any value from $(0, \frac{1}{2})$ when $n = 1$ and $[2\alpha] = 1$.

Proof. The proof of this lemma is pretty similar to that of Proposition 3.3 in [9]. The only difference is the index δ, which can be determined by Lemma 6.1. We therefore omit the details here. □

Now, we are ready to prove our main result.

Proof of Theorem 6.1. By scaling, it suffices to consider the case that $\|K\|_\infty + \|K\|_2 \le 1$, $\|u(0\|_2 \le 1$ and $T = 1$. We first verify the inequality (1.3), that is, we need to prove that there exists a constant $C = C(p)$ such that

$$\mathbb{E}\int_0^2 \|u(t)\|_2^p dt \le C \quad \text{and} \quad \mathbb{E}\|u\|^p_{\infty,[1,2]\times\mathbb{R}^n} \le C. \tag{2.7}$$

Let $p \ge 4$ and $\psi(t) = \|u(t)\|_2^2 + 1$. Then, applying Itô formula, we get

$$d\psi(t) = \psi(t)(F(t)dt + dG_t) \tag{2.8}$$

where

$$F(t) = \frac{-\langle \mathcal{D}^\alpha(u), u\rangle + \langle f(u), u\rangle + \|g(u)\|_2^2}{\|u(t)\|_2^2 + 1} \quad \text{and} \quad G_t = \int_0^t \frac{\langle g_i(u), u\rangle}{\|u(t)\|_2^2 + 1} dW_s^i$$

The solution of SDE (2.8) can be expressed as

$$\psi(t) = \psi(0)\exp\left[\int_0^t F(s)ds + G_t - \frac{1}{2}\langle G\rangle_t\right].$$

It follows from Assumption (A) that for all $t \le 2$

$$\langle G\rangle_t \le 2(\Lambda + 1)^2.$$

Then, the Novikov condition implies that

$$\exp\left[G_t - \frac{1}{2}\langle G\rangle_t\right]$$

is a martingale for $0 < t \le 2$. Using the assumption (A) again, we have

$$\mathbb{E}\psi^p(t) = \psi^p(0)\mathbb{E}\left\{p\left[\int_0^t F(s)ds + G_t - \frac{1}{2}\langle G\rangle_t\right]\right\} \le C\psi^p(0),$$

which implies that the first inequality of (2.7) holds. Next, we prove another inequality. Denote

$$X = \mathbb{E}\|u\|_{\infty,[1,2]\times\mathbb{R}^n} \quad \text{and} \quad \Theta = \left(\int_0^2 \|u(t)\|_2^4 dt\right)^{1/4}.$$

It follows from Lemma 6.3 that

$$\mathbb{P}\{X > a, \Theta \le \frac{a}{M}\} \le 2e^{-M^\delta},$$

where $a \ge 1$, $M \ge M_0$ and the M_0, δ are defined in Lemma 6.3. Here we use the fact $|u| = u^+ + u^-$ and we use Lemma 6.3 twice for u and $-u$. In particular, taking $M_0 \ge 1$ and $a = M^2$, we have

$$\mathbb{P}\{X > a, \Theta \le \sqrt{a}\} \le 2e^{-a^{\delta/2}}. \tag{2.9}$$

Moreover, the first inequality of (2.7) yields that

$$\mathbb{E}\Theta^{2p} \leq 2^{(p-2)/2} \int_0^2 \|u(s)\|_2^{2p} ds \leq C. \tag{2.10}$$

Combining (2.9) and (2.10), we have

$$\begin{aligned}
&\mathbb{E}\|u\|^p_{\infty,[1,2]\times\mathbb{R}^n} \\
&= p\int_0^\infty \mathbb{P}\{X > a\}a^{p-1}da \\
&\leq M_0^{2p} + p\int_{M_0^2}^\infty \mathbb{P}\{\Theta > \sqrt{a}\}a^{p-1}da + p\int_{M_0^2}^\infty \mathbb{P}\{X > a, \Theta \leq \sqrt{a}\}a^{p-1}da \\
&\leq M_0^{2p} + \mathbb{E}\Theta^{2p} + 2p\int_{M_0^2}^\infty e^{-a^{\delta/2}}a^{p-1}da \\
&\leq C.
\end{aligned}$$

We obtain the inequality (1.3).

Next, let us verify the inequality (1.4). To this end, we consider the solution v of the following SPDE

$$dv_t = \Delta^\alpha v + g_i(u)dW_t^i, \qquad v(2^{-1}) = 0. \tag{2.11}$$

Set $\phi = u - v$, then ϕ satisfies

$$\partial_t\phi = \mathcal{D}^\alpha\phi + f(\phi + v) + \mathcal{D}^\alpha v + \Delta^\alpha v, \quad \text{on } [2^{-1}, 2] \times \mathbb{R}^n. \tag{2.12}$$

From the assumption (A) and inequality (1.3), we have

$$\mathbb{E}\int_{2^{-1}}^2 \|g(u)\|_p^p dt \leq C.$$

Following Krylov's $W^{2,p}(\mathbb{R}^n)$-theory and utilising the estimates of the heat kernels (see [2, Corollary 3.5]), we have that there exists a positive constant β_1 such that $v \in C^{\beta_1}([2^{-1}, 2] \times \mathbb{R}^n)$, see Proposition 6.1. Moreover, we have the following estimates

$$\mathbb{E}\|v\|^p_{C^{\beta_1}([2^{-1},2]\times\mathbb{R}^n)} \leq \mathbb{E}\|g(u)\|_{L^p([2^{-1},2]\times\mathbb{R}^n)} \leq C \tag{2.13}$$

$$\mathbb{E}\int_{2^{-1}}^2 \|v\|^p_{W^{\frac{\alpha}{2},p}} dt \leq C. \tag{2.14}$$

The proof of (2.13) is similar to that of [20] and we omit the details here. It follows from the parabolic regularity result and (2.13) that there exists a positive constant β_2 such that $\phi \in C^{\beta_2}([1, 2] \times \mathbb{R}^n)$ and

$$\begin{aligned}
&\|\phi\|_{C^{\beta_2}([1,2]\times\mathbb{R}^n)} \\
&\leq C\left(\|\phi\|_{\infty,[2^{-1},2]\times\mathbb{R}^n} + \|\Delta^\alpha v\|_{L^p([1,2];W^{-\frac{\alpha}{2},p})}\right) \\
&\leq C\left(\|u\|_{\infty,[2^{-1},2]\times\mathbb{R}^n} + \|v\|_{\infty,[2^{-1},2]\times\mathbb{R}^n} + \|\Delta^\alpha v\|_{L^p([1,2];W^{-\frac{\alpha}{2},p})}\right).
\end{aligned}$$

Using (2.13) and $u = \phi + v$, we end up with the desired inequality (1.4) with $\beta = \min\{\beta_1, \beta_2\}$. We are done. □

Remark 6.1. Comparing the proof of Theorem 6.1 with that of Proposition 6.1, we have used the estimates of the heat kernels, instead of using the Bessel potential method.

3. Hölder continuity in the sense of almost sure continuity

In Section 2, we obtained the Hölder continuity with respect to the $L^p(\Omega)$ norm. This section concerns with another type of Hölder continuity, that is, the Hölder continuity holds almost surely. We first recall some estimates of the kernel function K satisfying $\partial_t K = \Delta^\alpha K$. We recall some properties of the kernel $K(t,s,x)$ (see, e.g., [1, 3, 10] for more details):

- for any $t > 0$,
$$\|K(t,\cdot)\|_{L^1(\mathbb{R}^n)} = 1 \text{ for all } t > 0:$$
- $K(t,x,y)$ is C^∞ on $(0,\infty)\times\mathbb{R}^n\times\mathbb{R}^n$ for each $t>0$;
- for $t > 0$, $x, y \in \mathbb{R}^n$, $x \neq y$, the sharp estimate of $K(t,x)$ is
$$K(t,x,y) \approx \min\left(\frac{t}{|x-y|^{n+2\alpha}}, t^{-n/(2\alpha)}\right);$$
- for $t > 0$, $x, y \in \mathbb{R}^n$, $x \neq y$, the estimate of the first order derivative of $K(t,x)$ is
$$|\nabla_x K(t,x,y)| \approx |y-x| \min\left\{\frac{t}{|y-x|^{d+2+2\alpha}}, t^{-\frac{d+2}{2\alpha}}\right\}. \tag{3.1}$$

The notation $f(x) \approx g(x)$ means that there is a number $0 < C < \infty$ independent of x, i.e., a constant, such that for every x we have $C^{-1}f(x) \leq g(x) \leq Cf(x)$. The estimate (3.1) for the first order derivative of $K(t,x)$ was derived in [1, Lemma 5]. Xie et al. [22] obtained the estimate of the m-th order derivative of $K(t,x)$ by induction.

Proposition 6.2. [22, Lemma 2.1] *For any $m \geq 0$, we have*
$$\partial_x^m K(t,x) = \sum_{l=0}^{l=\lfloor\frac{m}{2}\rfloor} C_l |x|^{m-2l} \min\left\{\frac{t}{|x|^{n+2\alpha+2(m-l)}}, t^{-\frac{n+2(m-l)}{2\alpha}}\right\}, \tag{3.2}$$
where $\lfloor\frac{m}{2}\rfloor$ means the largest integer that is less than $\frac{m}{2}$.

Proposition 6.3. [4, Theorem 3.4] *Let D be a bounded open set in $\mathbb{R}^n$ and let $X(t,x)$, $t \in [0,T]$, $x \in D$ be a family of random variables such that*
$$\mathbb{E}|X(t,x) - X(s,y)|^\delta \leq C\left(|t-s|^2 + |x-y|^2\right)^{\frac{n+1+\varepsilon}{2}}$$
for some positive constants $C > 0, \varepsilon > 0, \delta \geq 1$, and all $t, x \in [0,T]$, $x, y \in D$. Then, there exists a process $\bar{X}(t,x)$ such that
$$P(\bar{X}(t,x) = X(t,x)) = 1, \quad \forall (t,x) \in [0,T]\times D,$$
and the functions $\bar{X}(\cdot,\cdot)$ are Hölder continuous on $[0,T]\times D$ with an arbitrary exponent smaller than ε/δ.

The above property is called Kolmogorov Continuity Theorem. Some authors extended the above result to a general case, see for instance the monograph [19]. By using the above properties, we have the followings:

Theorem 6.2. *Let*

$$X(t,x) = \mathcal{K}g(t,x) = \int_0^t \int_{\mathbb{R}^n} K(t,r,y)g(r,x-y)dydW(r), \quad (t,x) \in [0,T] \times \mathbb{R}^n.$$

Assume that $g \in L^p(\Omega \times [0,T] \times \mathbb{R}^n)$ *for* $p > 2 + \frac{n}{\alpha}$ *and further assume that the kernel function* K *satisfies the equation* $\partial_t K = \Delta^\alpha K$. *Then, there exists a process* $\bar{X}(t,x)$ *such that*

$$P(\bar{X}(t,x) = X(t,x)) = 1, \quad \forall (t,x) \in [0,T] \times \mathbb{R}^n,$$

and the process $\bar{X}(\cdot,\cdot)$ *is Hölder continuous on* $[0,T] \times \mathbb{R}^n$.

Proof. Set $t \geq s$. In order to apply Proposition 6.3, we first have the following

$$\begin{aligned}
&\mathbb{E}|X(t,x) - X(s,y)|^\delta \\
&= \mathbb{E}\Big|\int_0^t \int_{\mathbb{R}^n} K(t-r,x-z)g(r,z)dzdW(r) \\
&\quad - \int_0^s \int_{\mathbb{R}^n} K(s-r,y-z)g(r,z)dzdW(r)\Big|^\delta \\
&\leq C\mathbb{E}\Big|\int_0^s \int_{\mathbb{R}^n} (K(t-r,x-z) - K(s-r,y-z))g(r,z)dzdW(r)\Big|^\delta \\
&\quad + C\mathbb{E}\Big|\int_s^t \int_{\mathbb{R}^n} K(t-r,x-z)g(r,z)dzdW(r)\Big|^\delta \\
&\leq C\mathbb{E}\Big|\int_0^s \Big(\int_{\mathbb{R}^n} (K(t-r,x-z) - K(s-r,y-z))g(r,z)dz\Big)^2 dr\Big|^{\frac{\delta}{2}} \\
&\quad + C\mathbb{E}\Big|\int_s^t \Big(\int_{\mathbb{R}^n} K(t-r,x-z)g(r,z)dz\Big)^2 dr\Big|^{\frac{\delta}{2}} \\
&\leq C\mathbb{E}\Big|\int_0^s \Big(\int_{\mathbb{R}^n} (K(t-r,x-z) - K(s-r,y-z))g(r,z)dz\Big)^2 dr\Big|^{\frac{\delta}{2}} \\
&\quad + C\mathbb{E}\Big|\int_s^t \Big(\int_{\mathbb{R}^n} K(t-r,x-z)g(r,z)dz\Big)^2 dr\Big|^{\frac{\delta}{2}} \\
&=: C(I_1 + I_2).
\end{aligned}$$

Estimate of I_1.

By using Proposition 6.2 and Hölder inequality, we have

$$\begin{aligned}
I_1 &= \mathbb{E}\Big|\int_0^s \Big(\int_{\mathbb{R}^n} (K(t-r,x-z) - K(s-r,y-z))g(r,z)dz\Big)^2 dr\Big|^{\frac{\delta}{2}} \\
&\le C\mathbb{E}\Big|\int_0^s \Big(\int_{\mathbb{R}^n} |K(t-r,x-z) - K(s-r,x-z)|\cdot|g(r,z)|dz\Big)^2 dr\Big|^{\frac{\delta}{2}} \\
&\quad + C\mathbb{E}\Big|\int_0^s \Big(\int_{\mathbb{R}^n} |K(s-r,x-z) - K(s-r,y-z)|\cdot|g(r,z)|dz\Big)^2 dr\Big|^{\frac{\delta}{2}} \\
&\le C(t-s)^\delta \mathbb{E}\Big|\int_0^s \Big(\int_{\mathbb{R}^n} |\frac{\partial K}{\partial t}(\xi-r,x-z)|^q dz\Big)^{\frac{2}{q}} \|g(r)\|^2_{L^p(\mathbb{R}^n)} dr\Big|^{\frac{\delta}{2}} \\
&\quad + C|x-y|^{\beta\delta}\mathbb{E}\Big|\int_0^s \Big(\int_{\mathbb{R}^n} |\frac{\partial^\beta K}{\partial x^\beta}(s-r,w-z)|^q dz\Big)^{\frac{2}{q}} \|g(r)\|^2_{L^p(\mathbb{R}^n)} dr\Big|^{\frac{\delta}{2}} \\
&\le C(t-s)^\delta \Big(\mathbb{E}\int_0^s \|g(t)\|_p^p dt\Big) \\
&\quad \times \Big[\int_0^s \Big(\int_{\mathbb{R}^n} |\frac{\partial K}{\partial t}(\xi-r,x-z)|^q dz\Big)^{\frac{2(p-1)}{p-2}} dr\Big]^{\frac{\delta}{2}} \\
&\quad + C|x-y|^{\beta\delta}\Big(\mathbb{E}\int_0^s \|g(t)\|_p^p dt\Big) \\
&\quad \times \Big[\int_0^s \Big(\int_{\mathbb{R}^n} |\frac{\partial^\beta K}{\partial x^\beta}(s-r,w-z)|^q dz\Big)^{\frac{2(p-1)}{p-2}} dr\Big]^{\frac{\delta}{2}} \\
&\le C((t-s)^\delta + |x-y|^{\beta\delta}),
\end{aligned}$$

where $q = p/(p-1)$, $\xi = \theta t + (1-\theta)s$, $w = \theta x + (1-\theta)y$, $0 < \theta < 1$, and we have used the following

$$\begin{aligned}
&\int_0^s \Big(\int_{\mathbb{R}^n} |\frac{\partial K}{\partial t}(\xi-r,x-z)|^q dz\Big)^{\frac{2(p-1)}{p-2}} dr \\
&\le C\int_0^s \Big(\int_0^{(\xi-r)^{\frac{1}{2\alpha}}} (\xi-r)^{-\frac{nq+2q\alpha}{2\alpha}} |z|^{n-1} d|z| \\
&\quad + \int_{(\xi-r)^{\frac{1}{2\alpha}}}^\infty |z|^{-(qn+2q\alpha)}|z|^{n-1}d|z|\Big)^{\frac{2(p-1)}{p-2}} dr \\
&\le C\Big[(\theta(t-s))^{\frac{(p-1)(n-nq-2q\alpha)}{\alpha(p-2)}+1} - \xi^{\frac{(p-1)(n-nq-2q\alpha)}{\alpha(p-2)}+1}\Big] \\
&\le C
\end{aligned}$$

due to the fact that $0 < \theta < 1$. Similarly, we have the following:

$$\begin{aligned}
&\int_0^s \left(\int_{\mathbb{R}^n} |\frac{\partial^\beta K}{\partial x^\beta}(s-r, w-z)|^q dz \right)^{\frac{2(p-1)}{p-2}} dr \\
&\leq C \int_0^s \left(\int_{(s-r)^{\frac{1}{2\alpha}}}^{\infty} |x|^{q\beta} \cdot |x|^{n-1} \frac{(s-r)^q}{|x|^{qn+2q\alpha+2q\beta}} d|x| \right. \\
&\qquad \left. + \int_0^{(s-r)^{\frac{1}{2\alpha}}} |x|^{q\beta} \cdot |x|^{n-1} (s-r)^{-\frac{nq+2q\beta}{2\alpha}} d|x| \right)^{\frac{2(p-1)}{p-2}} dr \\
&\leq C \int_0^s (s-r)^{-\frac{n+(p-1)\beta}{\alpha(p-2)}} dr \\
&\leq C
\end{aligned}$$

provided that $n + (p-1)\beta \leq \alpha(p-2)$. We note that if we can take p large enough and β small enough, then we have the inequality hold. In dealing with the estimates of I_1, we have also used the fractional mean value theorem. Actually, we can use the property of the kernel function that

$$\|K * f(t)\|_{C^\beta(\mathbb{R}^n)} \leq C t^{-\frac{\beta}{\alpha} - \frac{n}{p\alpha}} \|f(t)\|_p,$$

for any deterministic function $f \in L^p(\mathbb{R}^n)$.

Estimate of I_2.

Similar to the estimate of I_1, we have

$$\begin{aligned}
I_2 &= \mathbb{E}\left| \int_s^t \left(\int_{\mathbb{R}^n} K(t-r, x-z) g(r,z) dz \right)^2 dr \right|^{\frac{\delta}{2}} \\
&\leq \left(\mathbb{E} \int_0^s \|g(t)\|_p^p dt \right) \left[\int_s^t \left(\int_{\mathbb{R}^n} |K(t-r, x-z)|^q dz \right)^{\frac{2(p-1)}{p-2}} dr \right]^{\frac{\delta}{2}} \\
&\leq C \left[\int_s^t \left(\int_{\mathbb{R}^n} |K(t-r, x-z)|^q dz \right)^{\frac{2(p-1)}{p-2}} dr \right]^{\frac{\delta}{2}} \\
&\leq C \left[\int_s^t (t-r)^{-\frac{n}{(p-2)\alpha}} dr \right]^{\frac{\delta}{2}} \\
&\leq C(t-s)^{\delta \frac{(p-2)\alpha - n}{2(p-2)\alpha}}
\end{aligned}$$

provided that $\alpha(p-2) > n$. Combining the assumption on p, we have

$$I_2 \leq C(t-s)^{\frac{(p-1)\beta\delta}{2(p-2)\alpha}}.$$

From the estimates of I_1 and I_2, and applying the Proposition 6.3, we obtain the desired results. The proof is complete. □

Remark 6.2. It follows from Proposition 6.3 and Theorem 6.2, the Hölder continuity of the solutions of the following equation on a local bounded domain can also be obtained

$$du = \Delta^\alpha u dt + g(t,x)dW_t, \quad t > 0, x \in \mathbb{R}^n, \quad u|_{t=0} = 0.$$

We note here that the Hölder continuity holds almost surely. Moreover, if one wants to get the regularity of solutions in the form of $L^p(\Omega; C^{\alpha,\alpha/2}([0,T] \times D))$, for a bounded domain $D \subset \mathbb{R}^n$, then the tail estimates have to be used.

Acknowledgment

This work was supported in part by NSFC of China grants 11771123, 11171064.

References

[1] K. Bogdan and T. Jakubowski, Estimates of heat kernel of fractional Laplacian perturbed by gradient operators, Comm. Math. Phys., **271** (2007) 179-198.
[2] Z. Brzeźniak, On stochastic convolution in Banach spaces and applications, Stochastics Stochastics Rep., **61** (1997) 245-295.
[3] Z.-Q. Chen and E. Hu, Heat kernel estimates for $\Delta + \Delta^{\alpha/2}$ under gradient perturbation, Stochastic Process. Appl., **125** (2015) 2603-2642.
[4] G. Da Prato and J. Zabczyk, Stochastic differential equations in infinite dimensions, Encyclopedia Math. Appl. 44, Cambridge University Press, Cambridge, (1992).
[5] A. Debussche, S. de Moor and M. Hofmanová, A regularity result for quasilinear stochastic partial differential equations of parabolic type, SIAM J. Math. Anal., **47** (2015) 1590-1614.
[6] L. Denis, A. Matoussi and L. Stoica, L^p estimates for the uniform norm of solutions of quasilinear SPDE's, Probab. Theory Related Fields, **133** (2005) 437-463.
[7] K. Du and J. Liu, Schauder estimate for stochastic PDEs, C. R. Math. Acad. Sci. Paris, **354** (2016) 371-375.
[8] Q. Du, M. Gunzburger, R. B. Lehoucq and K. Zhou, A nonlocal vector calculus, nonlocal volume-constrained problems, and nonlocal balance laws, Math. Models Methods Appl. Sci., **23** (2013) 493-540.
[9] E. Hsu, Y. Wang and Z. Wang, Stochastic De Giorgi iteration and regularity of stochastic partial differential equations, arXiv: 1312.3311v3.
[10] C. Imbert, A non-local regularization of first order Hamilton–Jacobi equations, J. Differential Equations, **211** (2005) 218-246.
[11] I. Kim and K. Kim, An L_p-theory for stochastic partial differential equations driven by Lévy processes with pseudo-differential operators of arbitrary order, Stochastic Process. Appl., **126** (2016) 2761-2786.
[12] I. Kim, K. Kim and S. Lim, Parabolic Littlewood–Paley inequality for a class of time-dependent pseudo-differential operators of arbitrary order, and applications to high-order stochastic PDE, J. Math. Anal. Appl., **436** (2016) 1023-1047.
[13] K. Kim and P. Kim, An L_p-theory of a class of stochastic equations with the random fractional Laplacian driven by Lévy processes, Stochastic Process. Appl., **122** (2012) 3921-3952.
[14] K. H. Kim, $L_q(L_p)$ theory and Hölder estimates for parabolic SPDEs, Stochastic Process. Appl., **114** (2004) 313-330.

[15] K. H. Kim, An L_p-theory of SPDEs on Lipschitz domains, Potential Anal., **29** (2008) 303-326.
[16] S. B. Kuksin, N. S. Nadirashvili and A. L. Piatnitski, Hölder estimates for solutions of parabolic spdes, Theory Probab. Appl., **47** (2003) 157-164.
[17] N. V. Krylov, An analytic approach to SPDEs, in: Stochastic Partial Differential Equations: Six Perspectives, in: Math. Surveys Monogr. vol. 64, 1999, pp: 185-242.
[18] N. V. Krylov, On L_p-theory of stochastic partial differential equations in the whole space, SIAM J. Math. Anal., **27** (1996) 313-340.
[19] M. Ledoux and M. Talagrand, Probability in Banach spaces. Isoperimetry and Processes. Classics in Mathematics, Springer–Verlag, Berlin, (2011).
[20] G. Lv, H. Gao, J. Wei and J.-L. Wu, Morrey–Campanato estimates for the moments of stochastic integral operators and its application to SPDEs, arXiv: 1704.05580
[21] R. Tian, L. Ding and J. Wei, Hölder estimates of mild solutions for nonlocal SPDEs, Submitted.
[22] X. Xie, J. Duan, X. Li and G. Lv, A regularity result for the nonlocal Fokker–Planck equation with Ornstein–Uhlenbeck drift, arXiv: 1504.04631.
[23] X. Zhang, L_p-theory of semi-linear SPDEs on general measure spaces and applications, J. Funct. Anal., 239 (2006), no. 1, 44-75.

Chapter 7

Multiscale Modelling Couples Patches of Two-Layer Thin Fluid Flow

Meng Cao[1,*] and A. J. Roberts[2]

[1]*School of Mathematical Sciences, University of Adelaide, South Australia 5005*
**cao040610122@163.com*
[2]*School of Mathematical Sciences, University of Adelaide, South Australia 5005*

The multiscale gap-tooth scheme uses a given microscale simulator of complicated physical processes to enable macroscale simulations by computing only small sparse patches to significantly reduce large scale computing costs. This article develops the multiscale gap-tooth scheme to the case of nonlinear microscale simulations of thin fluid flow. The microscale model reduction is implemented by artificially assuming the fluid film flow having two artificial layers but no distinguishing physical feature. Centre manifold theory assures that there exists a slow manifold in the two-layer fluid film flow. Eigenvalue analysis confirms the stability of the microscale simulator. This article uses the gap-tooth scheme to numerically simulate the two-layer fluid film flow. Coupling conditions are developed by computing the values at the edges of patches by neighbouring macroscale values. We approximate the spatial derivatives using the numerical method of centre difference at staggered grids. Time derivatives are implemented using Matlab `ode15s`. Numerical eigenvalue analysis suggests that the gap-tooth scheme with the developed two-layer microscale simulator empowers feasible computation of large scale simulations of fluid film flows. We also implement numerical simulations of the fluid film flow by the gap-tooth scheme. Comparison between a numerical gap-tooth simulation and a numerical microscale simulation over the whole domain demonstrates that the gap-tooth scheme feasibly computes fluid film flow dynamics with computational savings.

1. Introduction

Mathematical equations describing geophysical fluid dynamics are typically written at the macroscale of kilometres. But the underlying turbulent flow and physics is best understood at the very much finer 'microscale', sub-metre scale. We aim to empower scientists and engineers to use brief bursts of a given microscale simulator of wave-like dynamics on small patches of the space-time domain in order to make efficient and accurate macroscale simulations without ever knowing a macroscale closure.

*Corresponding author.

Our modelling further develops the equation-free gap-tooth scheme [14, 33–35] to empower novel simulation of wave-like systems over large time and space scales from a *given* microscopic simulator. Previously most multiscale modelling techniques have been developed for dissipative systems [12, 15, 16, 27]. We suppose that the wave-like microscale simulator is computationally expensive so that only small time and spatial domain simulations are feasible: one example of future interest is direct numerical simulation of depth resolved turbulent fluid floods. The aim is for the microscale simulator to provide the necessary data for the macroscopic computation, so that whenever the microscale simulator improves, then the overall macroscale simulation will correspondingly improve.

[5] initiated the application of the gap-tooth scheme to linear wave equations with weak dissipation. [6] then developed and theoretically supported the gap-tooth method for more general dispersive and nonlinear wave-like systems, and as an indicative application and test, applied the methodology to a Smagorinski model of turbulent shallow water flow. The generic key is to use polynomial interpolation of macroscale quantities, across the unsimulated gap between patches, to provide coupling conditions on macroscale quantities on the edges of each patch of the microsimulator. However, in most applications the microscale simulator will have many internal modes. When there are such internal modes in the gap-tooth simulation, an outstanding issue is that at each time step we need to 'lift' the macroscale coupling data to an appropriate microscale configuration. As a first attempt to address this issue of lifting for wave-like dynamics, this paper uses the gap-tooth scheme to model viscous flow of a layer of fluid at moderate Reynolds number.

The flow of rainwater on the road, windscreen or other draining problems [7, 8], and paint and coating flows [38] are a few examples of fluid film flows. Dynamics of such thin film flows have been studied extensively [3, 24, 25, 30]. The first aim of this article is to construct a two-layer model for such thin film flow as the microscale simulator. The main reason to develop a two-layer model (Section 2) is that it has microscale modes requiring lifting, but without the full complexity of fully resolved vertical structures. However, the two-layer model is itself a novel model for fluid flows at moderate Reynolds number.

Consider a thin fluid flow of depth $h(x,t)$ on an inclined plate with the slope $\tan\theta$. Section 2 artificially assumes two layers in the thin fluid flow, which have no distinguishing physical feature, as shown in Figure 1. From the Navier–Stokes PDEs and the boundary conditions on the free surface and the flat substrate, Section 2.2 uses centre manifold theory [2, 21, 23, 26] to derive a semi-slow two-layer model in the flow fields of depth $h(x,t)$ and layer mean velocities $\bar{u}_1(x,t)$ in the lower layer and $\bar{u}_2(x,t)$ in the upper layer: nondimensionally, the main parts of the model are

$$\Delta t h = -\frac{1}{2}\left(\Delta x h\bar{u}_1 + \Delta x h\bar{u}_2\right), \tag{1.1a}$$

$$\Delta t\bar{u}_1 \approx 0.826\,(\tan\theta - \Delta x h) + \frac{1}{\mathrm{Re}}\left(-19.3\frac{\bar{u}_1}{h^2} + 6.98\frac{\bar{u}_2}{h^2}\right) - 1.48\bar{u}_1\Delta x\bar{u}_1\,, \tag{1.1b}$$

$$\Delta t \bar{u}_2 \approx 1.002\,(\tan\theta - \Delta x h) + \frac{1}{\mathrm{Re}}\left(6.98\frac{\bar{u}_1}{h^2} - 5.36\frac{\bar{u}_2}{h^2}\right) - 1.25\bar{u}_1 \Delta x \bar{u}_1\,, \tag{1.1c}$$

for Reynolds number Re, and the plate slope $\tan\theta$. The right-hand sides of (1.1), and the more refined version (2.16), include the effects of gravitational forcing, bed drag, nonlinear advection, and dispersion.

The stability analysis in section 2.4 shows instabilities at high wavenumber: consequently, section 2.4 introduces an asymptotically consistent regularising operator to stabilise the two-layer model.

Section 3 applies the gap-tooth scheme to simulate the fluid film flow with the two-layer model being the microscale simulator. The coupling conditions are extended by the novel proposed lifting of the one-layer mean velocity $\bar{u}(x,t)$ to the two-layer velocities $\bar{u}_1(x,t)$ and $\bar{u}_2(x,t)$. Numerical simulations are implemented using centre difference method at staggered grids for spatial derivatives and matlab `ode15s` for time derivatives. Comparisons between the gap-tooth simulation and the microscale simulation over the whole domain, in Section 4, indicate that the gap-tooth scheme successfully simulates the dynamics of thin fluid flow at moderate Reynolds numbers.

2. Construct the two-layer microscale simulator

This section describes the derivation of the two-layer model. First, section 2.1 lists the 2D continuity and Navier–Stokes equations of a layer thin fluid flow and the boundary conditions on the free surface and on the plate. Secondly, section 2.2 embeds these equations in a family of equations with modified surface stress and mid-depth continuity so that an emergent two-layer slow manifold exists. The computer algebra of Appendix 6 then constructs the two-layer manifold model as summarised in section 2.3.

2.1. *Governing equations and boundary conditions for two layer thin film flow*

Consider a thin fluid flow of depth $h(x,t)$ flowing down an inclined plate with slope $\tan\theta$. Denote the coordinate system by x and z, along and normal to the plate respectively. Assume the fluid film flow has two artificial layers, each of thickness $h(x,t)/2$, which have no distinguishing physical feature, as shown schematically in Figure 1. Denote each layer velocity field by $\boldsymbol{q}_j(x,z,t) = (u_j, w_j)$ and pressure field by $p_j(x,z,t)$, where $j=1$ is for lower layer and $j=2$ is for upper layer.

We nondimensionalise the system in terms of a typical film thickness H, a typical fluid velocity $U := \sqrt{gH}$, the constant density ρ, and the slope angle of the plate θ. Then the Reynolds number $\mathrm{Re} = HU/\nu_f$ with ν_f being the viscosity of the fluid. The nondimensional continuity and Navier–Stokes equations for the dynamics of the 2D fluid film flow are then

$$\boldsymbol{\nabla}\cdot\boldsymbol{q}_j = \Delta x u_j + \Delta z w_j = 0\,, \quad j = 1, 2\,, \tag{2.1}$$

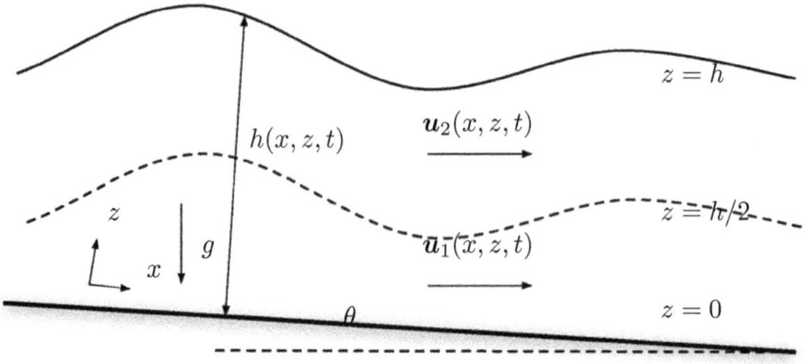

Fig. 1. Diagram of the two-layer modelling of thin film flow. Assume the fluid film flow has two artificial layers, which have no distinguishing physical feature. Each layer has thickness $h(x,t)/2$ and a mean layer velocity $u_j(x,z,t)$ with $j=1$ for lower layer and $j=2$ for upper layer. The plate has a slope $\tan\theta$. Then x and z are the lateral and normal coordinates, and g is gravity.

$$\Delta t\boldsymbol{q}_j + \boldsymbol{q}_j\cdot\boldsymbol{\nabla}\boldsymbol{q}_j = -\boldsymbol{\nabla}p_j + \frac{1}{\mathrm{Re}}\boldsymbol{\nabla}^2\boldsymbol{q}_j + \boldsymbol{g}\,,\quad j=1,2\,, \tag{2.2}$$

where the Reynolds number Re characterises the importance of the inertial terms compared to viscous dissipation, and where the vector $\boldsymbol{g}=(\tan\theta,-1)$ is the nondimensional forcing by gravity.

Well-known nondimensional boundary conditions hold on the plate and on the free surface.

- On the plate, prescribing no-slip requires

$$u_1 = w_1 = 0 \quad \text{on } z=0\,. \tag{2.3}$$

- On the free surface, the pressure is assumed to be zero and so the stress normal to the surface is zero for zero surface tension [24, 25, e.g.]:

$$\begin{aligned} 2&\left[\Delta z w_2 + (\Delta x h)^2\,\Delta x u_2 - \Delta x h\,(\Delta z u_2 + \Delta x w_2)\right] \\ &- \mathrm{Re}\left[1+(\Delta x h)^2\right] p_2 = 0 \quad \text{on } z=h\,, \end{aligned} \tag{2.4}$$

- The free surface having zero tangential stress results in

$$\left[1-(\Delta x h)^2\right](\Delta z u_2 + \Delta x w_2) + 2\,(\Delta x h)\,(\Delta z w_2 - \Delta x u_2) = 0 \quad \text{on } z=h\,. \tag{2.5}$$

- Also on the free surface, the kinematic condition is

$$\Delta t h = w_2 - u_2\Delta x h \quad \text{on } z=h\,. \tag{2.6}$$

- Lastly, on the artificial interface of the two artificial layers, continuity of the physical fields requires

$$p_1 = p_2 \quad \text{on } z=h/2\,, \tag{2.7}$$

$$u_1 = u_2 \quad \text{on } z = h/2\,, \tag{2.8}$$

$$w_1 = w_2 \quad \text{on } z = h/2\,, \tag{2.9}$$

$$\Delta z u_1 = \Delta z u_2 \quad \text{on } z = h/2\,. \tag{2.10}$$

The PDEs (2.1)–(2.2), together with the boundary conditions (2.3)–(2.10) describe the dynamics of fluid film flow on an inclined plate.

2.2. *Embed to support with centre manifold theory*

This section embeds the system of physical equations in a family of artificial problems parametrised by γ. This embedding empowers theoretical support for the two-layer model.

First modify the tangential stress surface condition (2.5) to have an artificial forcing proportional to the net shear in the upper layer velocity:

$$\begin{aligned}&\left[1-(\Delta x h)^2\right](\Delta z u_2 + \Delta x w_2) + 2(\Delta x h)^2(\Delta z w_2 - \Delta x u_2)\\ &= (1-\gamma)\frac{2}{h}\left[u_2 - u_2(x, h/2, t)\right] \quad \text{on } z = h\,.\end{aligned} \tag{2.11}$$

When evaluated at parameter $\gamma = 1$, the right-hand side of (2.11) vanishes and so this artificial boundary condition (2.11) reduces to the physical tangential stress conditions (2.5). Also modify the derivative continuity (2.10) on the interface of the two artificial layers to

$$\left(1-\frac{\gamma}{2}\right)\Delta z u_1 = \frac{\gamma}{2}\Delta z u_2 + 2(1-\gamma)\frac{u_1}{h} \quad \text{on } z = h/2\,. \tag{2.12}$$

When evaluated at $\gamma = 1$, this artificial interface condition (2.12) recovers the originally physical interface condition (2.10). These are the two modifications of the embedding.

The theoretical support is based upon a subspace of equilibria. In the absence of lateral variations, $\partial_x = 0$, for parameter $\gamma = 0$, and on a horizontal bed, $\theta = 0$, the fluid system (2.1)–(2.10) modified by equations (2.11) and (2.12) has two neutral modes of the layer shear flows $(u_1, u_2) \propto (z, h/2)$ and $(u_1, u_2) \propto (0, z - h/2)$. Conservation of fluid provides a third neutral mode in the dynamics. That is, under the assumptions $\gamma = \partial_x = \theta = 0$, there are three neutral modes, and hence a three dimensional 'slow' subspace of equilibria, corresponding to uniform shear flows on a fluid of any thickness h. In the state space $(h, u_1, u_2, w_1, w_2, p_1, p_2)$ the slow subspace is

$$\mathbb{E}_0 = \operatorname{span}\left\{\begin{bmatrix}1\\0\\0\\0\\0\\0\\0\end{bmatrix}, \begin{bmatrix}0\\z\\h/2\\0\\0\\0\\0\end{bmatrix}, \begin{bmatrix}0\\0\\z-h/2\\0\\0\\0\\0\end{bmatrix}\right\}.$$

Linearised about any equilibrium in the slow subspace $\mathbb{E}_0$ (together with the small parameters $\gamma = \partial_x = \theta = 0$), equations (2.1), (2.3) and (2.9) imply the normal velocity components $w_1 = w_2 = 0$. Then in the linearised system the normal component of the linearised PDE (2.2) with (2.4) and (2.7) establishes the pressures $p_1 = p_2 = 0$. Linearising the lateral component of the linearised PDE (2.2) together with the modified boundary conditions (2.8) and (2.11)–(2.12) gives a linear system for the lateral velocities u_j:

$$\Delta t u_j - \frac{1}{\mathrm{Re}}\frac{\partial^2 u_j}{\partial z^2} = 0\,, \tag{2.13a}$$

$$u_1 - u_2 = 0 \quad \text{on} \quad z = h/2\,, \tag{2.13b}$$

$$\Delta z u_2 - \frac{2}{h}\left[u_2 - u_2|_{z=h/2}\right] = 0 \quad \text{on} \quad z = h\,, \tag{2.13c}$$

$$\Delta z u_1 - \frac{2}{h}u_1 = 0 \quad \text{on} \quad z = h/2\,. \tag{2.13d}$$

The linearised PDE (2.13a) implies the eigenvalue $\lambda = -k^2/\mathrm{Re}$ by seeking the eigenvector of the lateral velocities $(u_1, u_2) \propto (0, \sin k(h - z))$. The boundary condition (2.13c) imposes the constraint $kh/2 = \tan kh/2$ with solutions $kh = 0, 8.986, 15.451, \ldots$. There exists a corresponding linearly independent generalised eigenvector with lower lateral velocity $u_1 \neq 0$. A spectral gap exists between the three zero eigenvalues of the slow subspace $\mathbb{E}_0$ and the non-zero eigenvalues headed by $\lambda \approx -80.763/(h^2\,\mathrm{Re})$.

Given the spectral gap, centre manifold theory for such 'infinite dimensional' systems [2, 21, 23, 26, e.g.] supports the existence, emergence and construction of a slow manifold model based upon the slow subspace shear modes. The slow manifold is constructed as a regular perturbation of the slow subspace [21, 23]. Importantly, the theory supports the model in a finite domain of the parameters γ, θ and ∂_x. Evaluating the resulting slow manifold model at the physical case of artificial parameter $\gamma = 1$ then provides the model for the physical flow dynamics. Table 1 [p. 119] shows evidence that the modelling converges at $\gamma = 1$.

2.3. *A low order model of the two layer flow*

Computer algebra (Appendix 6) constructs the semi-slow manifold of the two layer thin fluid flow: we call it 'semi-slow' because the model resolves two lateral shear modes, and because we use 'slow' to refer to a model that resolves only the gravest lateral shear mode. The computer algebra program derives the semi-slow model in the flow fields of depth $h(x,t)$, lower layer mean velocity $\bar{u}_1(x,t)$ and upper layer mean velocity $\bar{u}_2(x,t)$.

The order of errors in the construction is phrased in terms of small parameters. In the theoretical support, the subspace of equilibria are found for $\theta = \partial_x = \gamma = 0$ and so these are necessarily small parameters of the semi-slow manifold. Small lateral spatial derivative ∂_x corresponds to physical solutions varying slowly in x and in such a context has recently been made a rigorous approximation [26]. Because the theoretical support is based upon the 3D subspace of equilibria parametrised by h, $\bar{u}_1$ and $\bar{u}_2$, the semi-slow model is formally global in h, $\bar{u}_1$ and $\bar{u}_2$. Nonetheless, we typically discard high order terms in $\bar{u}_j$ as being insignificant in practical parameter regimes [25, e.g.]. Thus $\mathcal{O}(\bar{u}_1^p + \bar{u}_2^p + \partial_x^p + \theta^p)$ denotes the error terms for some exponent p, which means each term explicitly expressed in the model has in total less than p factors of these four parameters. The bigger the exponent number p, the higher the order of the modelling. The artificial small parameter γ has no physical meaning but is crucial to rigorously establish the semi-slow manifold. However, relatively high orders of the artificial parameter γ are required so that evaluating at $\gamma = 1$ is accurate.

Computer algebra in Appendix 6 derives the physical flow fields of pressures p_1 and p_2, and layer velocities u_1 and u_2 in terms of the film thickness h, layer mean velocities $\bar{u}_1$ and $\bar{u}_2$, and scaled local normal coordinate $Z = z/h$:

$$
\begin{aligned}
p_1 &= (1-Z)h \\
&\quad + h^2\frac{\partial^2 h}{\partial x^2}\left[(-0.422 + 0.104Z - 0.5Z^2) + \gamma(-0.0148 - 0.0107)\right] \\
&\quad + h\tan\theta\,\Delta x h\left[(0.0469 - 0.104Z) + \gamma(0.0375 + 0.0107Z)\right] \\
&\quad + \frac{2}{\mathrm{Re}}\left[(\Delta x\bar{u}_1 - \Delta x\bar{u}_2 - 2\Delta x\bar{u}_1 Z) + \frac{1}{h}\Delta x h\,(\bar{u}_1 + \bar{u}_2 - 2\bar{u}_1 Z)\right] \\
&\quad + \gamma\frac{1}{\mathrm{Re}}\left[(4.875\Delta x\bar{u}_1 - 1.125\Delta x\bar{u}_2 - 1.5\Delta x\bar{u}_1 Z + 0.5\Delta x\bar{u}_2 Z)\right. \\
&\quad \left. + \frac{1}{h}\Delta x h(-7.625\bar{u}_1 + 2.375\bar{u}_2 + 1.5\bar{u}_1 Z - 0.5\bar{u}_2 Z)\right] \\
&\quad + \mathcal{O}(\bar{u}_1^3 + \bar{u}_2^3 + \partial_x^3 + \theta^3, \gamma^2)\,, \qquad (2.14a) \\
p_2 &= (1-Z)h \\
&\quad + h^2\frac{\partial^2 h}{\partial x^2}\left[(-0.417 + 0.115Z - 0.5Z^2) + \gamma(-0.0178 - 0.0049Z)\right] \\
&\quad + h\tan\theta\,\Delta x h\left[(0.0521 - 0.115Z) + \gamma(0.0404 + 0.0049Z)\right] \\
&\quad + \frac{4}{\mathrm{Re}}\left[(-\Delta x\bar{u}_1 - \Delta x\bar{u}_2 + 2\Delta x\bar{u}_1 Z) + \frac{1}{h}\Delta x h\,(2\bar{u}_1 - \bar{u}_2 - 2\bar{u}_1 Z + \bar{u}_2 Z)\right] \\
&\quad + \gamma\frac{1}{\mathrm{Re}}\left[(11\Delta x\bar{u}_1 - 3\Delta x\bar{u}_2 - 13.75\Delta x\bar{u}_1 Z + 4.25\Delta x\bar{u}_2 Z)\right. \\
&\quad \left. + \frac{1}{h}\Delta x h(-13.75\bar{u}_1 + 4.25\bar{u}_2 + 13.75\bar{u}_1 Z - 4.25\bar{u}_2 Z)\right] \\
&\quad + \mathcal{O}(\bar{u}_1^3 + \bar{u}_2^3 + \partial_x^3 + \theta^3, \gamma^2)\,, \qquad (2.14b) \\
u_1(Z) &= \left[(4.0 + 1.5\gamma)\bar{u}_1 Z - 0.5\gamma\bar{u}_2 Z + (4.0\bar{u}_2 - 12.0\bar{u}_1)\gamma Z^3\right] \\
&\quad + \mathrm{Re}\,h^2\Delta x h\left[(-0.104 + 0.0107\gamma)Z + 0.5Z^2\right. \\
&\quad \left. - (0.5 + 0.123\gamma)Z^3 + 0.225\gamma Z^5\right] \\
&\quad + \mathrm{Re}\,h^2\tan\theta\,\left[(0.104 - 0.0107\gamma)Z - 0.5Z^2\right.
\end{aligned}
$$

$$\begin{aligned}
&+ (0.5 + 0.123\gamma)Z^3 - 0.225\gamma Z^5] \\
&+ \mathcal{O}(\bar u_1^3 + \bar u_2^3 + \partial_x^3 + \theta^3, \gamma^2)\,, \qquad (2.14c)\\
u_2(Z) = {}& [(6.0 - 11.9\gamma)\bar u_1 + (-2.0 + 4.13\gamma)\bar u_2 + (4.0 - 17.8\gamma)\bar u_2 Z \\
&+ (-8.0 + 48.3\gamma)\bar u_1 Z + (27.0\bar u_2 - 69.0\bar u_1)\gamma Z^2 \\
&+ (34.0\bar u_1 - 14.0\bar u_2)\gamma Z^3] \\
&+ \operatorname{Re} h^2 \Delta x h\,[(-0.089 + 0.10\gamma) + (0.45 - 0.68\gamma)Z \\
&+ (-0.63 + 1.8\gamma)Z^2 + (0.25 - 2.5\gamma)Z^3 + 1.8\gamma Z^4 - 0.49\gamma Z^5] \\
&+ \operatorname{Re} h^2 \tan\theta\,[(0.089 - 0.10\gamma) + (-0.45 + 0.68\gamma)Z \\
&+ (0.63 - 1.8\gamma)Z^2 + (-0.25 + 2.5\gamma)Z^3 - 1.8\gamma Z^4 + 0.49\gamma Z^5] \\
&+ \mathcal{O}(\bar u_1^3 + \bar u_2^3 + \partial_x^3 + \theta^3, \gamma^2)\,. \qquad (2.14d)
\end{aligned}$$

Equations (2.14) describe the low order shape of the manifold in the state space. Physically, upon setting the parameter $\gamma = 1$, these four equations (2.14) approximately describe the vertical structures of the pressures and layer velocities associated with the terms of depth $h(x,t)$, layer mean velocities $\bar u_j(x,t)$ and their lateral derivatives. One important feature of this approach is that we do not impose these vertical structures on the flow: instead we systematically solve the governing physical fluid equations to discover the structure functions appropriate for any suitable parameter regime.

The computer algebra in Appendix 6 also determines the evolutions of the depth $h(x,y)$ and layer mean velocities $\bar u_1(x,t)$ and $\bar u_2(x,t)$ on this semi-slow manifold (2.14) but now to high order in the parameter γ:

$$\begin{aligned}
\Delta t h = {}& -0.5\Delta x\,(h\bar u_1 + h\bar u_2)\,, \qquad (2.15a)\\
\Delta t \bar u_1 = {}& \big(0.75 + 0.0438\gamma + 0.0365\gamma^2 - 0.00439\gamma^3 + 0.0000522\gamma^4 \\
&- 0.000305\gamma^5 - 0.0000393\gamma^6\big)(\tan\theta - \Delta x h) \\
&+ \frac{1}{\operatorname{Re}}\Big[-\big(18\gamma + 1.35\gamma^2 + 0.0723\gamma^3 - 0.0869\gamma^4 - 0.0112\gamma^5 - 0.00637\gamma^6\big)\frac{\bar u_1}{h^2} \\
&+ \big(6.0\gamma + 1.05\gamma^2 - 0.038\gamma^3 - 0.022\gamma^4 - 0.008\gamma^5 - 0.00218\gamma^6\big)\frac{\bar u_2}{h^2}\Big] \\
&+ \mathcal{O}(\bar u_1^3 + \bar u_2^3 + \partial_x^3 + \theta^3, \gamma^7)\,, \qquad (2.15b)\\
\Delta t \bar u_2 = {}& \big(1.125 - 0.195\gamma + 0.0740\gamma^2 - 0.00126\gamma^3 - 0.0003062\gamma^4 \\
&- 0.000185\gamma^5 - 0.0000231\gamma^6\big)(\tan\theta - \Delta x h) \\
&+ \frac{1}{\operatorname{Re}}\Big[\big(15\gamma - 8.29\gamma^2 + 0.22\gamma^3 + 0.0278\gamma^4 + 0.0135\gamma^5 + 0.00351\gamma^6\big)\frac{\bar u_1}{h^2} \\
&+ \big(-9.0\gamma + 3.17\gamma^2 - 0.0456\gamma^3 - 0.0173\gamma^4 - 0.00559\gamma^5 - 0.00146\gamma^6\big)\frac{\bar u_2}{h^2}\Big] \\
&+ \mathcal{O}(\bar u_1^3 + \bar u_2^3 + \partial_x^3 + \theta^3, \gamma^7)\,, \qquad (2.15c)
\end{aligned}$$

Equation (2.15a) is a direct consequence of conservation of fluid. The momentum equations (2.15b)–(2.15c) include the effects of viscous drag and dissipation $\bar u_j/h^2$, and the gravity forcing $\tan\theta - \Delta x h$. The higher order physical effects of advection

Table 1. Partial sums from evaluating coefficients at $\gamma = 1$ of selected terms in equation (2.15b)–(2.15c) indicates that the power series in γ converges quickly.

	Δxh in (2.15b)	Δxh in (2.15c)	$\bar{u}_1/h^2/\mathrm{Re}$ in (2.15b)	$\bar{u}_1/h^2/\mathrm{Re}$ in (2.15c)	$\bar{u}_2/h^2/\mathrm{Re}$ in (2.15b)	$\bar{u}_2/h^2/\mathrm{Re}$ in (2.15c)
γ^0	−0.75	−1.125	0	0	0	0
γ^1	−0.7938	−0.930	−18.	15.	6.	-9.
γ^2	−0.8302	−1.004	−19.35	6.712	7.050	−5.288
γ^3	−0.8258	−1.002	−19.42	6.933	7.012	−5.333
γ^4	−0.8259	−1.002	−19.34	6.960	6.990	−5.350
γ^5	−0.8256	−1.002	−19.32	6.974	6.982	−5.356
γ^6	−0.8255	−1.002	−19.32	6.977	6.980	−5.357

and dispersion effects are included in the next equations (2.16b)–(2.16c) where we report on the model to errors $\mathcal{O}(\bar{u}_1^4 + \bar{u}_2^4 + \partial_x^4 + \theta^4, \gamma^7)$.

Equations (2.14a)–(2.15c) express the semi-slow manifold model in terms of the introduced artificial parameter γ. Every coefficient in these equations is a power series in γ. The partial sums in Table 1 indicate that these coefficient series in γ converges quickly for $\gamma = 1$. Thus, when the model is constructed to errors $\mathcal{O}(\gamma^7)$, it is apparent that all the shown digits are accurate. [24, 25] and [29] reported similar convergence in other related physical problems. Hereafter we calculate every coefficient in the model up to errors $\mathcal{O}(\gamma^7)$, and then evaluate at the artificial parameter $\gamma = 1$.

Truncating to errors $\mathcal{O}(\bar{u}_1^4 + \bar{u}_2^4 + \partial_x^4 + \theta^4, \gamma^7)$, omitting the intricate details of the derivation and upon setting the artificial parameter $\gamma = 1$, the evolution of the depth $h(x,t)$, the lower layer mean velocity $\bar{u}_1(x,t)$ and the upper layer mean velocity $\bar{u}_2(x,t)$ on the semi-slow manifold are described by the flow conservation equation and by effective lateral momentum equations

$$\Delta th = -0.5\left(\Delta xh\bar{u}_1 + \Delta xh\bar{u}_2\right), \tag{2.16a}$$

$$\begin{aligned}\Delta t\bar{u}_1 \approx{}& 0.826\left(\tan\theta - \Delta xh\right) + \frac{1}{\mathrm{Re}}\left(-19.3\frac{\bar{u}_1}{h^2} + 6.98\frac{\bar{u}_2}{h^2}\right)\\ &- 1.48\bar{u}_1\Delta x\bar{u}_1 - 0.225\bar{u}_2\Delta x\bar{u}_2 + 0.142\bar{u}_2\Delta x\bar{u}_1 + 0.0728\bar{u}_1\Delta x\bar{u}_2\\ &+ \frac{(\bar{u}_1 - \bar{u}_2)}{h}\left(-0.25\bar{u}_1 + 0.34\bar{u}_2\right)\Delta xh\\ &+ \frac{1}{\mathrm{Re}}\left(-3.84\frac{\partial^2\bar{u}_1}{\partial x^2} + 2.52\frac{\partial^2\bar{u}_2}{\partial x^2}\right),\end{aligned} \tag{2.16b}$$

$$\begin{aligned}\Delta t\bar{u}_2 \approx{}& 1.002\left(\tan\theta - \Delta xh\right) + \frac{1}{\mathrm{Re}}\left(6.98\frac{\bar{u}_1}{h^2} - 5.36\frac{\bar{u}_2}{h^2}\right)\\ &- 1.25\bar{u}_1\Delta x\bar{u}_1 - 1.57\bar{u}_2\Delta x\bar{u}_2 + 0.768\bar{u}_2\Delta x\bar{u}_1 + 0.930\bar{u}_1\Delta x\bar{u}_2\\ &+ \frac{(\bar{u}_1 - \bar{u}_2)}{h}\Delta xh\left(-0.78\bar{u}_1 + 0.38\bar{u}_2\right)\\ &+ \frac{1}{\mathrm{Re}}\left(-1.98\frac{\partial^2\bar{u}_1}{\partial x^2} + 5.23\frac{\partial^2\bar{u}_2}{\partial x^2}\right).\end{aligned} \tag{2.16c}$$

Equation (2.16a) is a direct consequence of conservation of fluid. The momentum equations (2.16b)–(2.16c) include the effects of gravity $\tan\theta - \Delta xh$, viscous drag $\bar{u}_j/h^2$, advection $\bar{u}_j\Delta x\bar{u}_i$, dispersion $\partial^2\bar{u}_j/\partial x^2$, and other viscous terms, such as $\Delta xh\Delta x\bar{u}_j$. Compared with the one-layer models [22, 24, 31, e.g.], the two-layer model (2.16b)–(2.16c) ensures more subtle effects and resolves more internal modes. These internal modes are necessary for a more generic microscale simulation in the gap-tooth scheme over that initiated for wave-like systems by [6].

2.4. *Eigenanalysis of the microscale model*

This section linearly analyses the two-layer model (2.16). Linear analysis indicates that an unphysical instability appears for high wavenumber. [4] explored three methods to avoid such instability: only resolving low wavenumbers; adding high order dissipation terms; and introducing a regularising operator. We implement, recommend and describe the last.

Linear analysis of the two-layer model Consider the modelled fluid film flow with two artificial layers on a flat plate with slope $\tan\theta$. The fluid shear flow has an equilibrium with thickness $h = 1$, without loss of generality. The model (2.16), assuming $\partial_t = \partial_x = 0$, simplifies to

$$\begin{aligned}0.826\tan\theta + \frac{1}{\mathrm{Re}}(-19.3\bar{u}_1 + 6.98\bar{u}_2) = 0\,,\\ 1.002\tan\theta + \frac{1}{\mathrm{Re}}(6.98\bar{u}_1 - 5.36\bar{u}_2) = 0\,,\end{aligned}$$

which predicts an equilibrium of layer mean velocities

$$\bar{u}_1 = 0.209\,\mathrm{Re}\tan\theta \quad\text{and}\quad \bar{u}_2 = 0.459\,\mathrm{Re}\tan\theta\,. \tag{2.17}$$

Impose small perturbations to this equilibrium and seek solutions in the form

$$\begin{aligned}h &= 1 + \hat{h}\exp(\lambda t + ikx)\,,\\ \bar{u}_1 &= 0.209\,\mathrm{Re}\tan\theta + \hat{u}_1\exp(\lambda t + ikx)\,,\\ \bar{u}_2 &= 0.459\,\mathrm{Re}\tan\theta + \hat{u}_2\exp(\lambda t + ikx)\,,\end{aligned} \tag{2.18}$$

for growth rate λ (possibly complex) and nondimensional wavenumber k.

Substitute the form (2.18) into the model (2.16), equate coefficients and derive the linear problem

$$\lambda\begin{bmatrix}\hat{h}\\ \hat{u}_1\\ \hat{u}_2\end{bmatrix} = M\begin{bmatrix}\hat{h}\\ \hat{u}_1\\ \hat{u}_2\end{bmatrix},$$

where the coefficient matrix M is

$$\begin{bmatrix} -0.334\tan\theta\operatorname{Re}ik & -0.5ik & -0.5ik \\ -(0.826 + 0.0259\tan\theta\operatorname{Re})ik & \frac{1}{\operatorname{Re}}(-19.3+3.84k^2) - 0.244\tan\theta\operatorname{Re}ik & \frac{1}{\operatorname{Re}}(6.98-2.52k^2) - 0.088\tan\theta\operatorname{Re}ik \\ -(1.002 + 0.0029\tan\theta\operatorname{Re})ik & \frac{1}{\operatorname{Re}}(6.98+1.98k^2) + 0.935\tan\theta\operatorname{Re}ik & \frac{1}{\operatorname{Re}}(-5.36-5.23k^2) - 0.526\tan\theta\operatorname{Re}ik \end{bmatrix}. \tag{2.19}$$

The coefficient matrix M has the characteristic equation

$$\begin{aligned} &\lambda^3 + \lambda^2\left[1.114\tan\theta\operatorname{Re}ik + \frac{1}{\operatorname{Re}}\left(1.39k^2+24.66\right)\right] \\ &\quad - \lambda\Big[0.476\tan^2\theta\operatorname{Re}^2k^2 + \tan\theta k(-2.265ik^2 - 14.03i - 0.0144\operatorname{Re}k) \\ &\quad + 15.09\frac{1}{\operatorname{Re}^2}k^4 - 0.914k^2 - 84.13\frac{1}{\operatorname{Re}^2}k^2 - 54.73\frac{1}{\operatorname{Re}^2}\Big] \\ &\quad - \Big[0.0725\tan^3\theta\operatorname{Re}^3ik^3 + \tan^2\theta\operatorname{Re}k^2(-0.0191\operatorname{Re}ik + 0.615k^2 + 1.908) \\ &\quad + \tan\theta\frac{1}{\operatorname{Re}}k(5.19ik^4 - 0.682\operatorname{Re}^2ik^2 - 28.94ik^2 - 18.83i - 0.841\operatorname{Re}k^3 \\ &\quad - 0.198\operatorname{Re}k) + \frac{1}{\operatorname{Re}}k^2(0.209k^2 - 18.26)\Big] = 0\,. \end{aligned} \tag{2.20}$$

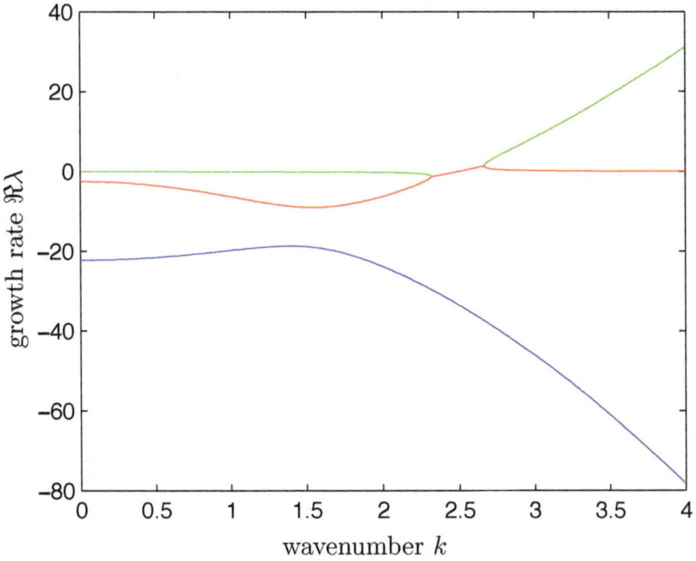

Fig. 2. The growth rate $\Re\lambda$ varies with the nondimensional wavenumber k from the characteristic equation (2.20) of the model (2.16). The slope of the plate is $\tan\theta = 0$. The Reynolds number is $\operatorname{Re} = 1$. An unphysical instability appears for high nondimensional wavenumber $k > 2.5$. The little interval near $k = 2.5$ is due to the numerical error.

Figure 2 plots the growth rates $\Re\lambda$ versus the nondimensional wavenumber k from equation (2.20). All the values represented by the blue curve are negative which nicely reflects viscous decay of lateral shear modes. When the nondimensional wavenumber $k < 2.5$, the values represented by the red curve are negative (viscous decay), and by the green curve are zeros (conservation of fluid). But when the nondimensional wavenumber $k > 2.5$, the green curve increases to positive which implies that instability arises in the system. This instability is nothing to do with physical instabilities, for example, in [11], who found the instability of the two liquid films down an inclined plate due to the different viscosity at the interface of the two layer flow and on the free surface. Here the instability arises at high wavenumber, a wavelength comparable to the thickness of the fluid, whereas our modelling is accurate for low wavenumber and long length scales.

Consistently avoid the instability We introduce a regularising operator to stabilise the unphysical instability. For the momentum equations (2.16b)–(2.16c), consider applying the regularising operator $\mathcal{L} = 1 - C\partial_x(h^2\partial_x)$ to both sides of both PDEs. The coefficient C is positive. The reason for using h^2 in the regularising operator $\mathcal{L}$ is to be dimensionally consistent which means we can cancel the h^{-2} in the drag terms in equations (2.16b)–(2.16c) and also ensure the regularising operator $\mathcal{L}$ is self-adjoint. This regularising operator generates dissipation effects to counteract the positive growth rates. The computer algebra of Appendix 6 applies the regularising operator to the momentum PDEs (2.16b)–(2.16c), and gives the PDEs, in term of the regularising parameter C,

$$\Delta th = -0.5\left(\Delta xh\bar{u}_1 + \Delta xh\bar{u}_2\right), \tag{2.21a}$$

$$\begin{aligned}
\mathcal{L}\Delta t\bar{u}_1 \approx{}& 0.826\left(\tan\theta - \Delta xh\right) \\
&+ \frac{1}{\mathrm{Re}}\left(-19.3\frac{\bar{u}_1}{h^2} + 6.98\frac{\bar{u}_2}{h^2}\right) \\
&- 1.48\bar{u}_1\Delta x\bar{u}_1 - 0.225\bar{u}_2\Delta x\bar{u}_2 + 0.142\bar{u}_2\Delta x\bar{u}_1 + 0.0728\bar{u}_1\Delta x\bar{u}_2 \\
&+ \frac{(\bar{u}_1 - \bar{u}_2)}{h}\left(-0.25\bar{u}_1 + 0.34\bar{u}_2\right)\Delta xh \\
&+ \frac{1}{\mathrm{Re}}\left(-3.84 + 19.3C\right)\frac{\partial^2\bar{u}_1}{\partial x^2} + \frac{1}{\mathrm{Re}}\left(2.52 - 6.98C\right)\frac{\partial^2\bar{u}_2}{\partial x^2},
\end{aligned} \tag{2.21b}$$

$$\begin{aligned}
\mathcal{L}\Delta t\bar{u}_2 \approx{}& 1.002\left(\tan\theta - \Delta xh\right) \\
&+ \frac{1}{\mathrm{Re}}\left(6.98\frac{\bar{u}_1}{h^2} - 5.36\frac{\bar{u}_2}{h^2}\right) \\
&- 1.25\bar{u}_1\Delta x\bar{u}_1 - 1.57\bar{u}_2\Delta x\bar{u}_2 + 0.768\bar{u}_2\Delta x\bar{u}_1 + 0.930\bar{u}_1\Delta x\bar{u}_2 \\
&+ \frac{(\bar{u}_1 - \bar{u}_2)}{h}\Delta xh\left(-0.78\bar{u}_1 + 0.38\bar{u}_2\right) \\
&+ \frac{1}{\mathrm{Re}}\left(-1.98 - 6.98C\right)\frac{\partial^2\bar{u}_1}{\partial x^2} + \frac{1}{\mathrm{Re}}\left(5.23 + 5.36C\right)\frac{\partial^2\bar{u}_2}{\partial x^2}.
\end{aligned} \tag{2.21c}$$

Now we linearise this system (2.21) and derive the characteristic equation, parametrised by the regularising C,

$$\begin{aligned}
&\lambda^3 + \lambda^2 \left[1.114 \tan\theta \operatorname{Re} ik + \frac{1}{\operatorname{Re}} \left(1.39k^2 + 24.66\right) + 24.66 \frac{1}{\operatorname{Re}} Ck^2 \right] \\
&- \lambda \left[0.476 \tan^2\theta \operatorname{Re}^2 k^2 + \tan\theta k(-2.265ik^2 - 14.03i - 0.0144 \operatorname{Re} k) \right. \\
&+ 15.09 \frac{1}{\operatorname{Re}^2} k^4 - 0.914k^2 - 84.13 \frac{1}{\operatorname{Re}^2} k^2 - 54.73 \frac{1}{\operatorname{Re}^2} \\
&\left. - 14.03C \tan\theta ik^3 - \frac{1}{\operatorname{Re}^2}(54.73C^2k^4 + 84.13Ck^4 + 109.46Ck^2) \right] \\
&- \left[0.0725 \tan^3\theta \operatorname{Re}^3 ik^3 + \tan^2\theta \operatorname{Re} k^2(-0.0191 \operatorname{Re} ik + 0.615k^2 + 1.908) \right. \\
&+ \tan\theta \frac{1}{\operatorname{Re}} k(5.19ik^4 - 0.682 \operatorname{Re}^2 ik^2 - 28.94ik^2 - 18.83i - 0.841 \operatorname{Re} k^3 \\
&- 0.198 \operatorname{Re} k) + \frac{1}{\operatorname{Re}} k^2(0.209k^2 - 18.26) + \frac{1}{\operatorname{Re}}(-18.83C^2 \tan\theta ik^4 \\
&- 28.94C \tan\theta ik^4 - 37.65C \tan\theta ik^2 - 18.26Ck^3) + 1.908C \tan^2\theta \operatorname{Re} k^3 \\
&\left. - 0.198C \tan\theta k^3 \right] = 0\,. \qquad (2.22)
\end{aligned}$$

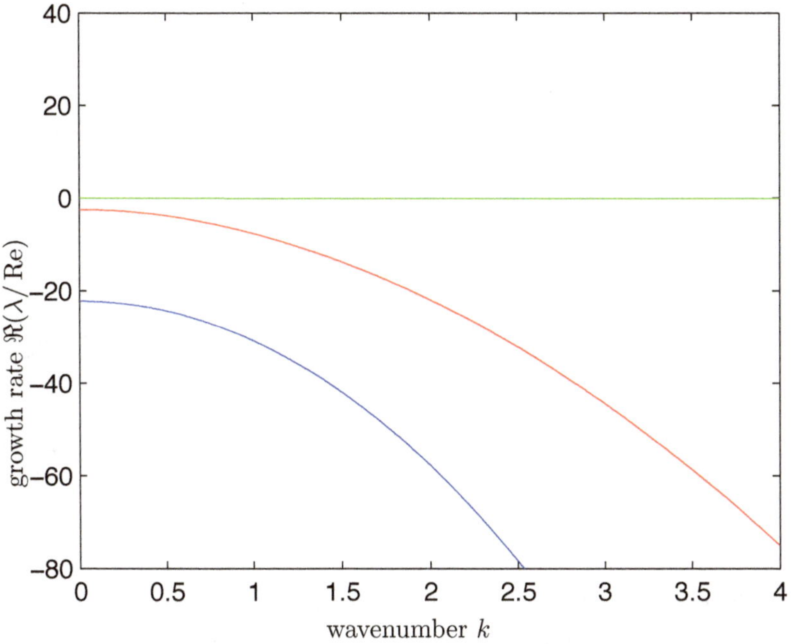

Fig. 3. Plot of the growth rate $\Re\lambda$ varying with the nondimensional wavenumber k. The coefficient $C = 0.5$ in the regularising operator $\mathcal{L}$. The slope of the plate is $\tan\theta = 0$. The Reynolds number is $\operatorname{Re} = 1$.

Figure 3 plots the growth rate $\Re\lambda$ varying with the nondimensional wavenumber k from the characteristic equation (2.22) for parameter $C = 0.5$ in the regularising operator $\mathcal{L}$. That there is no positive growth rate demonstrates that no instability occurs, even for the high wavenumber. The decay rates represented by the blue and red curves grow quickly for large wavenumber. Numerical checks indicate that the regularising coefficient $C > 0.17$ to eliminate the unphysical instabilities. This method is flexible through the wide range of choices of the positive coefficient C in the regularising operator $\mathcal{L}$. No high order derivatives, such as $h^2\partial^4\bar{u}_1/\partial x^4$, appear with this regularising method. Thus, introducing the regularising operator $\mathcal{L}$ usefully stabilises the model, and hereafter we implement the numerical simulations of the model (2.21) with the regularising coefficient $C = 0.5$.

3. Gap-tooth simulation of the two layer thin fluid flow

This section focuses on implementing the gap-tooth simulation of the thin fluid flow. This section uses the two-layer model (2.21) as the microscale simulator within patches. Coupling conditions (3.4a)–(3.4d) and (3.7) are developed to couple patches together.

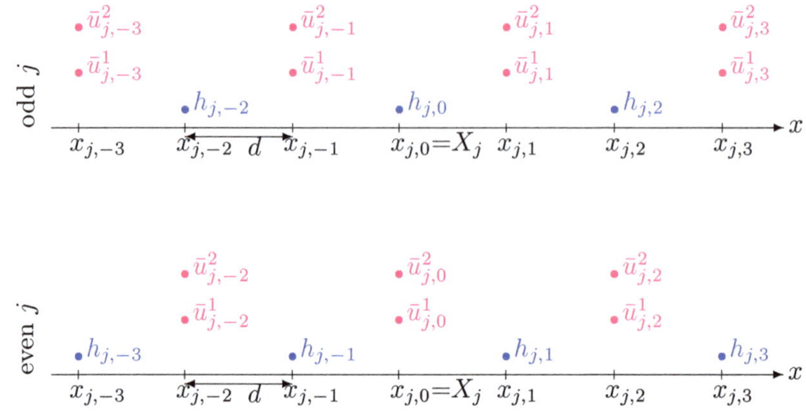

Fig. 4. Scheme of the staggered grid points of the depth $h_{j,i}$ (blue points) and the mean velocities $\bar{u}^1_{j,i}$ and $\bar{u}^2_{j,i}$ (magenta points) at the ith micro-grid point on the odd jth patch (top) and the even jth patch (bottom). This diagram shows the cases for $n = 5$ interior grid points in each patch ($n' = 3$).

Let us focus on one patch, the jth patch. Figure 4 shows the staggered grid for the depth $h_{j,i}$ (blue points) and the layer mean velocities $\bar{u}^1_{j,i}$ and $\bar{u}^2_{j,i}$ (magenta points) at the ith micro-grid point ($i = -n', \ldots, n'$ for $n' := (n+1)/2$) on the jth patch. The superscripts correspond to the two mean velocities $\bar{u}_1$ and $\bar{u}_2$ on a patch. Let each of m patches be centred on equi-spaced macroscale grid points $x = X_j = jD$, where $D = L/m$ is the macroscale spacing and L is the length of the whole domain. Each patch has relatively small width l. Assume each patch has a total of n microscale interior grid points, excluding the two edge grid points, so the microscale spatial step $d = l/(n+1)$. Let each patch around a macroscale grid point X_j execute a microscale simulator.

The microscale simulator (2.21) is a straightforward second order differential equation. Approximate the PDEs (2.21) on the jth patch with centred differences in microscale space of d shown in Figure 4. Use Matlab `ode15s` for continuous time integration. The dominant error comes from the macroscale coupling between patches, not the microscale discretisation [4, e.g.].

Define the mid patch point $x_{j,0} = X_j$. Let the macroscale value $H_j = h_j(X_j, t)$ for the odd j and $U_j = (\bar{u}_j^1(X_j, t) + \bar{u}_j^2(X_j, t))/2$ for even j. The inter-patch coupling uses only these macroscale values.

3.1. *Coupling conditions on the odd patches*

This subsection develops the coupling conditions on the odd patches. The velocity values $\bar{u}^1_{j,\pm n'}$ and $\bar{u}^2_{j,\pm n'}$ at the edges of the odd patches are approximated by interpolation of $U_{j\pm1}, U_{j\pm3}, \ldots$ from neighbouring patches, which is 'lifted' [17, e.g.] by requiring that the microscale dynamics lie on the slow manifold.

The dynamics in the interior of each patch is given by the microscale simulator (2.21). The regularising operator $\mathcal{L} = 1 - C\partial_x(h^2\partial_x)$ of the two layer mean velocities $\bar{u}_1(x,t)$ and $\bar{u}_2(x,t)$ on the odd jth patch requires more details. For the example of $n = 5$, Figure 4, for odd j the two momentum equations (2.21b)–(2.21c) are of the form

$$\mathcal{L}\begin{bmatrix} \Delta t\bar{u}^\ell_{j,-3} \\ \Delta t\bar{u}^\ell_{j,-1} \\ \Delta t\bar{u}^\ell_{j,1} \\ \Delta t\bar{u}^\ell_{j,3} \end{bmatrix} = \text{RHS}\,, \tag{3.1}$$

where the RHS refers to a finite difference discretisation of the right-hand sides of the two-layer PDEs (2.21), and the superscript $\ell = 1, 2$ for the lower and upper layer. The regularising operator $\mathcal{L}$ has the matrix form, for the example $n = 5$ case, of

$$\mathcal{L} = \frac{1}{4d^2}\begin{bmatrix} -Ch^2_{j,-2} & 1 + C(h^2_{j,-2} + h^2_{j,0}) & -Ch^2_{j,0} & 0 \\ 0 & -Ch^2_{j,0} & 1 + C(h^2_{j,0} + h^2_{j,2}) & -Ch^2_{j,2} \end{bmatrix}, \tag{3.2}$$

where d is the microscale spacial step. The values of $\bar{u}^\ell_{j,\pm n'}$ are at the edges of the jth patch. Recall that the macroscale mean velocity $U_j(t) = (\bar{u}^1_j(X_j, t) + \bar{u}^2_j(X_j, t))/2$ for the even j are known. Thus, this section completes the set of equations by finding the unknown microscale values $\Delta t\bar{u}^\ell_{j,\pm n'}$ and $\bar{u}^\ell_{j,\pm n'}$ from the known macroscale values U_j.

The challenge is to deduce microscale values appropriate to the macroscale structures. [12, 13] and [19] studied a heterogeneous multiscale method (HMM). The HMM contains two main components: an overall macroscale scheme for the macroscale variables and estimating the missing macroscale data by the microscale model. A compression operator Q and a reconstruction operator R are defined to satisfy $Q\bar{u}_j = U_j$, $RU_j = \bar{u}_j$ and $QRU_j = U_j$, where $\bar{u}_j$ is the microscale variable and U_j the macroscale variable. Such compression and reconstruction operators combine the microscale and macroscale variables. [16, 35] and [34] defined a coarse time-stepper by introducing a lifting operator and a corresponding restriction operator which transform between the microscale and macroscale variables. These works provide the methods to relate the microscale and macroscale.

This section analogously constructs a lifting operator to give the patch edge values of $\bar{u}^\ell_{j,\pm n'}$ and $\Delta t\bar{u}^\ell_{j,\pm n'}$ in terms of the macroscale mean velocity U_j by assuming the system lies on the one-layer slow manifold. Recall that the one-layer mean velocity $\bar{u}(X_j,t) = U_j = (\bar{u}^1_{j,0} + \bar{u}^2_{j,0})/2$. One constraint on the lifting is the coupling conditions [5, Eq. (11)] that give a linear, cubic or quintic approximation for the one layer mean velocity, such as the quintic

$$\begin{aligned}\bar{u}(X_j \pm rD,t) &= \frac{1}{2}(U_{j+1}+U_{j-1}) \pm \frac{r}{2}(U_{j+1}-U_{j-1})\\ &+ \frac{1}{16}(-1+r^2)(U_{j+2}-U_{j+1}-U_{j-1}+U_{j-2})\\ &\pm \frac{1}{48}(-r+r^3)(U_{j+2}-3U_{j+1}+3U_{j-1}-U_{j-2})\,. \qquad (3.3)\end{aligned}$$

The other requirement is that the patch lies on the slow manifold of macroscale waves. The following section 3.3 details the slow manifold of the two-layer model (2.21b)–(2.21c) in term of the mean velocity $\bar{u}$. For example and for simplicity, truncate the slow manifold description (3.9) and (3.11a)–(3.11b) to errors $\mathcal{O}(\epsilon^2)$ and obtain the values of $\bar{u}^\ell_{j,\pm n'}$ and $\Delta t\bar{u}^\ell_{j,\pm n'}$ on the odd jth patch as

$$\bar{u}^1_{j,\pm n'} = 0.587\bar{u}(X_j \pm rD,t) + 0.0129\,\mathrm{Re}\tan\theta\,, \qquad (3.4a)$$

$$\bar{u}^2_{j,\pm n'} = 1.413\bar{u}(X_j \pm rD,t) - 0.0129\,\mathrm{Re}\tan\theta\,, \qquad (3.4b)$$

$$\Delta t\bar{u}^1_{j,\pm n'} = -1.482\frac{1}{\mathrm{Re}}\frac{1}{h^2}\bar{u}(X_j \pm rD,t) + 0.489\tan\theta\,, \qquad (3.4c)$$

$$\Delta t\bar{u}^2_{j,\pm n'} = -3.526\frac{1}{\mathrm{Re}}\frac{1}{h^2}\bar{u}(X_j \pm rD,t) + 1.168\tan\theta\,, \qquad (3.4d)$$

where as a leading approximation we neglect the derivatives $\Delta x \approx 0$ in the slow manifold (3.11). Thus, equations (3.4a)–(3.4b) are the coupling conditions on the odd patches.

3.2. *Coupling conditions on the even patches*

This subsection develops the coupling conditions on the patches with even j. The values $h_{j,\pm n'}$ at the edges of the even patches are approximated from the neighbouring macroscale grid values to give coupling conditions.

Simulate on each patch by the discretisation of the microscale PDEs (2.21). The regularising operator $\mathcal{L} = 1 - C\partial_x(h^2\partial_x)$ of the two layer mean velocities $\bar{u}_1(x,t)$ and $\bar{u}_2(x,t)$ on the even jth patch requires more details. For example, we record here details for $n = 5$ as in Figure 4; other n is a direct generalisation. We need to approximate the second spatial derivatives ∂_x^2 in the left-hand and right-hand sides of the (2.21b)–(2.21c) at the positions $x_{j,\pm 2}$, which requires two virtual grid values $\bar{u}^\ell_{j,\pm 4}$. We set the values $\bar{u}^\ell_{j,\pm 4} = \bar{u}^\ell_{j,\pm 2}$ on the slow manifold. Thus, in Figure 4, for even j the momentum equations (2.21b)–(2.21c) are of the form

$$\mathcal{L}\begin{bmatrix}\Delta t\bar{u}^\ell_{j,-4}\\ \Delta t\bar{u}^\ell_{j,-2}\\ \Delta t\bar{u}^\ell_{j,0}\\ \Delta t\bar{u}^\ell_{j,2}\\ \Delta t\bar{u}^\ell_{j,4}\end{bmatrix} = \text{RHS}\,, \tag{3.5}$$

where the regularising operator $\mathcal{L}$ is discretised in the matrix form

$$\frac{1}{4d^2}\begin{bmatrix} -Ch^2_{j,-3} & 1+C(h^2_{j,-3}+h^2_{j,-1}) & -Ch^2_{j,-1} & 0 & 0\\ 0 & -Ch^2_{j,-1} & 1+C(h^2_{j,-1}+h^2_{j,1}) & -Ch^2_{j,1} & 0\\ 0 & 0 & -Ch^2_{j,1} & 1+C(h^2_{j,3}+h^2_{j,1}) & -Ch^2_{j,3}\end{bmatrix}, \tag{3.6}$$

and where the RHS refers to a discretisation of the right-hand sides of the momentum equations (2.21b)–(2.21c), the superscript $\ell = 1, 2$ for the lower and upper layer, and d is the microscale spatial step.

The regularising operator $\mathcal{L}$ and the RHSs need the values of $h_{j,\pm n'}$. Coupling conditions approximate the values of $h_{j,\pm n'}$ by interpolating the neighbouring macroscale values of $H_{j\pm 1}, H_{j\pm 2}, \ldots$. The coupling conditions [5, Eq. (11)] give a linear, cubic or quintic interpolation, such as the quintic

$$\begin{aligned} h_{j,\pm n'} &= h(X_j \pm rD, t) = \frac{1}{2}(H_{j+1} + H_{j-1}) \pm \frac{r}{2}(H_{j+1} - H_{j-1})\\ &+ \frac{1}{16}(-1 + r^2)(H_{j+2} - H_{j+1} - H_{j-1} + H_{j-2})\\ &\pm \frac{1}{48}(-r + r^3)(H_{j+2} - 3H_{j+1} + 3H_{j-1} - H_{j-2})\,, \end{aligned} \tag{3.7}$$

where r is the ratio of the macroscale step and half of the width of a patch. Thus the interpolation (3.7), together with equations (3.4a)–(3.4b), couple the patches together over the macroscale domain.

3.3. *The low order model of one layer flow*

This section derives a slow manifold of the two-layer model (2.21b)–(2.21c) in terms of the mean velocity $\bar{u}$. This slow manifold is used by sections 3.1–3.2 in order to lift the macroscale information to the microscale simulation on patches.

A fluid film model expressed in terms of the dynamics of both the fluid layer thickness and an overall lateral velocity (or momentum flux) is needed to resolve wave-like dynamics in many situations [28]: falling films [9, 20, p. 110]; wave transitions [10] to solitary waves [32]; higher Reynolds number flows [22, Eq. (19)]; in rising film flow and a slot coater [18, Eq. (37)]; rivulets under a sloping cylinder [1]. A slow manifold model of our two layer model corresponds to these earlier models of the fluid dynamics.

Indeed our construction here generates a slow manifold, one-layer model of the fluid dynamics which is the same as that of [24] (to the order of error of the analysis). The distinction is that here it is derived from the two-layer model, rather than the original fluid equations: this transitivity of modelling additionally validates the modelling process. For simplicity, we base the analysis upon a symmetric linear operator with slow eigenspace where the two layer velocities are in the ratio 1 : 2. To derive the slow manifold of the two-layer model, we embed the physical model (2.21b)–(2.21c) into a family of artificial problems by using another artificial parameter γ',

$$-\Delta t\bar{u}_1 + \text{RHS}_{(2.16\text{b})} + (1-\gamma')\frac{\text{Eu}}{\text{Re}}\frac{1}{h^2}\left(-4\bar{u}_1 + 2\bar{u}_2\right) = 0\,, \tag{3.8a}$$

$$-\Delta t\bar{u}_2 + \text{RHS}_{(2.16\text{c})} + (1-\gamma')\frac{\text{Eu}}{\text{Re}}\frac{1}{h^2}\left(2\bar{u}_1 - \bar{u}_2\right) = 0\,, \tag{3.8b}$$

where $\text{RHS}_{(2.16\text{b})}$ and $\text{RHS}_{(2.16\text{c})}$ are the right-hand sides of the PDEs (2.16b)–(2.16c). The variable Eu denotes an artificial Euler parameter used to enhance convergence [36, 37, e.g.]: computational experiments indicate that Eu $= 9/2$ delivers good convergence in the artificial parameter γ'. When the parameter $\gamma' = 1$, the PDEs (3.8) recover the original PDEs (2.16b)–(2.16c). When the artificial parameter $\gamma' = 0$, the linear operator in the PDEs (3.8) is

$$\frac{\text{Eu}}{h^2\,\text{Re}}\begin{bmatrix} -4 & 2 \\ 2 & -1 \end{bmatrix},$$

and it is this linear operator that guides effective recursive improvements of approximations to the original PDEs.

For a specific model we choose to truncate to errors $\mathcal{O}(\bar{u}^4 + \partial_x^4 + \theta^4, \gamma'^7)$. Then executing the computer algebra in Appendix 6 and evaluating at the artificial parameter $\gamma' = 1$, leads to the slow manifold where the two layer velocities are

$$\begin{aligned}\bar{u}_1 \approx{}& 0.587\bar{u} + 0.0129\,\text{Re}\,h^2\left(\tan\theta - \Delta x h\right)\\ &- 0.0468h^2\frac{\partial^2\bar{u}}{\partial x^2} - 0.205h\Delta x h\Delta x\bar{u} + 0.0700h\bar{u}\frac{\partial^2 h}{\partial x^2}\\ &+ \text{Re}\left(0.00465h^2\bar{u}\Delta x\bar{u} - 0.0115h\bar{u}^2\Delta x h - 0.0105h^3\Delta x h\frac{\partial^2 h}{\partial x^2}\right), \end{aligned}\tag{3.9a}$$

$$\begin{aligned}\bar{u}_2 \approx{}& 1.413\bar{u} - 0.0129\,\text{Re}\,h^2\left(\tan\theta - \Delta x h\right)\\ &+ 0.0468h^2\frac{\partial^2\bar{u}}{\partial x^2} + 0.205h\Delta x h\Delta x\bar{u} - 0.0700h\bar{u}\frac{\partial^2 h}{\partial x^2}\end{aligned}$$

$$- \operatorname{Re}\left(0.00465h^2\bar{u}\Delta x\bar{u} - 0.0115h\bar{u}^2\Delta xh - 0.0105h^3\Delta xh\frac{\partial^2 h}{\partial x^2}\right). \tag{3.9b}$$

Equations (3.9) are expressed in terms of the single layer depth-averaged velocity $\bar{u}(x,t)$ and the water depth $h(x,t)$.

On the slow manifold (3.9), conservation of mass requires

$$\Delta th = -\Delta xh\bar{u}\,. \tag{3.10}$$

The computer algebra in Appendix 6 differentiates equations (3.9) with respect to the time t, and gives the rate of change of the two layer velocities in term of the mean velocity $\bar{u}$ and fluid depth h as

$$\begin{aligned}
\Delta t\bar{u}_1 \approx{}& 0.489\left(\tan\theta - \Delta xh\right) - 1.482\frac{1}{\operatorname{Re}}\frac{\bar{u}}{h^2} - 0.904\bar{u}\Delta x\bar{u} \\
&+ \frac{1}{\operatorname{Re}}\left(2.552\frac{\partial^2\bar{u}}{\partial x^2} + 3.077\frac{1}{h}\Delta xh\Delta x\bar{u} - 0.650\frac{\bar{u}}{h}\frac{\partial^2 h}{\partial x^2}\right) \\
&+ \operatorname{Re}\left[0.0167h^3\frac{\partial^2\bar{u}}{\partial x^2} + 0.0438h^2\Delta xh\frac{\partial^2 h}{\partial x^2} + 0.0359h\bar{u}\left(\Delta xh\right)^2\right] \\
&- \operatorname{Re}\tan\theta\left(0.0298h\bar{u}\Delta xh + 0.0184h^2\Delta x\bar{u}\right), \qquad (3.11a) \\
\Delta t\bar{u}_2 \approx{}& 1.168\left(\tan\theta - \Delta xh\right) - 3.526\frac{1}{\operatorname{Re}}\frac{\bar{u}}{h^2} - 2.107\bar{u}\Delta x\bar{u} \\
&+ \frac{1}{\operatorname{Re}}\left(5.701\frac{\partial^2\bar{u}}{\partial x^2} + 6.962\frac{1}{h}\Delta xh\Delta x\bar{u} - 0.312\frac{\bar{u}}{h}\frac{\partial^2 h}{\partial x^2}\right) \\
&- \operatorname{Re}\left[0.00819h^3\frac{\partial^2\bar{u}}{\partial x^2} + 0.0482h^2\Delta xh\frac{\partial^2 h}{\partial x^2} + 0.0875h\bar{u}\left(\Delta xh\right)^2\right] \\
&+ \operatorname{Re}\tan\theta\left(0.0847h\bar{u}\Delta xh + 0.0355h^2\Delta x\bar{u}\right). \qquad (3.11b)
\end{aligned}$$

Equations (3.11) provide needed microscale information from the macroscale fields $\bar{u}$ and h.

The computer algebra in Appendix 6 simultaneously determines the evolution on the slow manifold of the one-layer model. The momentum equation is

$$\Delta t\bar{u} \approx 0.829\left(\tan\theta - \Delta xh\right) - 2.504\frac{1}{\operatorname{Re}}\frac{\bar{u}}{h^2} - 1.505\bar{u}\Delta x\bar{u} - 0.151\frac{\bar{u}^2}{h}\Delta xh\,. \tag{3.12}$$

This PDE (3.12) has the same terms as the low order model of [24, Eq. (11)]. The coefficients of these terms are the same to a relative error of less than 2.2%. This agreement partially verifies that the two-layer model is a reasonable model of the fluid film flow.

4. Numerical gap-tooth simulations of the two layer thin fluid flow

This section explores the numerical gap-tooth simulation with the two-layer microscale simulator (2.21) and the patch coupling conditions (3.7) and (3.3)–(3.4). Numerical eigenvalues and simulations show that the gap-tooth scheme reasonably simulates the macroscopic dynamics of the fluid film flow.

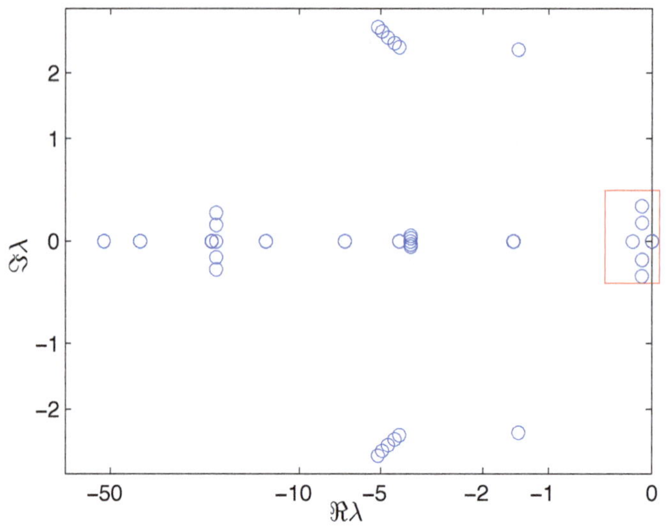

Fig. 5. Plots of the growth rate $\Re\lambda$ versus the frequency $\Im\lambda$ in the gap-tooth simulation of the thin film flow (with nonlinear axis scaling) over horizontal plate. There are $m = 10$ patches and $n = 9$ microscale grid points within a patch. The Reynolds number $\mathrm{Re} = 15$ and the coefficient in the regularising operator $\mathcal{L}$ is $C = 0.5$.

Consider the thin fluid flow on a horizontal plate: that is, the mean slope $\tan\theta = 0$. Distribute m patches in the macroscale domain of length $L = m\pi$, so the distance between the neighbouring patches is $D = L/m = \pi$. Divide each patch into $n+1$ equal microscale intervals by $n+2$ grid points, then the distance between neighbouring microscale points is $d = 2rD/(n+1)$, where r is the ratio between the half-width of each patch and the macroscale inter-patch distance D. We approximate the spatial derivatives in the right-hand side of the two-layer PDEs (2.21) by centred differences on the staggered microscale grids.

Figure 5 plots the growth rate $\Re\lambda$ versus the frequency $\Im\lambda$ in the gap-tooth simulation of the thin fluid flow. There are $m = 10$ patches and $n = 9$ microscale grids on a patch. The plate has a length $L = m\pi \approx 31.4$, so the distance between the neighbouring patches is $D = L/m = 3.14$. The ratio $r = 1/6$, so the microscale step on a patch is $d = 2rD/(n+1) \approx 0.17$. The Reynolds number $\mathrm{Re} = 15$ and the coefficient in the regularising operator $\mathcal{L}$ is $C = 0.5$. The negative growth rates imply the waves decay in time.

There are 90 pairs of eigenvalues for the system:

- the eigenvalues with large decay rates ($\Re\lambda < -1$) and zero imaginary parts predominantly represent the viscous decay of the lateral shear modes;
- the eigenvalues with large decay rates ($\Re\lambda < -1$) but with large imaginary parts ($|\Im\lambda| > 2$) correspond to microscale waves within each patch; and

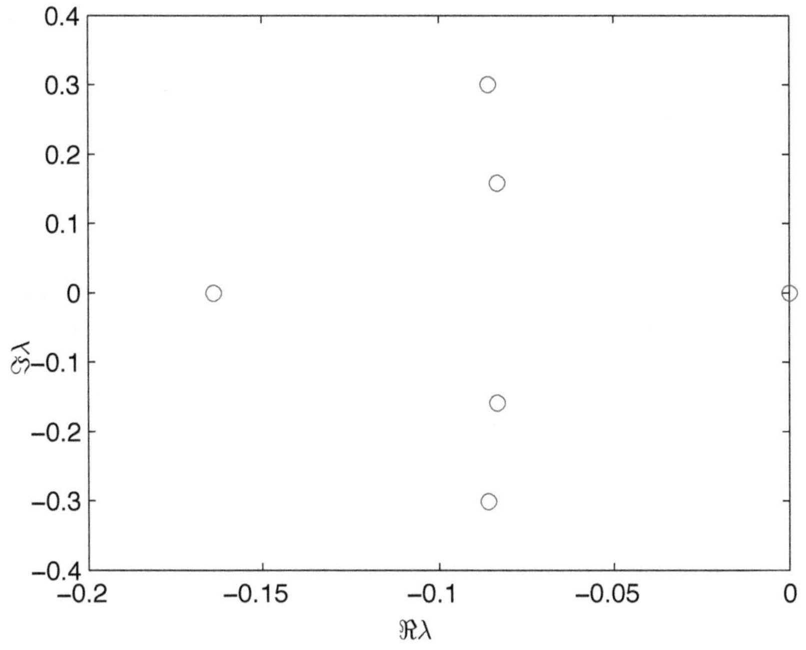

Fig. 6. Zoom into the values in the red rectangle in Figure 5. These eigenvalues which represent the macroscale modes.

- the eigenvalues with small growth rates ($|\Re\lambda| < 0.2$) in the red rectangle, zoomed in by Figure 6, represent the interesting macroscale wave-like dynamics. This set of ten eigenvalues includes two groups of four eigenvalues representing waves of wavenumbers 0.2 and 0.4, a zero eigenvalue representing conservation of water, and $\lambda \approx -0.16$ representing the decay of homogeneous shear.

The pattern of eigenvalues seen in Figure 5 is typical over a wide range of parameters. It shows the emergence of the macroscale wave modes from among the fast microscale modes within each patch.

Simulations confirm the nonlinear dynamics of the gap-tooth scheme are appropriate. Figure 7 plots and compares the free surface of the thin film flow in a gap-tooth simulation and in the microscale simulation over the whole domain. Time integration invoked the Matlab `ode15s` function. There are $m = 10$ patches and $n = 9$ microscale grids on a patch. The $t = 0$ graph shows that initially a perturbation $0.2\sin(2\pi/Lx)$ is imposed with small random noises to the equilibrium of fluid thickness one and initial layer mean velocities $\bar{u}_1 = 0$ and $\bar{u}_2 = 0.2$, so the mean layer velocity $\bar{u} = (\bar{u}_1 + \bar{u}_2)/2 = 0.1$. The $t = 2$ to $t = 10$ graphs show that the microscale modes on a patch smooth quickly and the macroscale waves propagate and slowly decay, which agrees with the large decay rates of the microscale waves and small decay rates of the macroscale modes in Figure 5. Further numerical simulation shows that the macroscale waves decay to the equilibrium of thickness one after time $t \approx 30$ for the Reynolds number Re $= 15$. When the Reynolds number Re becomes smaller, the macroscale waves decay faster to the equilibrium due to the stronger viscous effects. Figure 7 shows that the gap-tooth simulation agrees with the microscale simulation over the whole domain. The likely reason for the differences in $t = 10$ graph is that the gap-tooth simulation near the water bore involves the error $\mathcal{O}(D^2)$ for the macroscale step $D = L/m \approx 3.14$, while the microscale simulation on the whole macroscale domain involves the error $\mathcal{O}(d^2)$ for the microscale step $d = 2rD/(n+1) = 0.13:$ the ratio of these errors is $\mathcal{O}\big(D^2/d^2\big) \approx 500$, showing the macroscale simulation is limited largely by the interpolation between patches. As the number of patches increases, the error $\mathcal{O}(D^2)$ declines, and the gap-tooth simulation performs better. We show just $m = 10$ patches here for clarity.

Figure 8 plots the corresponded layer mean velocities $\bar{u}_1$ (blue) and $\bar{u}_2$ (red) in the gap-tooth simulation and in the microscale simulation on the whole domain.

5. Conclusion

This work applies the gap-tooth scheme to the thin fluid flow with a derived novel two-layer model (2.21) being the microscale simulator. The two-layer model (2.21) provides more microscale modes, but without the full complexity of fully resolved vertical structures.

Based on centre manifold theory, section 2–2.4 derive the two-layer model (2.21) from the continuity and modified Navier–Stokes equations. The model (2.21) includes the effects of gravity, drag, advection and dispersion. The slow manifold of the two-layer model agrees with the one-layer model of thin film flow by [24]. Eigenvalues analysis in section 2.4 indicates unphysical instability appear, but a consistent regularising operator $\mathcal{L}$ stabilises the model.

Then we simulate the thin fluid flow by new adaptations of the gap-tooth scheme with the two-layer model (2.21) being the microscale simulator. Section 3.3 algebraically derives the slow manifold of the two-layer model (2.21) in order to lift the macroscale information to the microscale simulation on patches. Then coupling conditions (3.7) and (3.4a)–(3.4b) are developed to couple the patches. Numerical eigenvalues in Figures 5–6 show that the macroscale modes decay slowly over the whole domain, while the microscale modes oscillate fast and decay quickly to quasi-equilibrium quickly. Non-zero frequencies $\Im\lambda$ in Figure 6 indicate waves are supported on the free surface over the whole domain. Numerical simulations in Figures 7–8 show that the gap-tooth scheme with the two-layer model (2.21) being microscale simulator work well enough for the thin film flow. Future work is planned to build on this base in order to apply the gap-tooth scheme to the complicated microscale physics of turbulent, possibly ice covered, water waves, with the aim of empowering effective computation over very large lateral scales.

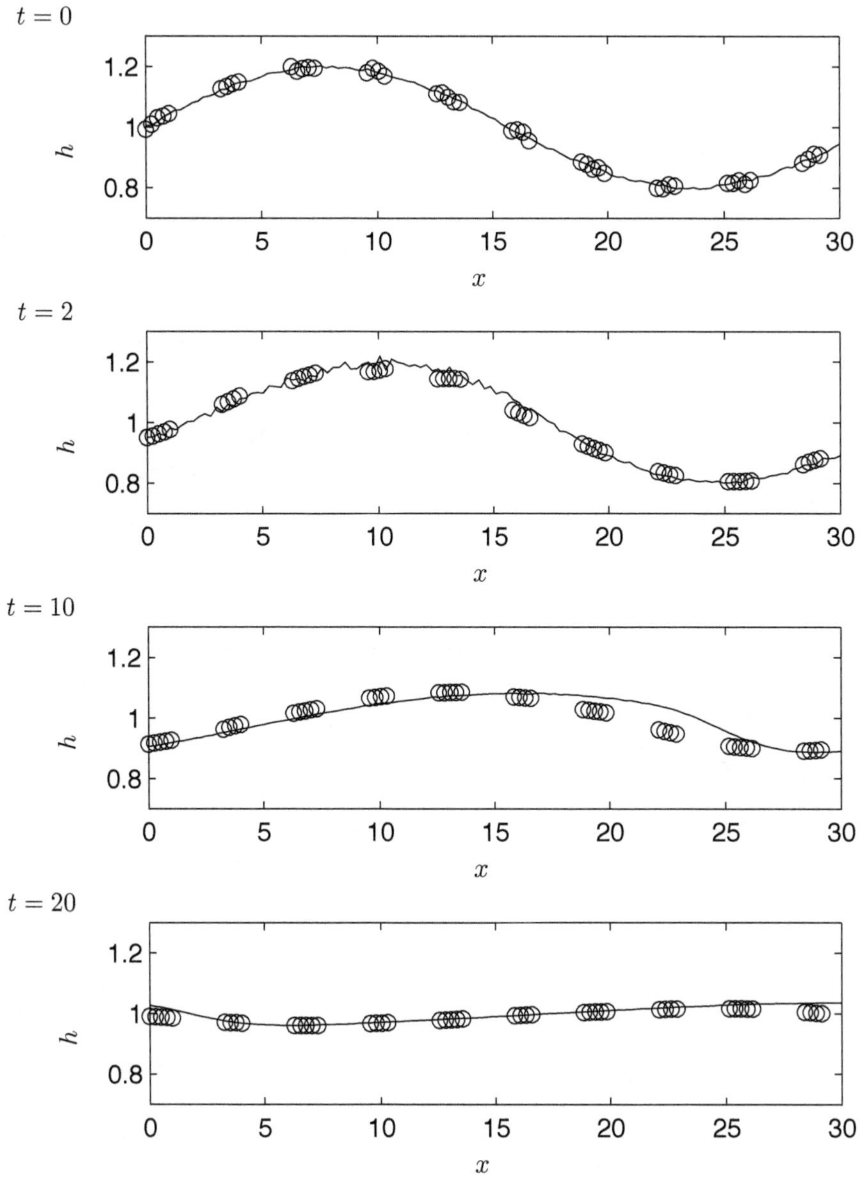

Fig. 7. Plots of the fluid thickness of the thin fluid flow on the domain $[0\ 10\pi]$: (circle) for the gap-tooth simulation; and (line) for microscale simulation on the whole domain. There are $m = 10$ patches and $n = 9$ microscale grids on a patch. The Reynolds number $\mathrm{Re} = 15$ and the regularising coefficient $C = 0.5$. The plate has a slope $\tan\theta = 0$. Figure 8 shows the corresponding fluid velocities.

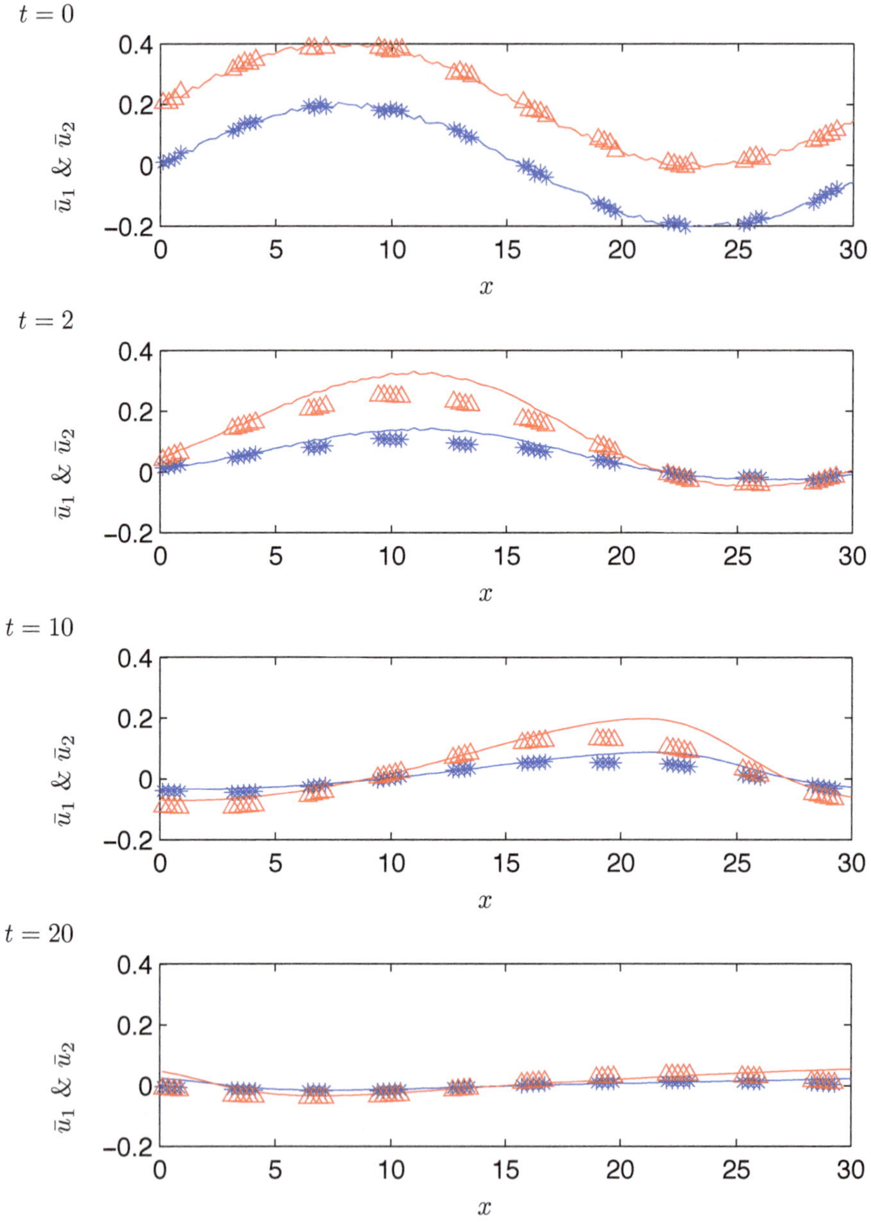

Fig. 8. Plots of the layer mean velocities: (blue line) represents $\bar{u}_1$ and (red line) represents $\bar{u}_2$ in the microscale simulation over the whole domain; and (blue star) represents $\bar{u}_1$ and (red triangle) represents $\bar{u}_2$ in the gap-tooth simulation of the fluid film flow on the domain $[0\ 10\pi]$. There are $m = 10$ patches and $n = 9$ microscale grids on a patch.

6. Ancillary computer algebra program

This appendix lists the computer algebra to derive a two-layer model of the viscous layer of fluid. Denote fluid thickness $h(x,t)$ by `h`, layer mean velocities $\bar{u}_j(x,t)$ by `uu`j for the lower layer $j=1$ and upper layer $j=2$, and their evolution $\Delta th =$ `gh` and $\Delta t\bar{u}_j =$ `gu`j. The Reynolds number is `re`, and the coefficients of lateral and normal gravitational forcing are `grx` and `grz:=1`.

Use the operator `h(m)` to denote m lateral derivatives of the fluid thickness h, $\partial_x^m h$, and similarly `uuj(m)` denotes m lateral derivatives of the layer mean velocity $\bar{u}_j$, $\partial_x^m \bar{u}_j$. Use `d` to count the number of lateral x derivatives so we can easily truncate the asymptotic expansion. These operators depend upon time and lateral space. Then the spatial derivative ∂_x`h(m)` $=$ `h(m+1)`, and the time derivative ∂_t`h(m)` $= \partial_x^m$`gh`.

References

[1] Alekseenko, S. V., Markovich, D. M. & Shtork, S. I., (1996), Wave flow of rivulets on the outer surface of an inclined cylinder, Phys. Fluids, 8, 3288–3299.

[2] Aulbach, B. & Wanner, T., (2000), The Hartman–Grobman theorem for Caratheodory-type differential equations in Banach spaces, Nonlinear Analysis, 40, 91–104. http://dx.doi.org/10.1016/S0362-546X(00)85006-3.

[3] Benjamin, T. B.,(1957), Wave formation in a laminar flow down an inclined plane, J. Fluid Mech., 2, 554–574. doi:10.1017/S0022112057000373.

[4] Cao, M., (2014), Modelling environmental turbulent fluids and multiscale modelling couples patches of wave-like system, PhD thesis, School of Mathematical Sciences, University of Adelaide.

[5] Cao, M. & Roberts, A. J., (2013), Multiscale modelling couples patches of wave-like simulations, in S. McCue, T. Moroney, D. Mallet & J. Bunder, eds, Proceedings of the 16th Biennial Computational Techniques and Applications Conference, CTAC-2012, Vol. 54 of ANZIAM J., pp. C153–C170. http://journal.austms.org.au/ojs/index.php/ANZIAMJ/article/view/6137.

[6] Cao, M. & Roberts, A. J., (2014), Multiscale modelling couples patches of nonlinear wave-like simulations, Technical report, [http://http://arxiv.org/abs/1404.6317].

[7] Chang, H. C., (1987), Evolution of nonlinear waves on vertically falling films normal form analysis, Chemical Engineering Science, 42, 515–533. doi:10.1016/0009-2509(87)80014-3.

[8] Chang, H. C., (1994a), Wave evolution on a falling film, Annu. Rev. Fluid Mech., 26, 103–136. doi:10.1146/annurev.fl.26.010194.000535.

[9] Chang, H. C., (1994b), Wave evolution on a falling film, Annu. Rev. Fluid Mech., 26, 103–136. http://dx.doi.org/10.1146/annurev.fl.26.010194.000535.

[10] Chang, H. C., Demekhin, E. & Saprikin, S. S., (2002), Noise driven wave transitions on a vertically falling film, J. Fluid Mech., 462, 255–283. http://dx.doi.org/10.1017/S0022112002008856.

[11] Chen, K., (1993), Wave formation in the gravity driven low reynolds number flow of two liquid films down an inclined plane, Physics of Fluids, 5, 30–38. doi:10.1063/1.858714.

[12] E., W. & Engquist, B., (2003), The heterogeneous multiscale methods, Comm. Math. Sci., 1(1), 87–132. http://projecteuclid.org/euclid.cms/1118150402.

[13] E., W., Engquist, B., Li, X., Ren, W. & Vanden-Eijnden, E., (2007), Heterogeneous Multiscale Methods: A Review, Communications in Computational Physics, 2(3), 367–450. doi:10.1.1.225.9038.

[14] Gear, C. W., Li, J. & Kevrekidis, I. G., (2003), The gap-tooth method in particle simulations, Phys. Lett. A, 316, 190–195. http://dx.doi.org/10.1016/j.physleta.2003.07.004.

[15] Hou, T. Y., Yang, D. & Ran, H., (2008), Multiscale analysis and computation for the three-dimensional incompressible Navier–Stokes equations, Multiscale Modelling and Simulation, 6(4), 1317–1346. doi:10.1137/070682046.

[16] Kevrekidis, I. G., Gear, G. W., Hyman, J. M., Kevrekidis, P. G., Runborg, O. & Theodoropoulos, C., (2003), Equation-free, coarsegrained multiscale computation: enabling microscopic simulators to perform system-level analysis, Comm. Math. Sci., 1(4), 715–762. http://projecteuclid.org/euclid.cms/1119655353.

[17] Kevrekidis, I. G. & Samaey, G., (2009), Equation-free multiscale computation: Algorithms and applications, Annu. Rev. Phys. Chem., 60, 321–44. doi:10.1146/annurev.physchem.59.032607.093610.

[18] Kheshgi, H. S., (1989), Profile equations for film flows at moderate Reynolds numbers, AIChE Journal, 35, 1719–1727. http://dx.doi.org/10.1002/aic.690351017.

[19] Malecha, Z., Chini, G. & Julien, K., (2013), A multiscale algorithm for simulating spatially-extended Langmuir circulation dynamics, J. Comput. Phys., doi:10.1016/j.jcp.2013.07.003.

[20] Nguyen, L. T. & Balakotaiah, V., (2000), Modeling and experimental studies of wave evolution on free falling viscous films, Phys. Fluids, 12, 2236–2256.

[21] Potzsche, C. & Rasmussen, M., (2006), Taylor approximation of integral manifolds, J. Dyn. Differ. Equ., 18(2), 427–460. doi:10.1007/s10884-006-9011-8.

[22] Prokopiou, T., Cheng, M. & Chang, H. C., (1991a), Long waves on inclined films at high reynolds number, J. Fluid Mech., 222, 665–691. doi:10.1017/S002211209100126X.

[23] Roberts, A. J., (1988), The application of centre-manifold theory to the evolution of systems which vary slowly in space, J. Austral. Math. Soc. Ser. B, 29, 480–500. doi:10.1017/S0334270000005968.

[24] Roberts, A. J., (1997), Low-dimensional models of thin film fluid dynamics, Phys. Letts. A, 212, 63–72. doi:10.1016/0375-9601(96)00040-0.

[25] Roberts, A. J., (1998), An accurate model of thin 2d fluid flows with inertia on curved substrates, in Free-surface flows with viscosity, Vol. 16 of Advances in Fluid Mechanics Series, Comput Mech Pub, chapter 3, pp. 69–88.

[26] Roberts, A. J., (2013), Macroscale, slowly varying, models emerge from the microscale dynamics in long thin domains, Technical report, [http://arxiv.org/abs/1310.1541].

[27] Roberts, A. J. & Kevrekidis, I. G., (2005), Higher order accuracy in the gap-tooth scheme for large-scale dynamics using microscopic simulators, ANZIAM Journal, 46, C637–C657. http://journal.austms.org.au/ojs/index.php/ANZIAMJ/article/view/981/847.

[28] Roberts, A. J. & Li, Z., (2006), An accurate and comprehensive model of thin fluid flows with inertia on curved substrates, J. Fluid Mech., 553, 33–73. http://dx.doi.org/10.1017/S0022112006008640.

[29] Roberts, A. J., Mei, Z. & Li, Z., (2002), Modelling the dynamics of turbulent floods, SIAM J. on Applied Mathematics, 63, 423–458. doi:10.1137/S0036139999358866.

[30] Roy, R. V., Roberts, A. J. & Simpson, M. E., (2002), A lubrication model of coating flows over a curved substrate in space, J. Fluid Mech., 454, 235–261. doi:10.1017/S0022112001007133.

[31] Ruyer-Quil, C. & Manneville, P., (1998), Modeling film flows down inclined planes, Eur. Phys. J. B, 6, 277–292. doi:10.1007/s100510050550.
[32] Ruyer-Quil, C. & Manneville, P., (2000), Improved modeling of flows down inclined planes, Eur. Phys. J. B, 15, 357–369. http://www.springerlink.com/link.asp?id=9ffp55pnhqkxe0lk.
[33] Samaey, G., Kevrekidis, I. G. & Roose, D., (2005), The gap-tooth scheme for homogenization problems, SIAM Multiscale Modeling and Simulation, 4, 278–306. http://dx.doi.org/10.1137/030602046.
[34] Samaey, G., Roberts, A. J. & Kevrekidis, I. G., (2009), Equation-free computation: an overview of patch dynamics, Vol. 8 of Multiscale Methods, Oxford Scholarship Online Monographs, doi:10.1093/acprof:oso/9780199233854.003.0008.
[35] Samaey, G., Roose, D. & Kevrekidis, I. G., (2005), The gap-tooth scheme for homogenization problems, Multiscale Modelling and Simulation, 4, 278–306. doi:10.1137/030602046.
[36] Van Dyke, M., (1964), Higher approximations in boundary-layer theory. Part 3. parabola in uniform stream, J. Fluid Mech., 19, 145–159.
[37] Van Dyke, M., (1984), Computer-extended series, Annu. Rev. Fluid Mech., 16, 287–310.
[38] Weinstein, S. J. & Ruschak, K. J., (2004), Coating flows, Annu. Rev. Fluid Mech., 36, 29–53. doi:10.1146/annurev.fluid.36.050802.122049.

Chapter 8

The Cauchy Problem for a Generalized Ostrovsky Equation with Positive Dispersion

Wei Yan

College of Mathematics and Information Science, Henan Normal University, Xinxiang, Henan 453007, China

yanwei19821115@sina.cn

This paper is devoted to studying the Cauchy problem for a generalized Ostrovsky equation with positive dispersion. Firstly, by using some Strichartz estimates, we establish an important bilinear estimate. Then, combining the bilinear estimate with the fixed point theorem, we prove that the Cauchy problem for a generalized Ostrovsky equation

$$u_t - \beta\partial_x^3 u + u^7 u_x - \gamma\partial_x^{-1} u = 0, \beta > 0, \gamma > 0$$

with data belonging to $H^s(\mathbb{R})$, $s > \frac{3}{14}$ on $\mathbb{R}$.

Keywords: Generalized Ostrovsky equation with positive dispersion; Cauchy problem

AMS Subject Classification: 35G25

1. Introduction

This paper is devoted to studying the Cauchy problem for a generalized Ostrovsky equation

$$u_t - \beta\partial_x^3 u + u^7 u_x - \gamma\partial_x^{-1} u = 0, \tag{1.1}$$

$$u(x, 0) = u_0(x) \tag{1.2}$$

with positive dispersion on $\mathbb{R}$.

Levandosky and Liu [9] investigated the stability of solitary waves of the generalized Ostrovsky equation

$$[u_t - \beta u_{xxx} + (f(u))_x]_x = \gamma u, x \in \mathbb{R} \tag{1.3}$$

where f is a C^2 function which is homogeneous of degree $p \geq 2$ in the sense that it satisfies $sf'(s) = f(s)$. Levandosky [10] investigated the stability of ground state solitary waves of (1.4) with $f(u) = c_1|u|^p + c_2|u|^{p-1}u$, $c_1, c_2 \in \mathbb{R}$.

When $f(u) = u^2$, (1.3) reduces to the Ostrovsky equation

$$\left[u_t - \beta u_{xxx} + (u^2)_x\right]_x = \gamma u, x \in \mathbb{R} \tag{1.4}$$

In this paper, we prove that the Cauchy problem for (1.1) is locally well-posed in $H^s(\mathbb{R})$ with $s > \frac{3}{14}$.

We always assume that $b = \frac{1}{2} + \frac{\epsilon}{24}$ and $c = \frac{1}{2} + \frac{\epsilon}{100}$. $\Phi(\xi) = \beta\xi^3 + \frac{\gamma}{\xi}$. $C > 0$ and $C' > 0$ denote two constants which may vary at different occurrence. Let $\eta(t)$ be a smooth cutoff function supported on $[-2, 2]$, $\eta \equiv 1$ on $[-1, 1]$, and $\eta_T(t) = \eta\left(\frac{t}{T}\right)$. For $x, \xi \in \mathbb{R}$, we denote by $\mathcal{F}_x u(\xi) = \frac{1}{\sqrt{2\pi}}\int_{\mathbb{R}} u(x)e^{-ix\xi}dx$ the Fourier transformation with respect to the space variable. We denote by $\mathcal{F}_t u(\tau) = \frac{1}{\sqrt{2\pi}}\int_{\mathbb{R}} u(t)e^{-it\tau}dt$ the Fourier transformation with respect to the time variable. We denote by $\mathcal{F}u(\xi,\tau) = \frac{1}{2\pi}\int_{\mathbb{R}^2} u(x,t)e^{-ix\xi-it\tau}dxdt$ the Fourier transformation with respect to the time and space variables. We define $\mathcal{F}_x J^s u(\xi) = \langle\xi\rangle^s \mathcal{F}_x u(\xi)$ and $S(t)f = \frac{1}{\sqrt{2\pi}}\int_{\mathbb{R}} e^{i(x\xi - t\Phi(\xi))}\mathcal{F}_x f(\xi)d\xi$.

The Sobolev space $H^s(\mathbb{R})$ is defined as follows:

$$H^s(\mathbb{R}) = \left\{u \in \mathcal{S}'(\mathbb{R}) : \|u\|_{H^s(\mathbb{R})} = \|\langle\xi\rangle^s \mathcal{F}_x u\|_{L^2_\xi} < \infty\right\},$$

and the space $X_{s,b}(\mathbb{R}^2)$ is defined as follows:

$$X_{s,b}(\mathbb{R}^2) = \left\{u \in \mathcal{S}'(\mathbb{R}^2) : \|u\|_{X_{s,b}(\mathbb{R}^2)} = \|\langle\xi\rangle^s\langle\tau - \Phi(\xi)\rangle^b \mathcal{F}u(\xi,\tau)\|_{L^2_\xi L^2_\tau} < \infty\right\}.$$

For the $X_{s,b}$ spaces, we refer the readers to [1, 2, 7, 11].

The main result of this paper is as follows.

Theorem 8.1. *Let $s > \frac{3}{14}$, $\phi \in H^s(\mathbb{R})$. Then the problems (1.1)–(1.2) are locally well-posed.*

The rest of the paper is arranged as follows. In Section 2, we give some preliminaries and some multilinear estimates. In Section 3, we prove the Theorem 8.1.

2. Preliminaries

In this section, we present some probabilistic lemmata and probabilistic Strichartz estimates, linear estimates needed in this paper. We always denote that $S(t)f = e^{t(\beta\partial_x^3 + \gamma\partial_x^{-1})}f = \frac{1}{\sqrt{2\pi}}\int_{\mathbb{R}} e^{i(x\xi - t\Phi(\xi))}\mathcal{F}_x f(\xi)d\xi$.

Lemma 8.1. *Let $T \in (0,1)$. Then, for $s \in \mathbb{R}$ and $-\frac{1}{2} < b' \leq 0 \leq b \leq b' + 1$, we have that*

$$\|\eta_T(t)S(t)f\|_{X_{s,b}(\mathbb{R}^2)} \leq CT^{\frac{1}{2}-b}\|f\|_{H^s(\mathbb{R})},$$

$$\left\|\eta_T(t)\int_0^t S(t-\tau)F(\tau)d\tau\right\|_{X_{s,b}(\mathbb{R}^2)} \leq CT^{1+b'-b}\|F\|_{X_{s,b'}(\mathbb{R}^2)}.$$

For the proof of Lemma 8.1, we refer the readers to [2, 3, 8].

Lemma 8.2. *Let $0 < \epsilon < \frac{1}{10^8}$ and $\mathcal{F}(P^a f)(\xi) = \chi_{\{|\xi| \geq a\}}(\xi)\mathcal{F}f(\xi)$ with $a \geq 2$, $q \geq 8$ and $\mathcal{F}(P_c f)(\xi) = \chi_{\{|\xi| \leq c\}}(\xi)\mathcal{F}f(\xi)$ with $0 < c \leq 8$ and $\mathcal{F}(D_x^d f)(\xi) = |\xi|^d\mathcal{F}f(\xi)$, $d \in \mathbb{R}$. Then, we have that*

$$\|u\|_{L^6_{xt}} \leq C\|u\|_{X_{0,\frac{1}{2}+\epsilon}}, \tag{2.1}$$

$$\|D_x^{\frac{1}{6}} P^a u\|_{L^6_{xt}} \leq C\|u\|_{X_{0,\frac{1}{2}+\epsilon}}, \tag{2.2}$$

$$\|u\|_{L_{xt}^{\frac{6}{1+\epsilon}}} \leq C\|u\|_{X_{0,\ \frac{6-\epsilon}{12-6\epsilon}}}, \tag{2.3}$$

$$\|\partial_x Pr^a u\|_{L_x^\infty L_t^2} \leq C\|u\|_{X_{0,\ \frac{1}{2}+\epsilon}}, \tag{2.4}$$

$$\|Pr_c u\|_{L_x^2 L_t^\infty} \leq C\|u\|_{X_{0,\ \frac{1}{2}+\epsilon}}, \tag{2.5}$$

$$\|Pr_c u\|_{L_x^\infty L_t^\infty} \leq C\|u\|_{X_{0,\ \frac{1}{2}+\epsilon}}. \tag{2.6}$$

Proof. From (2.8) of Lemma 2.7 of [5], we have that

$$\|S(t)u\|_{L^6_{xt}} \leq C\|u\|_{L^2_{xt}}. \tag{2.7}$$

Combining (2.7) with the standard proof of (2.13) of [6], we have that

$$\|u\|_{L^6_{xt}} \leq C\|u\|_{X_{0,\frac{1}{2}+\epsilon}}. \tag{2.8}$$

Combining (2.9) with the standard proof of (2.13) of [6], we have that (2.2) is valid. In particular, we have that

$$\|u\|_{L^6_{xt}} \leq C\|u\|_{X_{0,\ \frac{6-\epsilon}{12-6\epsilon}}}. \tag{2.9}$$

Interpolating (2.9) with

$$\|u\|_{L^2_{xt}} \leq C\|u\|_{X_{0,0}},$$

we have that (2.3) is valid.

From Lemma 2.9 in [4], we have that (2.5) is valid.

Combining the Sobolev embedding with (2.5), we have that

$$\begin{aligned}\|Pr_c u\|_{L_x^\infty L_t^\infty} &\leq C\left\|D_x^{\frac{1}{2}+\epsilon} Pr_c u\right\|_{L_x^2 L_t^\infty} \leq C\left\|D_x^{\frac{1}{2}+\epsilon} Pr_c u\right\|_{X_{0,\ \frac{1}{2}+\epsilon}} \\ &\leq C\|u\|_{X_{0,\ \frac{1}{2}+\epsilon}}.\end{aligned} \tag{2.10}$$

This completes the proof of Lemma 8.2.

Remark: Combining the Sobolev embeddings theorem with (2.2), we have that

$$\|P^a u\|_{L^8_{xt}} \leq C\|P^a D_x^{\frac{1}{6}} u\|_{L^6_{xt}} \leq C\|u\|_{X_{0,b}}. \tag{2.11}$$

Lemma 8.3. *We define I^s as follows:*

$$\mathcal{F}I^s_-(u_1,u_2) = \int_{\xi=\xi_1+\xi_2} \left|\Phi'(\xi_1)-\Phi'(\xi_2)\right|^s \mathcal{F}u_1(\xi_1,\tau_1)\mathcal{F}u_2(\xi_2,\tau_2)d\xi_1 d\tau_1.$$

Let $0 \leq s \leq \frac{1}{2}$ and $b > \frac{1}{2}$. Then we have that

$$\begin{aligned}&\left\|\int_{\substack{\xi=\sum_{j=1}^{2}\xi_j \\ \tau=\sum_{j=1}^{2}\tau_j}} |\xi_1^2-\xi_2^2|^s \mathcal{F}u_1(\xi_1,\tau_1)\mathcal{F}u_2(\xi_2,\tau_2)d\xi_1 d\tau_1\right\|_{L^2_{\xi\tau}} \\ &\leq C\left\|I^s_-(u_1,u_2)\right\|_{L^2_{xt}} \leq C\prod_{j=1}^{2}\|u_j\|_{X_{0,\ \frac{3+2s}{4}b}}.\end{aligned}$$

For the proof of Lemma 8.3, we refer the readers to (1.23) of [12].

2.1. *Multilinear estimates*

In this section, we present some crucial multilinear estimates which play an important role in establishing Theorem 8.1.

Lemma 8.4. *Let* $v_j \in X_{\frac{3}{14}+2\epsilon,\frac{1}{2}+\frac{\epsilon}{24}}$ *with* $(1 \leq j \leq 8, j \in N)$ *and* $b = \frac{1}{2} + \frac{\epsilon}{24}$. *Then, we have that*

$$\left\| \partial_x \left(\prod_{j=1}^{8} v_j \right) \right\|_{X_{\frac{3}{14}+2\epsilon,-\frac{1}{2}+\frac{\epsilon}{12}}} \leq C \prod_{j=1}^{8} \|v_j\|_{X_{\frac{3}{14}+2\epsilon,b}}. \tag{2.12}$$

Proof. By duality, to prove (2.12), it suffices to prove that

$$\begin{aligned} &\int_{\mathbb{R}^2} \left[\partial_x \left(\prod_{j=1}^{8} v_j \right) \right] v(x,t)dxdt \\ &\leq C \left[\prod_{j=1}^{8} \|v_j\|_{X_{\frac{3}{14}+2\epsilon,b}} \right] \|v\|_{X_{\frac{3}{14}+2\epsilon,\frac{1}{2}-\frac{\epsilon}{12}}}. \end{aligned} \tag{2.13}$$

By using the Plancherel's identity, to obtain (2.13), it suffices to prove that

$$\begin{aligned} &\int_{\mathbb{R}^2} \int_{\substack{\xi = \sum_{j=1}^{8} \xi_j \\ \tau = \sum_{j=1}^{8} \tau_j}} \left| \xi \mathcal{F}v(\tau,\xi) \prod_{j=1}^{8} \mathcal{F}v_j(\tau_j,\xi_j) \right| dV \\ &\leq C \left[\prod_{j=1}^{8} \|v_j\|_{X_{\frac{3}{14}+2\epsilon,b}} \right] \|v\|_{X_{\frac{3}{14}+2\epsilon,\frac{1}{2}-\frac{\epsilon}{12}}}, \end{aligned} \tag{2.14}$$

where $dV = d\xi_1 d\tau_1 d\xi_2 d\tau_2 d\xi_3 d\tau_3 d\xi_4 d\tau_4 d\xi_5 d\tau_5 d\xi_6 d\tau_6 d\xi_7 d\tau_7$. Let

$$\begin{aligned} &F_j(\tau_j,\xi_j) = \langle \xi_j \rangle^{\frac{3}{14}+2\epsilon} \langle \sigma_j \rangle^{\frac{1}{2}+\frac{\epsilon}{24}} \mathcal{F}v_j(\tau_j,\xi_j)(1 \leq j \leq 8, j \in N), \\ &F(\tau,\xi) = \langle \xi \rangle^{\frac{3}{14}+2\epsilon} \langle \sigma \rangle^{\frac{1}{2}-\frac{\epsilon}{12}} \mathcal{F}v(\tau,\xi). \end{aligned} \tag{2.15}$$

To prove (2.14), it suffices to prove that

$$\begin{aligned} &\int_{\mathbb{R}^2} \int_{\substack{\xi = \sum_{j=1}^{8} \xi_j \\ \tau = \sum_{j=1}^{8} \tau_j}} K(\tau_1,\xi_1,\tau_2,\xi_2,\tau_3,\xi_3,\tau_4,\xi_4,\tau_5,\xi_5,\tau_6,\xi_6,\tau_7,\xi_7,\tau,\xi) \\ &\qquad \times \left| F(\tau,\xi) \prod_{j=1}^{8} F_j(\tau_j,\xi_j) \right| dV \\ &\leq C \left[\prod_{j=1}^{8} \|F_j\|_{L^2_{\tau\xi}} \right] \|F\|_{L^2_{\tau\xi}}, \end{aligned} \tag{2.16}$$

where

$$K(\tau_1,\xi_1,\tau_2,\xi_2,\tau_3,\xi_3,\tau_4,\xi_4,\tau_5,\xi_5,\tau_6,\xi_6,\tau_7,\xi_7,\tau,\xi) = \frac{|\xi|\langle\xi\rangle^{\frac{3}{14}+2\epsilon}}{\langle\sigma\rangle^{\frac{1}{2}-\frac{\epsilon}{12}}\prod\limits_{j=1}^{8}\langle\xi_j\rangle^{\frac{3}{14}+2\epsilon}\langle\sigma_j\rangle^{b}}.$$

Without loss of generality, we assume that $|\xi_1| \geq |\xi_2| \geq \cdots \geq |\xi_8|$. We assume that

$$(\tau_1,\xi_1,\cdots,\tau_7,\xi_7,\tau,\xi)$$

is short for

$$(\tau_1,\xi_1,\tau_2,\xi_2,\tau_3,\xi_3,\tau_4,\xi_4,\tau_5,\xi_5,\tau_6,\xi_6,\tau_7,\xi_7,\tau,\xi).$$

Obviously,

$$\left\{(\tau_1,\xi_1,\cdots,\tau_7,\xi_7,\tau,\xi)\in\mathbb{R}^{16} : |\xi_1|\geq|\xi_2|\geq\cdots\geq|\xi_8|\right\} \subset \bigcup_{j=1}^{5}\Omega_j,$$

where

$$\Omega_1 = \left\{(\tau_1,\xi_1,\cdot\cdot,\tau_7,\xi_7,\tau,\xi)\in\mathbb{R}^{16} : |\xi_1|\geq|\xi_2|\geq\cdots\geq|\xi_8|, |\xi_1|\geq 80|\xi_2|\right\},$$
$$\Omega_2 = \left\{(\tau_1,\xi_1,\cdot\cdot,\tau_7,\xi_7,\tau,\xi)\in\mathbb{R}^{16} : |\xi_1|\geq|\xi_2|\geq\cdots\geq|\xi_8|, |\xi_1|\sim|\xi_2|\geq 80|\xi_3|\right\},$$
$$\Omega_3 = \left\{(\tau_1,\xi_1,\cdot\cdot,\tau_7,\xi_7,\tau,\xi)\in\mathbb{R}^{16} : |\xi_1|\geq|\xi_2|\geq\cdots\geq|\xi_8|, |\xi_1|\sim|\xi_3|\geq 80|\xi_4|\right\},$$
$$\Omega_4 = \left\{(\tau_1,\xi_1,\cdot\cdot,\tau_7,\xi_7,\tau,\xi)\in\mathbb{R}^{16} : |\xi_1|\geq|\xi_2|\geq\cdots\geq|\xi_8|, |\xi_1|\sim|\xi_4|\geq 80|\xi_5|\right\},$$
$$\Omega_5 = \left\{(\tau_1,\xi_1,\cdot\cdot,\tau_7,\xi_7,\tau,\xi)\in\mathbb{R}^{16} : |\xi_1|\geq|\xi_2|\geq\cdots\geq|\xi_8|, |\xi_1|\sim|\xi_5|\geq 80|\xi_6|\right\},$$
$$\Omega_6 = \left\{(\tau_1,\xi_1,\cdot\cdot,\tau_7,\xi_7,\tau,\xi)\in\mathbb{R}^{16} : |\xi_1|\geq|\xi_2|\geq\cdots\geq|\xi_8|, |\xi_1|\sim|\xi_6|\geq 80|\xi_7|\right\},$$
$$\Omega_7 = \left\{(\tau_1,\xi_1,\cdot\cdot,\tau_7,\xi_7,\tau,\xi)\in\mathbb{R}^{16} : |\xi_1|\geq|\xi_2|\geq\cdots\geq|\xi_8|, |\xi_1|\sim|\xi_7|\geq 80|\xi_8|\right\},$$
$$\Omega_8 = \left\{(\tau_1,\xi_1,\cdot\cdot,\tau_7,\xi_7,\tau,\xi)\in\mathbb{R}^{16} : |\xi_1|\geq|\xi_2|\geq\cdots\geq|\xi_8|, |\xi_1|\sim|\xi_8|\right\}.$$

We define

$$I_1 = \int_{\mathbb{R}^2}\int_{\substack{\xi=\sum\limits_{j=1}^{8}\xi_j\\ \tau=\sum\limits_{j=1}^{8}\tau_j}} K(\tau_1,\xi_1,\cdot\cdot,\tau_7,\xi_7,\tau,\xi)F(\tau,\xi)\prod_{j=1}^{8}F_j(\tau_j,\xi_j)dV,$$

$$f(x,t) = \mathcal{F}^{-1}\left(\frac{F}{\langle\sigma\rangle^{\frac{1}{2}-\frac{\epsilon}{12}}}\right), f_j = \mathcal{F}^{-1}\left(\frac{F}{\langle\sigma\rangle^{\frac{1}{2}+\frac{\epsilon}{24}}}\right)(1\leq j\leq 8).$$

(1) Ω_1. In this case, we consider $|\xi_1|\leq 8$, $|\xi_1|>8$, respectively.
When $|\xi_1|\leq 8$, we have that

$$K(\tau_1,\xi_1,\cdots,\tau_7,\xi_7,\tau,\xi) \leq \frac{C}{\langle\sigma\rangle^{\frac{1}{2}-\frac{\epsilon}{12}}\prod\limits_{j=1}^{8}\langle\sigma_j\rangle^{\frac{1}{2}+\frac{\epsilon}{24}}},$$

by using the Plancherel identity, the Hölder inequality, (2.4)–(2.5) and (2.10), we infer that

$$I_1 \leq C\left|\int_{\mathbb{R}^2} f(x,t)\prod_{j=1}^{8}f_j(x,t)dxdt\right|$$

$$\leq C\|f\|_{L_{xt}^{\frac{6}{1+\epsilon}}}\left(\prod_{j=1}^{2}\|f_j\|_{L_{xt}^{\infty}}\right)\left(\prod_{j=3}^{8}\|f_j\|_{L_{xt}^{6}}\right)$$

$$\leq C\|f\|_{X_{0,\frac{1}{2}-\frac{\epsilon}{12}}}\left(\prod_{j=1}^{2}\|f_j\|_{X_{\frac{1}{2}+\epsilon,\frac{1}{2}+\frac{\epsilon}{24}}}\right)\left(\prod_{j=3}^{8}\|f_j\|_{X_{0,\frac{1}{2}+\frac{\epsilon}{24}}}\right)$$

$$\leq C\left[\prod_{j=1}^{8}\|F_j\|_{L_{\tau\xi}^2}\right]\|F\|_{L_{\tau\xi}^2}.$$

When $|\xi_1|\geq 8$, we have that

$$K(\tau_1,\xi_1,\cdots,\tau_7,\xi_7,\tau,\xi)\leq C\frac{|\xi|}{\langle\sigma\rangle^{\frac{1}{2}-\frac{\epsilon}{12}}\prod\limits_{j=2}^{8}\langle\xi_j\rangle^{\frac{3}{14}+2\epsilon}\prod\limits_{j=1}^{8}\langle\sigma_j\rangle^{\frac{1}{2}+\frac{\epsilon}{24}}}$$

$$\leq C\frac{|\xi^2-\xi_3^2|^{\frac{1}{2}-\epsilon}|\xi|^{-1+2\epsilon}|\xi_1^2-\xi_2^2|^{\frac{1}{2}}}{\langle\sigma\rangle^{\frac{1}{2}-\frac{\epsilon}{12}}\prod\limits_{j=2}^{8}\langle\xi_j\rangle^{\frac{3}{14}+2\epsilon}\prod\limits_{j=1}^{8}\langle\sigma_j\rangle^{\frac{1}{2}+\frac{\epsilon}{24}}}.$$

By using the Plancherel identity, the Hölder inequality, Lemma 2.3 and the Sobolev embeddings theorem, we derive that

$$I_1\leq\left\|I_-^{\frac{1}{2}}(f_1,f_2)\right\|_{L_{xt}^2}\left(\prod_{j=4}^{8}\|J^{-\frac{1}{2}-\epsilon}f_j\|_{L_{xt}^{\infty}}\right)\|I_-^{\frac{1-\epsilon}{2}}(f_3,f)\|_{L_{xt}^2}$$

$$\leq C\left(\prod_{j=1}^{8}\|f_j\|_{X_{0,b}}\right)\|f\|_{X_{0,\frac{1}{2}-\frac{\epsilon}{12}}}\leq C\left[\prod_{j=1}^{8}\|F_j\|_{L_{\tau\xi}^2}\right]\|F\|_{L_{\tau\xi}^2}.$$

(2) Ω_2. In this case, we have that

$$K(\tau_1,\xi_1,\cdots,\tau_7,\xi_7,\tau,\xi)\leq C\frac{|\xi|\langle\xi\rangle^{\frac{3}{14}+2\epsilon}}{\langle\sigma\rangle^{\frac{1}{2}-\frac{\epsilon}{12}}\prod\limits_{j=1}^{8}\langle\xi_j\rangle^{\frac{3}{14}+2\epsilon}\prod\limits_{j=1}^{8}\langle\sigma_j\rangle^{\frac{1}{2}+\frac{\epsilon}{24}}}.$$

By using a delicate analysis, we have that

$$\|J^{-\frac{1}{2}-\epsilon}f_j\|_{L_{xt}^{\infty}}\leq C\|f_j\|_{X_{0,b}},5\leq j\leq 8.$$

Thus, by using the Plancherel identity and the Hölder inequality, from the above inequality, we derive that

$$I_1\leq C\|f_1\|_{L_{xt}^{\frac{28}{2-7\epsilon}}}\left\|I_-^{\frac{1}{2}}(f_2,f_3)\right\|_{L_{xt}^2}\left(\prod_{j=5}^{8}\|J^{-\frac{1}{2}-\epsilon}f_j\|_{L_{xt}^{\infty}}\right)\|f\|_{L_{xt}^{\frac{6}{1+\epsilon}}}$$

$$\leq C\left(\prod_{j=1}^{8}\|f_j\|_{X_{0,b}}\right)\|f\|_{X_{0,\frac{1}{2}-\frac{\epsilon}{12}}}\leq C\left[\prod_{j=1}^{8}\|F_j\|_{L_{\tau\xi}^2}\right]\|F\|_{L_{\tau\xi}^2}.$$

(3) Ω_3. Case $|\xi_1| \sim |\xi_3| \geq 80|\xi_4|$. We consider $|\xi_4| \leq 1, |\xi_4| \geq 1$, respectively. When $|\xi_4| \leq 1$, by using the Hölder inequality and (2.3), we have that

$$
\begin{aligned}
I_1 &\leq C \left\| I^{1/2} I_-^{1/2}(f_1, f_4) \right\|_{L^2_{xt}} \left(\prod_{j=2}^{3} \|J^{-\sigma} f_j\|_{L_{xt}^{\frac{28}{2-7\epsilon}}} \right) \\
&\quad \times \left(\prod_{j=5}^{8} \|f_j\|_{L_{xt}^{\frac{84}{4+7\epsilon}}} \right) \|f\|_{L_{xt}^{\frac{6}{1+\epsilon}}} \leq C \left(\prod_{j=1}^{8} \|f_j\|_{X_{0,b}} \right) \|f\|_{X_{0,\frac{1}{2}-\frac{\epsilon}{12}}} \\
&\leq C \left(\prod_{j=1}^{8} \|f_j\|_{X_{0,b}} \right) \|f\|_{X_{0,\frac{1}{2}-\frac{\epsilon}{12}}} \leq C \left[\prod_{j=1}^{8} \|F_j\|_{L^2_{\tau\xi}} \right] \|F\|_{L^2_{\tau\xi}}.
\end{aligned}
$$

Here, we use the fact that

$$
\|J^{-\sigma} f_j\|_{L_{xt}^{\frac{28}{2-7\epsilon}}} \leq C \|f_j\|_{X_{0,b}}, j = 2, 3.
$$

When $|\xi_4| \geq 1$, we consider $|\xi_5| \leq 1$,$|\xi_5| \geq 1$, respectively.
When $|\xi_5| \leq 1$, by using the Hölder inequality and (2.3), we have that

$$
\begin{aligned}
I_1 &\leq C \left\| I^{1/2} I_-^{1/2}(f_1, f_5) \right\|_{L^2_{xt}} \left(\prod_{j=2}^{4} \|J^{-\sigma} f_j\|_{L_{xt}^{\frac{28}{2-7\epsilon}}} \right) \\
&\quad \times \left(\prod_{j=6}^{8} \|f_j\|_{L_{xt}^{\frac{84}{10+49\epsilon}}} \right) \|f\|_{L_{xt}^{\frac{6}{1+\epsilon}}} \leq C \left(\prod_{j=1}^{8} \|f_j\|_{X_{0,b}} \right) \|f\|_{X_{0,\frac{1}{2}-\frac{\epsilon}{12}}} \\
&\leq C \left[\prod_{j=1}^{8} \|F_j\|_{L^2_{\tau\xi}} \right] \|F\|_{L^2_{\tau\xi}}.
\end{aligned}
$$

Here, we use the fact that

$$
\|f_j\|_{L_{xt}^{\frac{84}{10+49\epsilon}}} \leq C \|f_j\|_{X_{0,b}}, 6 \leq j \leq 8.
$$

When $|\xi_5| \geq 1$, we consider $|\xi_6| \leq 1, |\xi_6| \geq 1$, respectively.
When $|\xi_6| \leq 1$, we can deal with this case as before.
When $|\xi_6| \geq 1$, we consider $|\xi_7| \leq 1, |\xi_7| \geq 1$, respectively.
When $|\xi_7| \leq 1$, by using the Hölder inequality and (2.3), we have that

$$
\begin{aligned}
I_1 &\leq C \left\| I^{1/2} I_-^{1/2}(f_1, f_6) \right\|_{L^2_{xt}} \left(\prod_{j=2}^{5} \|J^{-\sigma} f_j\|_{L_{xt}^{\frac{28}{2-7\epsilon}}} \right) \\
&\quad \times \left(\prod_{j=7}^{8} \|f_j\|_{L_{xt}^{\frac{84}{2+35\epsilon}}} \right) \|f\|_{L_{xt}^{\frac{6}{1+\epsilon}}} \leq C \left(\prod_{j=1}^{8} \|f_j\|_{X_{0,b}} \right) \|f\|_{X_{0,\frac{1}{2}-\frac{\epsilon}{12}}} \\
&\leq C \left[\prod_{j=1}^{8} \|F_j\|_{L^2_{\tau\xi}} \right] \|F\|_{L^2_{\tau\xi}}.
\end{aligned}
$$

Here, we use the fact that

$$\|f_j\|_{L_{xt}^{\frac{84}{2+35\epsilon}}} \le C\|f_j\|_{X_{0,b}}, 6 \le j \le 8.$$

When $|\xi_7| \ge 1$, we consider $|\xi_8| \le 8, |\xi_8| \ge 8$, respectively.
When $|\xi_8| \le 8$, by using the Hölder inequality and (2.3), we have that

$$I_1 \le C \left\| I^{1/2} I_{-}^{1/2}(f_1, f_6) \right\|_{L_{xt}^2} \left(\prod_{j=2}^{5} \|J^{-\sigma} f_j\|_{L_{xt}^{\frac{28}{2-7\epsilon}}} \right)$$

$$\times \|J^{-2\sigma} f_7\|_{L_{xt}^{\frac{42}{2+35\epsilon}}} \|f_8\|_{L_{xt}^{\infty}} \|f\|_{L_{xt}^{\frac{6}{1+\epsilon}}} \le C \left(\prod_{j=1}^{8} \|f_j\|_{X_{0,b}} \right) \|f\|_{X_{0,\frac{1}{2}-\frac{\epsilon}{12}}}$$

$$\le C \left[\prod_{j=1}^{8} \|F_j\|_{L_{\tau\xi}^2} \right] \|F\|_{L_{\tau\xi}^2}.$$

Here we use the fact that

$$\|J^{-2\sigma} f_7\|_{L_{xt}^{\frac{42}{2+35\epsilon}}} \le C\|f_7\|_{X_{0,b}}.$$

When $|\xi_8| \ge 8$, we consider $|\xi| \le 1, |\xi| \ge 1$, respectively.
When $|\xi| \le 1$, by using Lemma 8.3, we have that

$$I_1 \le C \left\| I^{\frac{1}{2}-2\epsilon} I_{-}^{\frac{1}{2}-2\epsilon}(f_1, f) \right\|_{L_{xt}^2} \left(\prod_{j=2}^{8} \|J^{-\sigma} f_j\|_{L_{xt}^{14}} \right)$$

$$\le C \left(\prod_{j=1}^{8} \|f_j\|_{X_{0,b}} \right) \|f\|_{X_{0,\frac{1}{2}-\frac{\epsilon}{12}}} \le C \left[\prod_{j=1}^{8} \|F_j\|_{L_{\tau\xi}^2} \right] \|F\|_{L_{\tau\xi}^2}.$$

Here we use the fact that

$$\|J^{-\sigma} f_j\|_{L_{xt}^{14}} \le C\|f_j\|_{X_{0,b}}, 2 \le j \le 8.$$

When $|\xi| \ge 1$, by using the Hölder inequality, we have that

$$I_1 \le C \left(\prod_{j=1}^{8} \|J^{-\frac{1+28\epsilon}{16}} f_j\|_{L_{xt}^{\frac{64}{7-\epsilon}}} \right) \|f\|_{L_{xt}^{\frac{8}{1+\epsilon}}} \le C \left(\prod_{j=1}^{8} \|f_j\|_{X_{0,b}} \right) \|f\|_{X_{0,\frac{1}{2}-\frac{\epsilon}{12}}}$$

$$\le C \left[\prod_{j=1}^{8} \|F_j\|_{L_{\tau\xi}^2} \right] \|F\|_{L_{\tau\xi}^2}.$$

(4) Ω_4. Case $|\xi_1| \sim |\xi_4| \ge 80|\xi_5|$. This case can be proved similarly to case $|\xi_1| \sim |\xi_3| \ge 80|\xi_4|$.
(5) Ω_5. Case $|\xi_1| \sim |\xi_5| \ge 80|\xi_6|$. This case can be proved similarly to case $|\xi_1| \sim |\xi_3| \ge 80|\xi_4|$.
(6) Ω_6. Case $|\xi_1| \sim |\xi_6| \ge 80|\xi_6|$. This case can be proved similarly to case $|\xi_1| \sim |\xi_3| \ge 80|\xi_4|$.

(7) Ω_7. Case $|\xi_1| \sim |\xi_7| \geq 80|\xi_8|$. This case can be proved similarly to case $|\xi_1| \sim |\xi_3| \geq 80|\xi_4|$.
(8) Ω_8. When $|\xi_1| \sim |\xi_8|$, by using the Hölder inequality, we have that

$$\begin{aligned} I_1 &\leq C\left(\prod_{j=1}^{8} \|J^{-\frac{1+28\epsilon}{16}} f_j\|_{L_{xt}^{\frac{64}{7-\epsilon}}}\right) \|f\|_{L_{xt}^{\frac{8}{1+\epsilon}}} \leq C\left(\prod_{j=1}^{8} \|f_j\|_{X_{0,b}}\right) \|f\|_{X_{0,\frac{1}{2}-\frac{\epsilon}{12}}} \\ &\leq C\left[\prod_{j=1}^{8} \|F_j\|_{L^2_{\tau\xi}}\right] \|F\|_{L^2_{\tau\xi}}. \end{aligned}$$

This completes the proof of Lemma 8.4.

3. Proof of Theorem 8.1

In this section, we are in a position to prove Theorem 8.1.

Note that (1.1)–(1.2) are equivalent to the following integral equation

$$v(t) = S(t)u_0 + \int_0^t S(t-\tau)v\partial_x(v^7)d\tau. \tag{3.1}$$

Obviously, v satisfies (3.1) on $[-T, T]$ if v satisfies

$$v(t) = \eta(t)u_0 + \eta_T(t)\int_0^t S(t-\tau)v\partial_x(v^7)d\tau \tag{3.2}$$

for some $T \ll 1$. We define

$$\Gamma v(t) = \eta(t)u_0 + \eta_T(t)\int_0^t S(t-\tau)v\partial_x(v^7)d\tau. \tag{3.3}$$

We define $B = \left\{\|u\|_{X_{\sigma,b}} \leq 2C\|u_0\|_{H^s}\right\}$.

By using Lemmas 8.1 and 8.4, for sufficiently small $T > 0$, we get that

$$\begin{aligned} \|\Gamma v\|_{X_{\sigma,b}} &\leq \left\|\eta(t)u_0 + \eta_T(t)\int_0^t S(t-\tau)v\partial_x(v^7)d\tau\right\|_{X_{\sigma,b}} \\ &\leq C\|\eta(t)u_0\|_{X_{\sigma,b}} + \left\|\eta_T(t)\int_0^t S(t-\tau)v\partial_x(v^7)d\tau\right\|_{X_{\sigma,b}} \\ &\leq C\|u_0\|_{H^s} + CT^{\frac{\epsilon}{24}}\left\|v\partial_x(v^7)\right\|_{X_{\sigma,-\frac{1}{2}+\frac{\epsilon}{12}}} \\ &\leq C\|u_0\|_{H^s} + CT^{\frac{\epsilon}{24}}\|u\|^8_{X_{\sigma,b}} \\ &\leq C\|u_0\|_{H^s} + CT^{\frac{\epsilon}{24}}(2C\|u_0\|_{H^s})^8 \leq 2C\|u_0\|_{H^s}. \end{aligned} \tag{3.4}$$

Similarly, by using Lemma 8.1 and 8.4, we conclude that

$$\begin{aligned} \|\Gamma v_1 - \Gamma v_2\|_{X_{\sigma,b}} &\leq CT^{\frac{\epsilon}{24}}\|v_1 - v_2\|_{X_{\sigma,b}}\left(\|v_1\|^7_{X_{\sigma,b}} + \|v_2\|^7_{X_{\sigma,b}} + R^7\right) \\ &\leq CT^{\frac{\epsilon}{24}}(2C\|u_0\|_{H^s})^7\|v_1 - v_2\|_{X_{\sigma,b}} \\ &\leq \frac{1}{2}\|v_1 - v_2\|_{X_{\sigma,b}}. \end{aligned} \tag{3.5}$$

Thus, there exists u such that $\Gamma u = u$. The proof of uniqueness is standard.

This completes the proof of Theorem 8.1.

References

[1] M. Beals, Self-spreading and strength of singularities for solutions to semilinear wave equations, *Ann. of Math.*, 118(1983), 187-214.

[2] J. Bourgain, Fourier transform restriction phenomena for certain lattice subsets and applications to nonlinear evolution equations, I and II, *Geom. Funct. Anal.*, 3(1993), 107-156, 209-262.

[3] A. Grünrock, New applications of the Fourier restriction norm method to wellposedness problems for nonlinear Evolution Equations, Ph.D. Universität Wuppertal, 2002, Germany, Dissertation.

[4] B. L. Guo and Z. H. Huo, The global attractor of the damped forced Ostrovsky equation, *J. Math. Anal. Appl.*, 329(2007), 392-407.

[5] G. L. Gui and Y. Liu, On the Cauchy problem for the Ostrovsky equation with positive dispersion, *Comm. Partial Diff. Eqns.*, 32(2007), 1895-1916.

[6] C. E. Kenig, G. Ponce and L. Vega, The Cauchy problem for the Korteweg-de Vries equation in Sobolev spaces of negative indices, *Duke Math.*, 71(1993), 1-21.

[7] S. Klainerman and M. Machedon, Smoothing estimates for null forms and applications, *Internat. Math. Res. Notices*, (9):383ff. approx. 7 pp. (electronic), 1994.

[8] C. E. Kenig, G. Ponce and L. Vega, The Cauchy problem for the Korteweg-de Vries equation in Sobolev spaces of negative indices, *Duke Math. J.*, 71(1993), 1-21.

[9] S. Levandosky and Y. Liu, Stability of solitary waves of a generalized Ostrovsky equation, *SIAM J. Math. Anal.*, 38(2006), 985-1011.

[10] S. Levandosky, On the stability of solitary waves of a generalized Ostrovsky equation, *Anal. Math.Phys.*, 2(4)(2012), 407-437.

[11] J. Rauch and M. Reed, Nonlinear microlocal analysis of semilinear hyperbolic systems in one space dimension, *Duke Math. J.*, 49(1982), 397-475.

[12] W. Yan, Y. S. Li, J. H. Huang and J. Q. Duan, The Cauchy problem for the Ostrovsky equation with positive dispersion, *Nonlinear Differential Equations Appl.*, 25(2018), 25:22.

Chapter 9

Well-Posedness for Stochastic Two Component Dullin–Gottwald–Holm System with Lévy Noise

Yong Chen[1,*] and Jingyun Luo[2]

[1]*School of Science, Zhejiang Sci-Tech University, Hangzhou 310018, P.R. China*
**youngchen329@126.com*

[2]*School of Science, Zhejiang Sci-Tech University, Hangzhou 310018, P.R. China*
nature19951115@163.com

In this paper, the well-posedness of the stochastic DGH system with Lévy noise is established. The method is adding dissipation term firstly, then deriving the local well-posedness of stochastic dissipative DGH system in the $L^2(\Omega; \mathcal{D}([0,T]; H^s(\mathbb{R}) \times H^{s-1}(\mathbb{R})))$ by contracting mapping principle. Continually, by regularization and stopping time skill, we prove that the solution of dissipation system is a Cauchy sequence, whose limitation is the solution of the original stochastic DGH system.

1. Introduction

Initially, Dullin et al. [Dullin et al. (2001)] studied the following DGH equation:

$$m_t - Au_x + um_x + 2u_x m + \gamma u_{xxx} = 0, \tag{1.1}$$

where $m = u - \alpha^2 u_{xx}$, by using asymptotic expansion directly in the Hamiltonian for Euler's equation in the shallow water regime. Additionally, the equation above can be rewritten as:

$$u_t - \alpha^2 u_{xxt} - Au_x + 3uu_x + \gamma u_{xxx} = \alpha^2(bu_x u_{xx} + uu_{xxx}).$$

If $\gamma \to 0$, the equation turns out to be the Camassa–Holm (CH) equation [Camassa and Holm (1993)]. What is more, the Korteweg–de Vries (KdV) equation can be obtained from the DGH equation with $\alpha = 0$. Tian et al. [Tian et al. (2005)] and Yin [Yin (2004)] studied the existence of strong solutions of DGH equation (1.1). Zhang et al. [Zhang and Yin (2010)] proved the existence of global weak solutions to (1.1) provided that the initial data satisfies some certain conditions.

The following two-component DGH system:

$$\begin{cases} m_t - Au_x + \gamma u_{xxx} + 2mu_x + um_x + \rho\rho_x = 0, & x \in \mathbb{R},\ t > 0, \\ m = u - \alpha^2 u_{xx}, \\ \rho_t + (\rho u)_x = 0 \quad x \in \mathbb{R},\ t > 0, \end{cases} \tag{1.2}$$

*Corresponding author.

was derived in [Constantin and Ivanov (2010), Guo and Gao (2012)]. In [Guo and Gao (2012)], Guo et al. obtained the well-posedness, wave-breaking criteria and global strong solutions of (4.2). Using the bi-linear estimate technique to the approximate solutions, Chen et al. [Chen et al. (2013)] improved the local well-posedness result for System (4.2) in $H^s \times H^{s-1}$ with $s > 3/2$. In [Liu and Yin (2013)], Liu et al. established the local well-posedness in Besov space and stability of solitary waves for System (4.2).

In this article, we consider the following stochastic two component DGH system with Lévy noise:

$$\begin{cases} du + uu_x dt = f(u,\rho)dt + \int_{\mathbb{Z}} g_1(u,\rho,z)\tilde{N}_1(dt,dz), \ x \in \mathbb{R}, t > 0 \\ d\rho + (u\rho)_x dt = \int_{\mathbb{Z}} g_2(u,\rho,z)\tilde{N}_2(dt,dz) \end{cases} \tag{1.3}$$

with initial value:

$$u(0,x) = u_0(x), \rho(0,x) = \rho_0(x). \tag{1.4}$$

Here $f(u,\rho) = (1-\partial_x^2)^{-1}\partial_x[u^2 + \frac{1}{2}u_x^2 + (\gamma - A)u + \frac{1}{2}\rho^2 + \gamma u_x]$, and $\tilde{N}_i(i = 1,2)$ are compensated Poisson random measure (cPrm) corresponding to a Prm N_i with σ-finite intensity measure $\nu_i(dz)$ on a measurable space $\mathbb{Z}$, i.e., $\tilde{N}_i(dt,dz) = N_i(dt,dz) - \nu_i(dt,dz)$. For $u, v \in H^s, \rho_1, \rho_2 \in H^{s-1}$, suppose the function g_i satisfies:

$$\int_{\mathbb{Z}} \|g_i(u,\rho,z)\|_{H^s}^2 \nu_i(dz) \le C\|u\|_{H^s}^2 + \|\rho\|_{H^{s-1}}^2, \tag{1.5}$$

$$\int_{\mathbb{Z}} \|g_i(u,\rho_1,z) - g_i(v,\rho_2,z)\|_{H^s}^2 \nu(dz) \le C\|u-v\|_{H^s}^2 + \|\rho_1 + \rho_2\|_{H^{s-1}}^2. \tag{1.6}$$

To get the well-posedness of (1.3), let us come to the following equations:

$$\begin{cases} du + uu_x dt = \varepsilon u_{xx} dt + f(u,\rho)dt + \int_{\mathbb{Z}} g_1(u,\rho,z)\tilde{N}_1(dt,dz), \ x \in \mathbb{R}, t > 0, \\ d\rho = \varepsilon \rho_{xx} dt - (u\rho)_x dt + \int_{\mathbb{Z}} g_2(u,\rho,z)\tilde{N}_2(dt,dz), \ x \in \mathbb{R}, t > 0. \end{cases} \tag{1.7}$$

The first integral equivalent form of the system (1.7) is as follow:

$$\begin{aligned} u(t) = e^{\varepsilon t\partial_x^2}u_0 + \int_0^t e^{\varepsilon(t-\tau)\partial_x^2} f(u,\rho)d\tau - \int_0^t e^{\varepsilon(t-\tau)\partial_x^2} uu_x d\tau \\ + \int_0^t \int_{\mathbb{Z}} e^{\varepsilon(t-\tau)\partial_x^2} g_1(u,\rho,z)\tilde{N}_1(d\tau,dz), \end{aligned} \tag{1.8}$$

and the second equation:

$$\rho(t) = e^{\varepsilon t\partial_x^2}\rho_0 - \int_0^t e^{\varepsilon(t-\tau)\partial_x^2} u\rho_x d\tau + \int_0^t \int_{\mathbb{Z}} e^{\varepsilon(t-\tau)\partial_x^2} g_2(u,\rho,z)\tilde{N}_2(d\tau,dz). \tag{1.9}$$

Then, by the contracting mapping principle, the local well-posedness of (1.8) and (1.9) in $L^2(\Omega;\mathcal{D}[0,T];H^s \times H^{s-1}), s > 1$ can be obtained as follows.

Theorem 9.1. *Let $s > 1, u_0 \in H^s, \rho_0 \in H^{s-1}$. Then, there exists $T > 0$ which makes $(u_\varepsilon,\rho_\varepsilon) \in L^2(\Omega;\mathcal{D}[0,T];H^s \times H^{s-1})$ is a unique solution of (1.7).*

Then, letting $\varepsilon \to 0$, we will prove the solution $(u_\varepsilon, \rho_\varepsilon)$ of (1.7) converges to the solution (u, ρ) of (1.3). However, we should consider the equations (1.7) with the following regularized initial values:

$$u_\varepsilon(0, x) = u_{0\varepsilon}, \quad \rho_\varepsilon(0, x) = \rho_{0\varepsilon}, \tag{1.10}$$

where $u_{0\varepsilon} = u_0 * \lambda_\varepsilon$, $\rho_{0\varepsilon} = \rho_0 * \lambda_\varepsilon$, λ is the Friedrichs mollifier in the space variable $\lambda_\varepsilon(x) = \varepsilon^{-\frac{1}{4}}\lambda(\varepsilon^{-\frac{1}{4}}x)$ and the Fourier transform $\hat{\lambda}$ of λ satisfies $\hat{\lambda} \in C_0^\infty, \hat{\lambda}(\xi) \geq 0$ and $\hat{\lambda}(\xi) = 1$ for any $\xi \in (-1, 1)$. We still let u_ε and ρ_ε be the solution of (1.7) and (1.10). We will prove $(u_\varepsilon, \rho_\varepsilon)$ is a Cauchy sequence and the limit of it is the solution of (1.3). The main result of the paper is given below.

Theorem 9.2. *Assume $(u_0, \rho_0) \in H^s \times H^{s-1}$ with $s > 3/2$. Then there exists $T > 0$ such that for almost all $\omega \in \Omega$, there is a unique solution $(u, \rho) \in \mathcal{D}([0, T]; H^s \times H^{s-1})$ of system (1.3).*

2. The proof of Theorem 9.1

In this section, the proof of Theorem 9.1 is obtained with Lemmas 9.2–9.4 by the contracting mapping principle.

Lemma 9.1. *(Moser-type estimates) [Chemin (1998), Lemarie-Rieusset (2002)] Let $s > 0, \epsilon > 0$. Then*

(i) for $0 < s < 1/2$, $\|uv\|_{H^{2s-\frac{1}{2}}} \leq C\|u\|_{H^s}\|v\|_{H^s}$,
(ii) for $s > 1/2$, $\|uv\|_{H^s} \leq C\|u\|_{H^s}\|v\|_{H^s}$,
(iii) for $s = 1/2$, $\|uv\|_{H^{s-\epsilon}} \leq C\|u\|_{H^s}\|v\|_{H^s}$.

Lemma 9.2. *Assume $u, v \in \mathcal{D}([0, T]; H^s)$ with $s > 0$. Then, it has*

$$\|\int_0^t e^{\varepsilon(t-\tau)\partial_x^2}(uv)_x d\tau\|_{H^s} \leq CT^\alpha \|u\|_{\mathcal{D}([0,T];H^s)}\|v\|_{\mathcal{D}([0,T];H^s)},$$

where $T^\alpha = \max\{T^{\frac{1}{2}}, T^{\frac{1-\varepsilon}{2}}, T^{\frac{1+2s}{4}}\}$.

Proof. For $s > 1/2$, we have

$$\begin{aligned}\|\int_0^t e^{\varepsilon(t-\tau)\partial_x^2}(uv)_x d\tau\|_{H^s} &= \int_0^t \|e^{\varepsilon(t-\tau)\partial_x^2}\partial_x(1-\partial_x^2)^{\frac{s}{2}}uv\|_{L^2}d\tau \\ &\leq C\int_0^t (t-\tau)^{-\frac{1}{2}}\|(1-\partial_x^2)^{\frac{s}{2}}uv\|_{L^2}d\tau \\ &\leq Ct^{\frac{1}{2}}\|u\|_{\mathcal{D}([0,T];H^s)}\|v\|_{\mathcal{D}([0,T];H^s)}.\end{aligned}$$

For $s = 1/2$, we have

$$\begin{aligned}\|\int_0^t e^{\varepsilon(t-\tau)\partial_x^2}(uv)_x d\tau\|_{H^s} &= \int_0^t \|e^{\varepsilon(t-\tau)\partial_x^2}\partial_x(1-\partial_x^2)^{\frac{\varepsilon}{2}}(1-\partial_x^2)^{\frac{s-\varepsilon}{2}}uv\|_{L^2}d\tau \\ &\leq C\int_0^t (t-\tau)^{-\frac{\varepsilon+1}{2}}\|(1-\partial_x^2)^{\frac{s-\varepsilon}{2}}uv\|_{L^2}d\tau \\ &\leq Ct^{\frac{1-\varepsilon}{2}}\|u\|_{\mathcal{D}([0,T];H^s)}\|v\|_{\mathcal{D}([0,T];H^s)}.\end{aligned}$$

For $0 < s < 1/2$, we have

$$
\begin{aligned}
&\|\int_0^t e^{\varepsilon(t-\tau)\partial_x^2}(uv)_x d\tau\|_{H^s}\\
=&\frac{1}{2}\int_0^t \|e^{\varepsilon(t-\tau)\partial_x^2}\partial_x(1-\partial_x^2)^{\frac{1}{2}(\frac{1}{2}-s)}(1-\partial_x^2)^{\frac{1}{2}(2s-\frac{1}{2})}uv\|_{L^2}d\tau\\
\leq&C\int_0^t (t-\tau)^{-\frac{3-2s}{4}}\|(1-\partial_x^2)^{\frac{1}{2}(2s-\frac{1}{2})}uv\|_{L^2}d\tau\\
\leq&Ct^{\frac{1+2s}{4}}\|u\|_{\mathcal{D}([0,T];H^s)}\|v\|_{\mathcal{D}([0,T];H^s)}.
\end{aligned}
$$

The proof is complete.

Lemma 9.3. *Assume that* $(u,\rho)\in\mathcal{D}([0,T];H^s\times H^{s-1})$ *with* $s>1$*:*

$$
\|\int_0^t e^{\varepsilon(t-\tau)\partial_x^2}f(u,\rho)d\tau\|_{H^s}\leq CT^{\theta}(\|u\|^2_{\mathcal{D}([0,T];H^s)}+\|\rho\|^2_{\mathcal{D}([0,T];H^{s-1})}+1),
$$

where $T^\theta=\max\{T,T^{1-\frac{\varepsilon}{2}},T^{\frac{1+2s}{4}}\}$.

Proof. The proof is similar to that of Lemma 9.2. For $s>3/2$, we have

$$
\begin{aligned}
&\|\int_0^t e^{\varepsilon(t-\tau)\partial_x^2}f(u,\rho)d\tau\|_{H^s}\\
\leq&C\int_0^t \|e^{\varepsilon(t-\tau)\partial_x^2}(1-\partial_x^2)^{\frac{s-1}{2}}(u^2+u_x^2+u+\rho^2+u_x)\|_{L^2}d\tau\\
\leq&CT(\|u\|^2_{\mathcal{D}([0,T];H^{s-1})}+\|u_x\|^2_{\mathcal{D}([0,T];H^{s-1})}+\|u\|_{\mathcal{D}([0,T];H^{s-1})}\\
&+\|\rho\|^2_{\mathcal{D}([0,T];H^{s-1})}+\|u_x\|_{\mathcal{D}([0,T];H^{s-1})})\\
\leq&CT(\|u\|^2_{\mathcal{D}([0,T];H^s)}+\|\rho\|^2_{\mathcal{D}([0,T];H^{s-1})}+1).
\end{aligned}
$$

For $s=3/2$,we can use the same way to get:

$$
\begin{aligned}
&\|\int_0^t e^{\varepsilon(t-\tau)\partial_x^2}f(u,\rho)d\tau\|_{H^s}\\
\leq&C\int_0^t \|e^{\varepsilon(t-\tau)\partial_x^2}\partial_x(1-\partial_x^2)^{\frac{\varepsilon-1}{2}}(1-\partial_x^2)^{\frac{s-\varepsilon-1}{2}}(u^2+u_x^2+u+\rho^2+u_x)\|_{L^2}d\tau\\
\leq&Ct^{\frac{3-\varepsilon}{2}}(\|u\|^2_{\mathcal{D}([0,T];H^{s-1})}+\|u_x\|^2_{\mathcal{D}([0,T];H^{s-1})}+\|u\|_{\mathcal{D}([0,T];H^{s-1})}\\
&+\|\rho\|^2_{\mathcal{D}([0,T];H^{s-1})}+\|u_x\|_{\mathcal{D}([0,T];H^{s-1})})\\
\leq&Ct^{\frac{3-\varepsilon}{2}}(\|u\|^2_{\mathcal{D}([0,T];H^s)}+\|\rho\|^2_{\mathcal{D}([0,T];H^{s-1})}+1).
\end{aligned}
$$

For $1<s<3/2$, it can be:

$$
\begin{aligned}
&\|\int_0^t e^{\varepsilon(t-\tau)\partial_x^2}f(u,\rho)d\tau\|_{H^s}\\
\leq&C\int_0^t \|e^{\varepsilon(t-\tau)\partial_x^2}\partial_x(1-\partial_x^2)^{\frac{1-2s}{4}}(1-\partial_x^2)^{\frac{2(s-1)-\frac{1}{2}}{2}}(u^2+u_x^2+u+\rho^2+u_x)\|_{L^2}d\tau
\end{aligned}
$$

$$\leq Ct^{\frac{1+2s}{4}}(\|u\|^2_{\mathcal{D}([0,T];H^{s-1})} + \|u_x\|^2_{\mathcal{D}([0,T];H^{s-1})} + \|u\|_{\mathcal{D}([0,T];H^{s-1})}$$
$$+ \|\rho\|^2_{\mathcal{D}([0,T];H^{s-1})} + \|u_x\|_{\mathcal{D}([0,T];H^{s-1})})$$
$$\leq CT^{\frac{1+2s}{4}}(\|u\|^2_{\mathcal{D}([0,T];H^{s})} + \|\rho\|^2_{\mathcal{D}([0,T];H^{s-1})} + 1).$$

The proof is complete.

Lemma 9.4. *Assume* $(u,\rho) \in \mathcal{D}([0,T]; H^s \times H^{s-1})$, *then it has:*

$$\mathbb{E} \sup_{t\in[0,T]} \| \int_0^t \int_{\mathbb{Z}} e^{\varepsilon(t-\tau)\partial_x^2} g_i(u,\rho,z)\tilde{N}_i(d\tau,dz)\|^2_{H^s} \leq C\int_0^T \mathbb{E} \sup_{r\in[0,\tau\wedge\mathcal{T}_R]} \|u\|^2_{H^s} d\tau.$$

Proof. By Barkholder–Davis–Gundy (B–D–G) inequality [Peszat and Zabczyk (2007)] and condition (1.5), there is:

$$\mathbb{E} \sup_{t\in[0,T]} \| \int_0^t \int_{\mathbb{Z}} e^{\varepsilon(t-\tau)\partial_x^2} g_i(u,\rho,z)\tilde{N}_i(d\tau,dz)\|^2_{H^s}$$
$$\leq C\mathbb{E}\int_0^T \int_{\mathbb{Z}} \|g_i(u,\rho,z)\|^2_{H^s} \nu_i(d\tau,dz)$$
$$\leq C\int_0^T \mathbb{E} \sup_{r\in[0,\tau]} (\|u(r)\|^2_{H^s} + \|\rho(r)\|^2_{H^{s-1}})d\tau. \quad (2.1)$$

The proof is complete.

3. The proof of Theorem 9.2

In this section, we will prove the solution of (1.7) and (1.10) is a Cauchy sequence, and the limit is the solution of (1.3).

Additionally, under the assumption of Theorem 9.1, there exists a unique solution $(u_\varepsilon,\rho_\varepsilon) \in \mathcal{D}([0,T]; H^s \times H^{s-1})$ for (1.7) and (1.10). To continue, the following lemmas are given in order to solve the problem.

Lemma 9.5. *[Bona and Smith (1975)] Under the above assumptions, the following estimates need any* ε *satisfying* $0 < \varepsilon < 1/4$ *and* $s > 0, q > 0$, *what is more,*

$$\|u_{0\varepsilon}\|^2_{H^q} \leq c\varepsilon^{\frac{s-q}{2}} \quad \text{and} \quad \|u_{0\varepsilon} - u_0\|^2_{H^q} \leq c\varepsilon^{\frac{s-q}{2}}.$$

Lemma 9.6. *[Kato and Ponce (1988)] (i) If* $r > 0$, *then* $H^r \cap L^\infty$ *is an algebra, and*

$$\|uv\|_{H^r} \leq C(\|u\|_{L^\infty}\|v\|_{H^r} + \|u\|_{H^r}\|v\|_{L^\infty}).$$

(ii) Let $r > 0$, *if* $u \in H^r \cap W^{1,\infty}$ *and* $v \in H^{r-1} \cap L^\infty$, *then we get*

$$\|[\Lambda^r, u]v\|_{L^2} \leq C(\|u_x\|_{L^\infty}\|\Lambda^{r-1}v\|_{L^2} + \|\Lambda^r u\|_{L^2}\|v\|_{L^\infty}),$$

where $[\Lambda^r,u]v = \Lambda^r(uv) - u\Lambda^r v, \Lambda = (1-\partial_x^2)^{1/2}$.

(iii) Let $q \geq 0$, $u \in H^q$ *and* $\|u_x\|_{L^\infty} < \infty$. *Then there is*

$$(\Lambda^q u, \Lambda^q(uu_x)) \leq C\|u_x\|_{L^\infty}\|u\|^2_{H^q}.$$

Lemma 9.7. *Let $s > 3/2, q > 1$. Then we have*

$$\lim_{R\to+\infty} \ln P(\sup_{0\le t\le T} \|u_\varepsilon(t)\|_{H^s} + \|\rho_\varepsilon(t)\|_{H^{s-1}} > R) = -\infty, \tag{3.1}$$

$$\lim_{R\to+\infty} \ln P(\sup_{0\le t\le T} \|u_\varepsilon(t)\|^2_{H^q} + \|\rho_\varepsilon(t)\|_{H^{q-1}} > R\varepsilon^{\frac{s-q}{2}}) = -\infty. \tag{3.2}$$

Proof. Let $\Phi_{i\varepsilon}(t) = \int_0^t \int_{\mathbb{Z}} e^{\varepsilon(t-\tau)\partial_x^2} g_i(u_\varepsilon, \rho, z)\tilde{N}_i(d\tau, dz), v(t) = u_\varepsilon(t) - \Phi_{1\varepsilon}(t), w(t) = \rho_\varepsilon(t) - \Phi_{2\varepsilon}(t)$. Then $v(t)$ and $w(t)$ are mild solutions of the deterministic equations with random coefficients.

$$v' = \varepsilon v_{xx} - (v + \Phi_{1\varepsilon})(v + \Phi_{1\varepsilon})_x + f(v + \Phi_{1\varepsilon}, (w + \Phi_{1\varepsilon})), \quad v_0(x) = u_{0\varepsilon}(x), \tag{3.3}$$

$$w' = \varepsilon w_{xx} - [(v + \Phi_{2\varepsilon})(w + \Phi_{2\varepsilon})]_x, \quad w_0(x) = \rho_{0\varepsilon}(x), \tag{3.4}$$

By B–D–G inequality, we can gain

$$\mathbb{E} \sup_{t\in[0,T]} (\|\Phi_{1\varepsilon}(t)\|^2_{H^s} + \|\Phi_{2\varepsilon}(t)\|^2_{H^{s-1}}) \le C\mathbb{E} \sup_{t\in[0,T]} (\|u_\varepsilon\|^2_{H^s} + \|\rho_\varepsilon\|^2_{H^{s-1}}) < +\infty, \tag{3.5}$$

$$\mathbb{E} \sup_{t\in[0,T]} (\|\Phi_{1\varepsilon}(t)\|^2_{H^{s+1}} + \|\Phi_{2\varepsilon}(t)\|^2_{H^s}) \le C\mathbb{E} \sup_{t\in[0,T]} (\|u_\varepsilon\|^2_{H^{s+1}} + \|\rho_\varepsilon\|^2_{H^s}) < +\infty. \tag{3.6}$$

For $R > 0$, define the following stopping times

$$\tau_R^1 = \inf\{t \in [0,T]; \|\Phi_{1\varepsilon}(t)\|^2_{H^{s+1}} + \|\Phi_{2\varepsilon}(t)\|^2_{H^s} > R\},$$
$$\tau_R^2 = \inf\{t \in [0,T]; \|\Phi_{1\varepsilon}(t)\|^2_{H^s} + \|\Phi_{2\varepsilon}(t)\|^2_{H^{s-1}} > R\}.$$

Let $\tau_R = \tau_R^1 \wedge \tau_R^2$. Then by (3.5)–(3.6), we can gain

$$\begin{aligned} &P(\tau_R \le T) \\ \le& \frac{\mathbb{E}\sup_{t\in[0,T]}(\|\Phi_{1\varepsilon}(t)\|^2_{H^s} + \|\Phi_{1\varepsilon}(t)\|^2_{H^{s+1}} + \|\Phi_{2\varepsilon}(t)\|^2_{H^s} + \|\Phi_{2\varepsilon}(t)\|^2_{H^{s-1}})}{R} \\ \le& C\frac{\mathbb{E}\sup_{t\in[0,T]}(\|u_\varepsilon\|^2_{H^s} + \|u_\varepsilon\|^2_{H^{s+1}} + \|\rho_\varepsilon\|^2_{H^s} + \|\rho_\varepsilon\|^2_{H^{s-1}})}{R}. \end{aligned} \tag{3.7}$$

Multiplying both sides of (3.3) with $\Lambda^s v \Lambda^s$, then we can get an equation:

$$\begin{aligned} \|v\|^2_{H^s} + \varepsilon \int_0^{t\wedge\tau_R} \|v\|^2_{H^{s+1}} dr =& \|v_0\|^2_{H^s} - \int_0^{t\wedge\tau_R} (\Lambda^s v, \Lambda^s((v + \Phi_\varepsilon)(v + \Phi_\varepsilon)_x))dr \\ &+ \int_0^{t\wedge\tau_R} (\Lambda^s v, \Lambda^s f(v + \Phi_\varepsilon, w + \Phi_\varepsilon))dr. \end{aligned} \tag{3.8}$$

Since $(v + \Phi_{1\varepsilon})(v + \Phi_{1\varepsilon})_x = vv_x + v\Phi_{1\varepsilon x} + v_x\Phi_{1\varepsilon} + \Phi_{1\varepsilon}\Phi_{1\varepsilon x}$, by Lemma 9.6 (i), (iii) and Young inequality, we have part one that:

$$\begin{aligned} &\int_0^{t\wedge\tau_R} (\Lambda^s v, \Lambda^s(vv_x + v\Phi_{1\varepsilon x} + \Phi_{1\varepsilon}\Phi_{1\varepsilon x}))dr \\ \le& C\int_0^{t\wedge\tau_R} \|v\|_{H^s}(\|v\|_{H^s}\|v_x\|_{L^\infty} + \|v\|_{H^s}\|\Phi_{1\varepsilon x}\|_{L^\infty} + \|\Phi^2_{1\varepsilon}\|_{H^{s+1}})dr \end{aligned}$$

$$\leq C\int_0^{t\wedge\tau_R}\|v\|_{H^s}(\|v\|_{H^s}^2+R)dr$$
$$\leq C\int_0^{t\wedge\tau_R}(R^{\frac{3}{2}}+\|v\|_{H^s}^3)dr. \tag{3.9}$$

And by Lemma 9.6 (ii), we can get part two like:

$$\int_0^{t\wedge\tau_R}(\Lambda^s v,\Lambda^s(v_x\Phi_{1\varepsilon}))dr$$
$$=\int_0^{t\wedge\tau_R}(\Lambda^s v,[\Lambda^s,\Phi_{1\varepsilon}]v_x)dr+\int_0^{t\wedge\tau_R}(\Lambda^s v,\Phi_{1\varepsilon}\Lambda^s v_x)dr$$
$$\leq C\int_0^{t\wedge\tau_R}(\|v\|_{H^s}^2(\|\Phi_{1\varepsilon}\|_{H^s}+\|\Phi_{1\varepsilon}\|_{H^{s+1}}))dr$$
$$\leq C\int_0^{t\wedge\tau_R}(\|v\|_{H^s}^3+R)dr. \tag{3.10}$$

Moreover, by Hölder and Young inequalities, it has

$$\int_0^{t\wedge\tau_R}(\Lambda^s v,\Lambda^s f(v+\Phi_{1\varepsilon},w+\Phi_{2\varepsilon}))dr$$
$$\leq C\int_0^{t\wedge\tau_R}\|v\|_{H^s}(\|v+\Phi_{1\varepsilon}\|_{H^{s-1}}^2+\|(v+\Phi_{1\varepsilon})_x\|_{H^{s-1}}^2+\|v+\Phi_{1\varepsilon}\|_{H^{s-1}}$$
$$+\|w+\Phi_{2\varepsilon}\|_{H^{s-1}}^2+\|(v+\Phi_{1\varepsilon})_x\|_{H^{s-1}})dr$$
$$\leq C\int_0^{t\wedge\tau_R}\|v\|_{H^s}(\|v\|_{H^s}^2+\|v\|_{H^s}+\|w\|_{H^{s-1}}^2+R)dr$$
$$\leq C(R)\int_0^{t\wedge\tau_R}(\|v\|_{H^s}^3+\|w\|_{H^{s-1}}^3)dr. \tag{3.11}$$

It follows from (3.9)–(3.11) that

$$\|v\|_{H^s}^2\leq\|v_0\|_{H^s}^2+C(R)\int_0^{t\wedge\tau_R}(\|v\|_{H^s}^3+\|w\|_{H^{s-1}}^3)dr. \tag{3.12}$$

To be continued, as for ρ, we can also multiply both sides of (3.4) with $\Lambda^{s-1}w\Lambda^{s-1}$, integrating with respect to x and t and then lead to the equation:

$$\|w\|_{H^{s-1}}^2+\varepsilon\int_0^{t\wedge\tau_R}\|w\|_{H^s}^2dr$$
$$=\|w_0\|_{H^{s-1}}^2-\int_0^{t\wedge\tau_R}(\Lambda^{s-1}w,\Lambda^s[(v+\Phi_{1\varepsilon})(w+\Phi_{2\varepsilon}n)])dr. \tag{3.13}$$

Since $(v+\Phi_{1\varepsilon})(w+\Phi_{2\varepsilon})=vw+v\Phi_{2\varepsilon}+w\Phi_{2\varepsilon}+\Phi_{1\varepsilon}\Phi_{2\varepsilon}$, by the Lemma 9.6 (i), (iii) and Young inequality, we can gain the first part as:

$$\int_0^{t\wedge\tau_R}(\Lambda^{s-1}w,\Lambda^s(v\Phi_{2\varepsilon}+w\Phi_{1\varepsilon}+\Phi_{1\varepsilon}\Phi_{2\varepsilon}))dr$$
$$\leq C\int_0^{t\wedge\tau_R}\|w\|_{H^{s-1}}(\|v\|_{H^s}\|\Phi_{2\varepsilon}\|_{H^s}+\|w\|_{H^s}\|\Phi_{1\varepsilon}\|_{H^s}+\|\Phi_{1\varepsilon}\|_{H^s}\|\Phi_{2\varepsilon}\|_{H^s})dr$$

$$\leq C(R)\int_0^{t\wedge\tau_R}(\|w\|_{H^{s-1}}^3+\|v\|_{H^s}^3)dr. \tag{3.14}$$

The second part is:

$$\begin{aligned}
&\int_0^{t\wedge\tau_R}(\Lambda^{s-1}w,\Lambda^s(vw))dr\\
=&\int_0^{t\wedge\tau_R}[(\Lambda^{s-1}w,[\Lambda^s,v]w)+(\Lambda^{s-1}w,v\Lambda^s w)]dr\\
\leq& C\int_0^{t\wedge\tau_R}[\|w\|_{H^{s-1}}(\|v_x\|_{L^\infty}\|\partial^{s-1}w\|_{L^2}+\|\partial^s v\|_{L^2}\|w\|_{L^\infty})\\
&+\frac{1}{2}\int_{\mathbb{R}}\frac{\partial(\Lambda^{s-1}w)^2}{\partial_x}vdx]dr\\
\leq& C\int_0^{t\wedge\tau_R}[\|w\|_{H^{s-1}}(\|v\|_{H^s}\|w\|_{H^{s-1}}+\|v\|_{H^s}\|w\|_{H^{s-1}})\\
&-\frac{1}{2}\int_{\mathbb{R}}(\Lambda^{s-1}w)^2v_xdx]dr\\
\leq& C\int_0^{t\wedge\tau_R}\|w\|_{H^{s-1}}^2\|v\|_{H^s}dr\\
\leq& C\int_0^{t\wedge\tau_R}(\|w\|_{H^{s-1}}^3+\|v\|_{H^s}^3)dr.
\end{aligned} \tag{3.15}$$

Add part one and part two together then we have:

$$\|w\|_{H^{s-1}}^2\leq\|w_0\|_{H^{s-1}}^2+C(R)\int_0^{t\wedge\tau_R}(\|w\|_{H^{s-1}}^3+\|v\|_{H^s}^3)dr. \tag{3.16}$$

As following, we add (3.12)–(3.16), then we have:

$$\begin{aligned}
&\sup_{t\in[0,T\wedge\tau_R]}(\|v(t\wedge\tau_R)\|_{H^s}^2+\|w(t\wedge\tau_R)\|_{H^{s-1}}^2)\\
&\leq(\|v_0\|_{H^s}^2+\|w_0\|_{H^{s-1}}^2)+C(R)\int_0^{t\wedge\tau_R}(\|v\|_{H^s}^3+\|w\|_{H^{s-1}}^3)dt\\
&=Y(T).
\end{aligned} \tag{3.17}$$

Let $(\|v(T\wedge\tau_R)\|_{H^s}^2+\|w(T\wedge\tau_R)\|_{H^{s-1}}^2)=A$, then $\frac{dY(T)}{dT}=C\sup_{t\in[0,T\wedge\tau_R]}A^2\leq CY^2(T)$, and from which we can gain

$$\sup_{t\in[0,T\wedge\tau_R]}(\|v\|_{H^s}^2+\|w\|_{H^{s-1}}^2)^2\lesssim Y(T)\leq\frac{C}{1-CT}. \tag{3.18}$$

By (3.5)–(3.7) and (3.18), it has

$$\begin{aligned}
\sup_{t\in[0,T\wedge\tau_R]}(\|u_\varepsilon\|_{H^s}^2+\|\rho_\varepsilon\|_{H^{s-1}}^2)\leq&\sup_{t\in[0,T\wedge\tau_R]}(\|v\|_{H^s}^2+\|\Phi_\varepsilon\|_{H^s}^2)\\
&+\sup_{t\in[0,T\wedge\tau_R]}(\|w\|_{H^{s-1}}^2+\|\Phi_\varepsilon\|_{H^{s-1}}^2)\\
\leq&\frac{C}{1-CT}+R.
\end{aligned} \tag{3.19}$$

By (3.7), for any $\delta > 0$, there exists R that fits:

$$P(\tau_R \le T) < \frac{\delta}{2}. \tag{3.20}$$

For such R, by (3.19), there is:

$$\begin{aligned}
&\ln P(\sup_{t\in[0,T\wedge\overline{\tau}_R]}(\|u_\varepsilon\|^2_{H^s} + \|\rho_\varepsilon\|^2_{H^{s-1}}) > R_1) \\
\le &\ln \frac{\mathbb{E}\sup_{t\in[0,T\wedge\overline{\tau}_R]}(\|u_\varepsilon\|^2_{H^s} + \|\rho_\varepsilon\|^2_{H^{s-1}})}{R_1} \\
\le &\ln(\frac{C}{1-CT} + R) - \ln(R_1) \to -\infty, \text{ as } R_1 \to +\infty.
\end{aligned} \tag{3.21}$$

Hence, there exists R_1 such that

$$P(\sup_{t\in[0,T\wedge\tau_R]}(\|u_\varepsilon\|^2_{H^s} + \|\rho_\varepsilon\|^2_{H^{s-1}}) > R_1) < \frac{\delta}{2}. \tag{3.22}$$

By (3.20) and (3.22), we can get that

$$\begin{aligned}
&P(\sup_{t\in[0,T]}(\|u_\varepsilon\|^2_{H^s} + \|\rho_\varepsilon\|^2_{H^{s-1}}) > R_1) \\
\le &P(\sup_{t\in[0,T\wedge\tau_R]}(\|u_\varepsilon\|^2_{H^s} + \|\rho_\varepsilon\|^2_{H^{s-1}}) > R_1) + P(\tau_R \le T) \\
< &\delta,
\end{aligned} \tag{3.23}$$

from which (3.1) is obtained.

For $R > 0$, define the stopping time $\overline{\tau}_R = \inf\{t \in [0,T]; (\|u_\varepsilon\|^2_{H^s} + \|\rho_\varepsilon\|^2_{H^{s-1}}) > R\}$. Applying Itô formula, then we have

$$\begin{aligned}
&\|u_\varepsilon(t\wedge\overline{\tau}_R)\|^2_{H^q} + 2\varepsilon\int_0^{t\wedge\tau_R}\|u_\varepsilon(t\wedge\overline{\tau}_R)\|^2_{H^{q+1}}dr \\
= &\|u_{0\varepsilon}\|^2_{H^q} + 2\int_0^{t\wedge\tau_R}(\Lambda^q u_\varepsilon, \Lambda^q(f(u_\varepsilon,\rho_\varepsilon) - u_\varepsilon u_{\varepsilon x}))dr &(3.24)\\
&+ 2\int_0^{t\wedge\tau_R}\int_{\mathbb{Z}}(\Lambda^q u_\varepsilon, \Lambda^q g_1(u_\varepsilon,\rho,z))\tilde{N}_1(dr,dz) &(3.25)\\
&+ \int_0^{t\wedge\tau_R}\int_{\mathbb{Z}}\|g_1(u_\varepsilon,\rho,z)\|^2_{H^q}\nu_1(dr,dz).
\end{aligned}$$

and

$$\begin{aligned}
&\|\rho_\varepsilon(t\wedge\overline{\tau}_R)\|^2_{H^{q-1}} + 2\varepsilon\int_0^{t\wedge\tau_R}\|\rho_\varepsilon(t\wedge\overline{\tau}_R)\|^2_{H^q}dr \\
= &\|\rho_{0\varepsilon}\|^2_{H^{q-1}} + 2\int_0^{t\wedge\tau_R}(\Lambda^{q-1}\rho_\varepsilon, \Lambda^q(u\rho))dr &(3.26)\\
&+ 2\int_0^{t\wedge\tau_R}\int_{\mathbb{Z}}(\Lambda^{q-1}\rho_\varepsilon, \Lambda^{q-1}g_2(u_\varepsilon,\rho,z))\tilde{N}_2(dr,dz) &(3.27)\\
&+ \int_0^{t\wedge\tau_R}\int_{\mathbb{Z}}\|g_2(u_\varepsilon,\rho,z)\|^2_{H^{q-1}}\nu_2(dr,dz).
\end{aligned}$$

By Lemma 3.2 (i), (iii) and embedding theorem, we can gain

$$\begin{aligned}
&\int_0^{t\wedge\tau_R} (\Lambda^q u_\varepsilon, \Lambda^q(f(u_\varepsilon,\rho_\varepsilon) - u_\varepsilon u_{\varepsilon x}))dr \\
\leq & C\int_0^{t\wedge\tau_R} \|u_\varepsilon\|_{H^q}(\|u_\varepsilon\|_{H^q}\|\rho_\varepsilon\|_{L^\infty} + \|u_\varepsilon^2\|_{H^{q-1}} + \|u_{\varepsilon x}^2\|_{H^{q-1}})dr \\
\leq & C\int_0^{t\wedge\tau_R} \|u_\varepsilon\|_{H^q}\|u_\varepsilon\|_{H^q}^2 dr \\
\leq & C\sqrt{R}\int_0^{t\wedge\tau_R} \|u_\varepsilon\|_{H^q}^2 dr.
\end{aligned} \tag{3.28}$$

By condition (1.5), B–D–G and Young inequalities, we can have

$$\begin{aligned}
&\mathbb{E}\sup_{t\in[0,T\wedge\overline{\tau}_R]}\int_0^{t\wedge\tau_R}\int_{\mathbb{Z}}(\Lambda^q u_\varepsilon, \Lambda^q g_1(u_\varepsilon,\rho_\varepsilon,z)\tilde{N}_1(dr,dz)) \\
\leq & C\mathbb{E}(\int_0^{T\wedge\overline{\tau}_R}\|u_\varepsilon\|_{H^q}^2\int_{\mathbb{Z}}\|g_i(u_\varepsilon,\rho_\varepsilon,z)\|_{H^q}^2\nu_i(dr,dz))^{1/2} \\
\leq & \frac{1}{2}\mathbb{E}\sup_{t\in[0,T\wedge\overline{\tau}_R]}\|u_\varepsilon\|_{H^q}^2 + C\int_0^T\mathbb{E}\sup_{r\in[0,\tau\wedge\overline{\tau}_R]}(\|u_\varepsilon(r)\|_{H^q}^2 + \|\rho_\varepsilon(r)\|_{H^{q-1}}^2)d\tau,
\end{aligned} \tag{3.29}$$

and

$$\begin{aligned}
&\mathbb{E}\sup_{t\in[0,T\wedge\overline{\tau}_R]}\int_0^{t\wedge\tau_R}\int_{\mathbb{Z}}(\Lambda^{q-1}\rho_\varepsilon, \Lambda^{q-1} g_2(u_\varepsilon,\rho_\varepsilon,z)\tilde{N}_2(dr,dz)) \\
\leq & \frac{1}{2}\mathbb{E}\sup_{t\in[0,T\wedge\overline{\tau}_R]}\|\rho_\varepsilon\|_{H^{q-1}}^2 + C\int_0^T\mathbb{E}\sup_{r\in[0,\tau\wedge\overline{\tau}_R]}(\|u_\varepsilon(r)\|_{H^{q-1}}^2 + \|\rho_\varepsilon(r)\|_{H^{q-2}}^2)d\tau.
\end{aligned} \tag{3.30}$$

It follows from (3.26)–(3.30) and Lemma 3.1 that

$$\begin{aligned}
&\mathbb{E}\sup_{t\in[0,T\wedge\overline{\tau}_R]}(\|u_\varepsilon\|_{H^q}^2 + \|\rho_\varepsilon\|_{H^{q-1}}^2) \\
\leq & 4\varepsilon^{\frac{s-q}{2}} + C(1+\sqrt{R})\int_0^T\mathbb{E}\sup_{r\in[0,t\wedge\overline{\tau}_R]}(\|u_\varepsilon(r)\|_{H^q}^2 + \|\rho_\varepsilon(r)\|_{H^{q-1}}^2)dt.
\end{aligned} \tag{3.31}$$

Then by Gronwall inequality, there is

$$\mathbb{E}\sup_{t\in[0,T\wedge\overline{\tau}_R]}(\|u_\varepsilon\|_{H^q}^2 + \|\rho_\varepsilon\|_{H^{q-1}}^2) \lesssim \varepsilon^{\frac{s-q}{2}}e^{C(1+\sqrt{R})T}. \tag{3.32}$$

By (3.1), for any $\delta > 0$, there exists R that fits

$$P(\sup_{t\in[0,T]}(\|u_\varepsilon\|_{H^s}^2 + \|\rho_\varepsilon\|_{H^{s-1}}^2) > R) < \frac{\delta}{2} \tag{3.33}$$

then by (3.32), it has

$$\ln P(\sup_{t\in[0,T\wedge\overline{\tau}_R]}(\|u_\varepsilon\|_{H^q}^2 + \|\rho_\varepsilon\|_{H^{q-1}}^2) > R_1\varepsilon^{\frac{s-q}{2}})$$

$$\leq \ln \frac{\mathbb{E}\sup_{t\in[0,T\wedge\tilde{\tau}_R]}(\|u_\varepsilon\|^2_{H^q} + \|\rho_\varepsilon\|^2_{H^{q-1}})}{R_1\varepsilon^{\frac{s-q}{2}}}$$
$$\leq C(1+\sqrt{R})T - \ln(R_1) \to -\infty, \text{ as } R_1 \to +\infty. \tag{3.34}$$

Hence, there exists R_1 that fits

$$P(\sup_{t\in[0,T\wedge\tilde{\tau}_R]}(\|u_\varepsilon\|^2_{H^q} + \|\rho_\varepsilon\|^2_{H^{q-1}}) > R_1\varepsilon^{\frac{s-q}{2}}) < \frac{\delta}{2}. \tag{3.35}$$

So by (3.33) and (3.35), we can get

$$P(\sup_{t\in[0,T]}(\|u_\varepsilon\|^2_{H^q} + \|\rho_\varepsilon\|^2_{H^{q-1}}) > R_1\varepsilon^{\frac{s-q}{2}})$$
$$\leq P(\sup_{t\in[0,T]}(\|u_\varepsilon\|^2_{H^q} + \|\rho_\varepsilon\|^2_{H^{q-1}}) > R_1\varepsilon^{\frac{s-q}{2}}, \sup_{t\in[0,T]}(\|u_\varepsilon\|^2_{H^s} + \|\rho_\varepsilon\|^2_{H^{s-1}}) \leq R)$$
$$+ P(\sup_{t\in[0,T]}(\|u_\varepsilon\|^2_{H^s} + \|\rho_\varepsilon\|^2_{H^{s-1}}) > R)$$
$$\leq P(\sup_{t\in[0,T\wedge\tilde{\tau}_R]}(\|u_\varepsilon\|^2_{H^q} + \|\rho_\varepsilon\|^2_{H^{q-1}}) > R_1\varepsilon^{\frac{s-q}{2}}) \tag{3.36}$$
$$+ P(\sup_{t\in[0,T]}(\|u_\varepsilon\|^2_{H^s} + \|\rho_\varepsilon\|^2_{H^{s-1}}) > R)$$
$$< \delta, \tag{3.37}$$

from which (3.2) is obtained.

Now, we prove that $(u_\varepsilon, \rho_\varepsilon)$ is a Cauchy sequence. Let $(u_\varepsilon, \rho_\varepsilon)$ and (u_η, ρ_η) be solutions of problem (1.7), corresponding to the parameters ε and η, with $0 < \eta < \varepsilon < 1/4$, and let $h = (u_\varepsilon, \rho_\varepsilon) - (u_\eta, \rho_\eta) = (u_\varepsilon - u_\eta, \rho_\varepsilon - \rho_\eta) = (h_1, h_2), v = (u_\varepsilon, \rho_\varepsilon) + (u_\eta, \rho_\eta) = (u_\varepsilon + u_\eta, \rho_\varepsilon + \rho_\eta) = (v_1, v_2)$. Then h satisfies

$$dh_1 = (\varepsilon u_{\varepsilon xx} - \eta u_{\eta xx})dt - \frac{1}{2}(h_1 v_1)_x dt + (f(u_\varepsilon, \rho_\varepsilon) - f(u_\eta, \rho_\eta))dt$$
$$+ \int_{\mathbb{Z}} (g_1(u_\varepsilon, \rho_\varepsilon, z) - g_1(u_\eta, \rho_\eta, z))\tilde{N}_1(dt, dz), \tag{3.38}$$

and

$$dh_2 = (\varepsilon \rho_{\varepsilon xx} - \eta \rho_{\eta xx})dt - \frac{1}{2}(h_2 v_2)_x dt \tag{3.39}$$
$$+ \int_{\mathbb{Z}} (g_2(u_\varepsilon, \rho_\varepsilon, z) - g_2(u_\eta, \rho_\eta, z))\tilde{N}_2(dt, dz), \tag{3.40}$$

with the initial value $h_{10} = u_{0\varepsilon} - u_{0\eta}, h_{20} = \rho_{0\varepsilon} - \rho_{0\eta}$.

Proposition 9.1. *For P-a.e. $\omega \in \Omega$, there exists a $T > 0$ such that $(u_\varepsilon, \rho_\varepsilon)$ is a Cauchy sequence in the space $\mathcal{D}([0,T]; H^s \times H^{s-1})$ with $s > 3/2$.*

Proof. For $1/2 < q < min(1, s-1), R > 0$, defining the stopping times

$$\tilde{\tau}^1_R = \inf\{t \in [0,T]; \|u_\varepsilon\|^2_{H^s} + \|u_\eta\|^2_{H^s} + \|\rho_\varepsilon\|^2_{H^{s-1}} + \|\rho_\eta\|^2_{H^{s-1}} > R\},$$
$$\tilde{\tau}^2_R = \inf\{t \in [0,T]; \|u_\varepsilon\|^2_{H^{q+2}} + \|\rho_\varepsilon\|^2_{H^{q+1}} > R\varepsilon^{\frac{s-q-2}{2}}\}, \text{ and}$$

$$\tilde{\tau}_R^3 = \inf\{t \in [0,T]; \|u_\eta\|_{H^{q+2}}^2 + \|\rho_\eta\|_{H^{q+1}}^2 > R\eta^{\frac{s-q-2}{2}}\}.$$

Let $\tau_R^\varepsilon = \tilde{\tau}_R^1 \wedge \tilde{\tau}_R^2 \wedge \tilde{\tau}_R^3$. Applying Itô formula on $\|h_1(t\wedge\tau_R^\varepsilon)\|_{H^q}^2$ and $\|h_2(t\wedge\tau_R^\varepsilon)\|_{H^{q-1}}^2$, we get

$$\begin{aligned}
&\|h_1(t\wedge\tau_R^\varepsilon)\|_{H^q}^2 \\
&= \|h_{10}\|_{H^q}^2 + 2\int_0^{t\wedge\tau_R^\varepsilon} (\Lambda^q h_1, \Lambda^q(\varepsilon u_{\varepsilon xx} - \eta u_{\eta xx}))d\tau \\
&\quad - \int_0^{t\wedge\tau_R^\varepsilon} (\Lambda^q h_1, \Lambda^q(h_1 v_1)_x)d\tau \\
&\quad + 2\int_0^{t\wedge\tau_R^\varepsilon} (\Lambda^q h_1, \Lambda^q(f(u_\varepsilon,\rho_\varepsilon) - f(u_\eta,\rho_\eta)))d\tau \\
&\quad + 2\int_0^{t\wedge\tau_R^\varepsilon}\int_{\mathbb{Z}} (\Lambda^q h_1, \Lambda^q(g_1(u_\varepsilon,\rho_\varepsilon,z) - g_1(u_\eta,\rho_\eta,z)))\tilde{N}_1(d\tau,dz) \\
&\quad + \int_0^{t\wedge\tau_R^\varepsilon}\int_{\mathbb{Z}} \|g_1(u_\varepsilon,\rho_\varepsilon,z) - g_1(u_\eta,\rho_\eta,z)\|_{H^q}^2 \nu_1(d\tau,dz).
\end{aligned}$$

and

$$\begin{aligned}
&\|h_2(t\wedge\tau_R^\varepsilon)\|_{H^{q-1}}^2 \\
&= \|h_{20}\|_{H^{q-1}}^2 + 2\int_0^{t\wedge\tau_R^\varepsilon} (\Lambda^{q-1} h_2, \Lambda^{q-1}(\varepsilon \rho_{\varepsilon xx} - \eta \rho_{\eta xx}))d\tau \\
&\quad - \int_0^{t\wedge\tau_R^\varepsilon} (\Lambda^{q-1} h_2, \Lambda^{q-1}(h_2 v_2)_x)d\tau \\
&\quad + 2\int_0^{t\wedge\tau_R^\varepsilon}\int_{\mathbb{Z}} (\Lambda^{q-1} h_2, \Lambda^{q-1}(g_2(u_\varepsilon,\rho_\varepsilon,z) - g_2(u_\eta,\rho_\eta,z)))\tilde{N}_2(d\tau,dz) \\
&\quad + \int_0^{t\wedge\tau_R^\varepsilon}\int_{\mathbb{Z}} \|g_2(u_\varepsilon,\rho_\varepsilon,z) - g_2(u_\eta,\rho_\eta,z)\|_{H^{q-1}}^2 \nu_2(d\tau,dz).
\end{aligned}$$

Using Hölder and Young inequalities, we get

$$\begin{aligned}
&\int_0^{t\wedge\tau_R^\varepsilon} (\Lambda^q h_1, \Lambda^q(\varepsilon u_{\varepsilon xx} - \eta u_{\eta xx}))d\tau \\
&\le C\Big(\int_0^{t\wedge\tau_R^\varepsilon} \|h_1\|_{H^q}^2 d\tau + \int_0^{t\wedge\tau_R^\varepsilon} \varepsilon^2\|u_\varepsilon\|_{q+2}^2 + \eta^2\|u_\eta\|_{q+2}^2 d\tau\Big) \\
&\le C\varepsilon^{\frac{s-q+2}{2}}R + C\int_0^{t\wedge\tau_R^\varepsilon} \|h_1\|_{H^q}^2 d\tau,
\end{aligned} \tag{3.41}$$

and similarly,

$$\begin{aligned}
&\int_0^{t\wedge\tau_R^\varepsilon} (\Lambda^{q-1} h_2, \Lambda^{q-1}(\varepsilon \rho_{\varepsilon xx} - \eta \rho_{\eta xx}))d\tau \\
&\le C\varepsilon^{\frac{s-q+2}{2}}R + C\int_0^{t\wedge\tau_R^\varepsilon} \|h_2\|_{H^{q-1}}^2 d\tau.
\end{aligned} \tag{3.42}$$

By Lemma 3.2, we can gain

$$\begin{aligned}
&\int_0^{t\wedge\tau_R^\varepsilon}(\Lambda^q h_1,\Lambda^q(h_1v_1)_x)d\tau\\
&=\int_0^{t\wedge\tau_R^\varepsilon}(\Lambda^q h_1,\Lambda^q(h_{1x}v_1))d\tau+\int_0^{t\wedge\tau_R^\varepsilon}(\Lambda^q h_1,\Lambda^q(h_1v_{1x}))d\tau\\
&=\int_0^{t\wedge\tau_R^\varepsilon}(\Lambda^q h_1,[\Lambda^q,v_1]h_{1x})d\tau+\int_0^{t\wedge\tau_R^\varepsilon}(\Lambda^q h_1,v_1\Lambda^q h_{1x})d\tau \qquad (3.43)\\
&\quad+\int_0^{t\wedge\tau_R^\varepsilon}(\Lambda^q h_1,\Lambda^q(h_1v_{1x}))d\tau\\
&\le C\int_0^{t\wedge\tau_R^\varepsilon}(1+\|u_\varepsilon\|_{H^s}+\|u_\eta\|_{H^s})\|h_1\|_{H^q}^2d\tau\\
&\le C(R)\int_0^{t\wedge\tau_R^\varepsilon}\|h_1\|_{H^q}^2ds, \qquad (3.44)
\end{aligned}$$

and

$$\begin{aligned}
&\int_0^{t\wedge\tau_R^\varepsilon}(\Lambda^q h_1,\Lambda^q(f(u_\varepsilon,\rho_\varepsilon)-f(u_\eta,\rho_\eta)))d\tau\\
&\le C\int_0^{t\wedge\tau_R^\varepsilon}\|h_1\|_{H^q}(\|h_1v_1\|_{H^{q-1}}+\|h_{1x}v_{1x}\|_{H^{q-1}}+\|h_1\|_{H^q}+\|h_2v_2\|_{H^{q-1}})d\tau\\
&\le C\int_0^{t\wedge\tau_R^\varepsilon}(\|v_1\|_{H^s}+\|v_2\|_{H^{s-1}})(\|h_1\|_{H^q}^2+\|h_2\|_{H^{q-1}}^2)d\tau\\
&\le C(R)\int_0^{t\wedge\tau_R^\varepsilon}(\|h_1\|_{H^q}^2+\|h_2\|_{H^{q-1}}^2)ds. \qquad (3.45)
\end{aligned}$$

By B–D–G inequality and condition (1.6), then we get

$$\begin{aligned}
&\mathbb{E}\sup_{t\in[0,T\wedge\tau_R^\varepsilon]}\int_0^{t\wedge\tau_R^\varepsilon}\int_{\mathbb{Z}}(\Lambda^q h_1,\Lambda^q(g_1(u_\varepsilon,\rho_\varepsilon,z)-g_1(u_\eta,\rho_\eta,z)))\tilde{N}_1(d\tau,dz)\\
&\le C\mathbb{E}\Big(\int_0^{T\wedge\tau_R^\varepsilon}\|h_1\|_{H^q}^2\|g_1(u_\varepsilon,\rho_\varepsilon,z)-g_1(u_\eta,\rho_\eta,z)\|_{H^q}^2\nu_1(d\tau,dz)\Big)^{1/2}\\
&\le\frac{1}{2}\mathbb{E}\sup_{t\in[0,T\wedge\tau_R^\varepsilon]}\|h_1\|_{H^q}^2+C\int_0^T\mathbb{E}\sup_{r\in[0,\tau\wedge\tau_R^\varepsilon]}(\|h_1(r)\|_{H^q}^2+\|h_2(r)\|_{H^{q-1}}^2)d\tau, \qquad (3.46)
\end{aligned}$$

and similarly

$$\begin{aligned}
&\mathbb{E}\sup_{t\in[0,T\wedge\tau_R^\varepsilon]}\int_0^{t\wedge\tau_R^\varepsilon}\int_{\mathbb{Z}}(\Lambda^{q-1} h_2,\Lambda^{q-1}(g_2(u_\varepsilon,\rho_\varepsilon,z)-g_2(u_\eta,\rho_\eta,z)))\tilde{N}_2(d\tau,dz)\\
&\le\frac{1}{2}\mathbb{E}\sup_{t\in[0,T\wedge\tau_R^\varepsilon]}\|h_2\|_{H^{q-1}}^2+C\int_0^T\mathbb{E}\sup_{r\in[0,\tau\wedge\tau_R^\varepsilon]}(\|h_1(r)\|_{H^{q-1}}^2+\|h_2(r)\|_{H^{q-2}}^2)d\tau. \qquad (3.47)
\end{aligned}$$

It follows from above inequalities that

$$\mathbb{E}\sup_{t\in[0,T\wedge\tau_R^\varepsilon]}(\|h_1(r)\|_{H^q}^2+\|h_2(r)\|_{H^{q-1}}^2)$$

$$\le \|h_{01}\|^2_{H^q} + \|h_{02}\|^2_{H^{q-1}} + \varepsilon^{\frac{s-q+2}{2}} R$$
$$+ C(R)\int_0^T \mathbb{E} \sup_{r\in[0,\tau\wedge\tau_R^\varepsilon]} (\|h_1(r)\|^2_{H^q} + \|h_2(r)\|^2_{H^{q-1}})d\tau.$$

Moreover, the Gronwall inequality also implies

$$\mathbb{E} \sup_{t\in[0,T\wedge\tau_R^\varepsilon]} (\|h_1(r)\|^2_{H^q} + \|h_2(r)\|^2_{H^{q-1}}) \le Ce^{C\sqrt{R}T}\varepsilon^{\frac{s-q}{2}}. \tag{3.48}$$

Then, similar to (3.1)–(3.2), we can get

$$\lim_{R\to+\infty} \ln P(\sup_{0\le t\le T} \|h(t)\|_{H^s} > R\varepsilon^{\frac{s-q}{2}}) = -\infty. \tag{3.49}$$

Now again, defining the stopping times

$$\tilde{\tau}_R^4 = \inf\{t\in[0,T]; (\|u_\varepsilon\|^2_{H^{s+k}} + \|\rho_\varepsilon\|^2_{H^{s+k-1}}) > R\varepsilon^{-\frac{k}{2}},\ k=1,2\},$$
$$\tilde{\tau}_R^5 = \inf\{t\in[0,T]; (\|u_\eta\|^2_{H^{s+k}} + \|\rho_\eta\|^2_{H^{s+k-1}}) > R\eta^{-\frac{k}{2}},\ k=1,2\}, \text{ and}$$
$$\tilde{\tau}_R^6 = \inf\{t\in[0,T]; \|h_1(r)\|^2_{H^q} + \|h_2(r)\|^2_{H^{q-1}} > R\varepsilon^{\frac{s-q}{2}}\}.$$

Let $\tilde{\tau}_R = \tilde{\tau}_R^1\wedge\tilde{\tau}_R^4\wedge\tilde{\tau}_R^5\wedge\tilde{\tau}_R^6$. Using Itô formula on $\|h_1(t\wedge\tilde{\tau}_R)\|^2_{H^s}$, there is:

$$\begin{aligned}
&\|h_1(t\wedge\tilde{\tau}_R)\|^2_{H^s}\\
&= \|h_{10}\|^2_{H^s} + 2\int_0^{t\wedge\tilde{\tau}_R} (\Lambda^s h_1, \Lambda^s(\varepsilon u_{\varepsilon xx} - \eta u_{\eta xx}))d\tau\\
&\quad - \int_0^{t\wedge\tilde{\tau}_R} (\Lambda^s h_1, \Lambda^s(h_1 v_1)_x)d\tau + 2\int_0^{t\wedge\tilde{\tau}_R} (\Lambda^s h_1, \Lambda^s(f(u_\varepsilon,\rho_\varepsilon) - f(u_\eta,\rho_\eta)))d\tau\\
&\quad + 2\int_0^{t\wedge\tilde{\tau}_R}\int_{\mathbb{Z}} (\Lambda^s h_1, \Lambda^s(g_1(u_\varepsilon,\rho_\varepsilon,z) - g_1(u_\eta,\rho_\eta,z)))\tilde{N}_1(d\tau,dz)\\
&\quad + \int_0^{t\wedge\tilde{\tau}_R}\int_{\mathbb{Z}} \|g_1(u_\varepsilon,\rho_\varepsilon,z) - g_1(u_\eta,\rho_\eta,z)\|^2_{H^s}\nu_1(d\tau,dz).
\end{aligned} \tag{3.50}$$

Similar to (3.41), (3.45) and (3.46), we have

$$\int_0^{t\wedge\tilde{\tau}_R} (\Lambda^s h_1, \Lambda^s(\varepsilon u_{\varepsilon xx} - \eta u_{\eta xx}))d\tau \le CR\varepsilon + C\int_0^{t\wedge\tilde{\tau}_R} \|h_1\|^2_{H^s}d\tau, \tag{3.51}$$

$$\begin{aligned}
&\int_0^{t\wedge\tilde{\tau}_R} (\Lambda^s h_1, \Lambda^s(f(u_\varepsilon,\rho_\varepsilon) - f(u_\eta,\rho_\eta)))d\tau\\
&\qquad \le C(R)\int_0^{t\wedge\tilde{\tau}_R} (\|h_1\|^2_{H^s} + \|h_2\|^2_{H^{s-1}})ds,
\end{aligned} \tag{3.52}$$

and

$$\begin{aligned}
&\mathbb{E} \sup_{t\in[0,T\wedge\tilde{\tau}_R]} \int_0^{t\wedge\tilde{\tau}_R}\int_{\mathbb{Z}} (\Lambda^s h_1, \Lambda^s(g_1(u_\varepsilon,\rho_\varepsilon,z) - g_1(u_\eta,\rho_\eta,z))\tilde{N}_1(d\tau,dz))\\
&\le \frac{1}{2}\mathbb{E} \sup_{t\in[0,T\wedge\tilde{\tau}_R]} \|h_1\|^2_{H^s} + C\int_0^T \mathbb{E} \sup_{r\in[0,\tau\wedge\tilde{\tau}_R]} (\|h_1(r)\|^2_{H^s} + \|h_2(r)\|^2_{H^{s-1}})d\tau.
\end{aligned} \tag{3.53}$$

Finally, by Lemma 3.2 and Cauchy inequality, we can gain

$$\begin{aligned}
&\int_0^{t\wedge\tilde{\tau}_R} (\Lambda^s h_1, \Lambda^s(h_1 v_1)_x) d\tau \\
=&\int_0^{t\wedge\tilde{\tau}_R} (\Lambda^s h_1, \Lambda^s(h_{1x} v_1)) d\tau + \int_0^{t\wedge\tilde{\tau}_R} (\Lambda^s h_1, \Lambda^s(h_1 v_{1x})) d\tau \\
=&\int_0^{t\wedge\tilde{\tau}_R} (\Lambda^s h_1, [\Lambda^s, v_1] h_{1x}) d\tau + \int_0^{t\wedge\tilde{\tau}_R} (\Lambda^s h_1, v_1 \Lambda^s h_{1x}) d\tau \\
&+ \int_0^{t\wedge\tilde{\tau}_R} (\Lambda^s h_1, [\Lambda^s, h_1] v_{1x}) d\tau + \int_0^{t\wedge\tilde{\tau}_R} (\Lambda^s h_1, h_1 \Lambda^s v_{1x}) d\tau \\
\leq& C\int_0^{t\wedge\tilde{\tau}_R} \|h_1\|_{H^s}(\|h_1\|_{H^s}\|v_1\|_{H^s} + \|h_1\|_{L^\infty}\|v_1\|_{H^{s+1}}) d\tau \\
\leq& C\varepsilon^{\frac{s-q-1}{2}} + C(R)\int_0^{t\wedge\tilde{\tau}_R} \|h_1\|_{H^s}^2 d\tau.
\end{aligned} \tag{3.54}$$

It follows from (3.50)–(3.54) that

$$\begin{aligned}
\mathbb{E}\sup_{t\in[0,T\wedge\tilde{\tau}_R]} \|h_1\|_{H^s}^2 \leq& 2\|h_{01}\|_{H^s}^2 + C\varepsilon^{\frac{s-q-1}{2}} \\
&+ C\int_0^T \mathbb{E}\sup_{r\in[0,\tau\wedge\tilde{\tau}_R]} (\|h_1(r)\|_{H^s}^2 + \|h_2(r)\|_{H^{s-1}}^2) d\tau,
\end{aligned}$$

and similarly, it has

$$\begin{aligned}
\mathbb{E}\sup_{t\in[0,T\wedge\tilde{\tau}_R]} \|h_2\|_{H^{s-1}}^2 \leq& 2\|h_{02}\|_{H^{s-1}}^2 + C\varepsilon^{\frac{s-q-1}{2}} \\
&+ C\int_0^T \mathbb{E}\sup_{r\in[0,\tau\wedge\tilde{\tau}_R]} (\|h_1(r)\|_{H^s}^2 + \|h_2(r)\|_{H^{s-1}}^2) d\tau.
\end{aligned}$$

Hence, we can get that

$$\mathbb{E}\sup_{t\in[0,T\wedge\tilde{\tau}_R]} (\|h_1\|_{H^s}^2 + \|h_2\|_{H^{s-1}}^2) \to 0 \quad as \quad \varepsilon \to 0. \tag{3.55}$$

Then, $\exists\ \eta > 0$, by Lemma 3.3 that there exists a constant $R > 0$ such that

$$\begin{aligned}
&P(\tilde{\tau}_R \leq T) \\
\leq& P(\sup_{t\in[0,T]} (\|u_\varepsilon\|_{H^s}^2 + \|\rho_\varepsilon\|_{H^{s-1}}^2 + \|u_\eta\|_{H^s}^2 + \|u_\eta\|_{H^{s-1}}^2) > R) \\
&+ P(\sup_{t\in[0,T]} \|u_\varepsilon\|_{H^{s+k}}^2 + \|\rho_\varepsilon\|_{H^{s+k-1}}^2 > R\varepsilon^{-\frac{k}{2}}) \\
&+ P(\sup_{t\in[0,T]} \|u_\eta\|_{H^{s+k}}^2 + \|\rho_\eta\|_{H^{s+k-1}}^2 > R\eta^{-\frac{k}{2}}) \\
&+ P(\sup_{t\in[0,T]} \|h_1\|_{H^q}^2 + \|h_2\|_{H^{q-1}}^2 > R\eta^{\frac{s-q}{2}}) \\
\leq& \frac{\eta}{2}.
\end{aligned}$$

To these R and $\forall\lambda > 0$, because of (3.55), there must exist ε_0 that fits $\varepsilon < \varepsilon_0$,

$$P(\sup_{t\in[0,T]} \|h_1(t\wedge\tau_R)\|_{H^s} + \|h_2(t\wedge\tau_R)\|_{H^{s-1}} > \lambda) \leq \frac{\eta}{2}.$$

Therefore

$$P(\sup_{t\in[0,T]} \|h(t)\|_{H^s} > \lambda) \leq P(\tilde{\tau}_R \leq T) + P(\sup_{t\in[0,T]} \|h(t\wedge\tilde{\tau}_R)\|_{H^s} > \lambda) \leq \eta.$$

The whole article above proves that h converges to 0 in probability in $\mathcal{D}([0,T]; H^s \times H^{s-1})$. The proof of Proposition 9.1 is completed.

According to Proposition 9.1, from which we can gain the well-posedness of (1.3).

Acknowledgment

This work is partially supported by China NSF Grant Nos. 11531006, 11401532, 11671359, and Zhejiang Provincial NSF of China under Grant No. LQ15A010012, LY18A010027.

References

[Dullin et al. (2001)] R. Dullin, G. Gottwald and D. Holm, An integrable shallow water equation with linear and nonlinear dispersion, Phys. D. Rev. Lett., 87(9), 4501-4504, 2001.

[Camassa and Holm (1993)] R. Camassa and D. Holm, An integrable shallow water equation with peaked solitons, Phys. Rev. Lett., 71, 1661-1664, 1993.

[Tian et al. (2005)] L. X. Tian, G. L. Gui and Y. Liu, On the well-posedness problem and the scattering problem for DGH equation, Comm. Math. Phys., 257(3), 667-701, 2005.

[Yin (2004)] Z. Yin, Well-posedness, global existence and blowup phenomena for an integrable shallow water equation, Discrete Contin. Dyn. Syst., 10 393-411, 2004.

[Zhang and Yin (2010)] S. Zhang and Z. Yin, Global weak solutions for the Dullin–Gottwald–Holm equation, Nonlinear Anal. TMA, 72 1690-1700, 2010.

[Constantin and Ivanov (2010)] A. Constantin and R. Ivanov, On an integrable two-component Camassa–Holm shallow water system, Phys. Lett. A, 372, 7129-7132, 2008.

[Guo and Gao (2012)] F. Guo, H. Gao and Y. Liu, On the wave-breaking phenomena for the two-component Dullin–Gottwald–Holm system, J. Lond. Math. Soc., 86, 810-834, 2012.

[Chen et al. (2013)] Y. Chen, H. Gao and Y. Liu, On the Cauchy problem for the two-component Dullin–Gottwald–Holm system, Discrete Contin. Dyn. Syst., 33, 3407-3441, 2013.

[Liu and Yin (2013)] X. X. Liu and Z. Y. Yin, Local well-posedness and stability of solitary waves for the two-component Dullin–Gottwald–Holm system, Nonlinear Analysis, 88, 1-15, 2013.

[Chemin (1998)] J. Y. Chemin, Perfect Incompressible Fluids, Oxford Univ. Press, New York, 1998.

[Lemarie-Rieusset (2002)] P. G. Lemarie-Rieusset, Recent Developments in the Navier–Stokes Problem, Res. Notes Math., Chapman and Hall/CRC, 2002.

[Peszat and Zabczyk (2007)] S. Peszat and J. Zabczyk, Stochastic Partial Differential Equations with Lévy Noise, Encyclopedia of Mathematics and its Applications, Cambridge: Cambridge University Press, 2007.

[Bona and Smith (1975)] J. L. Bona and R. Smith, The initial value problem for the Korteweg–de Vries equation, Phil. Trans. Roy. Soc. London A, 278, 555-601, 1975.

[Kato and Ponce (1988)] T. Kato and G. Ponce, Commutator estimayes and the Euler and Navier–Stokes equation, Commun. Pure Appl. Math., 41, 891-907, 1988.

Chapter 10

On Smooth Approximation of Lévy Processes in Skorokhod Space

Lingyu Feng* and Xianming Liu†
School of Mathematics and Statistics, Huazhong University of Sciences and Technology, Wuhan, 430074, P.R. China
**lyfeng@hust.edu.cn; †xmliu@hust.edu.cn*

In this paper, we study the smooth approximation of arbitrary Lévy processes. Such approximation process is known as a time average approximation, and is suitable for dealing with stochastic flow problems. Our first result shows that the smooth approximation processes converges to Lévy processes in Skorokhod space. Next, we explore that if the drift term of a stochastic differential equation (abbreviated as SDE) satisfies usual Lipschitz constrains and a linear growth condition, then the solution of the SDE converges in Skorokhod space. As a corollary, the first passage time for the solution of approximation system converges to that of original system in law.
Keywords. Skorokhod topology; Tightness; Smooth Approximation of Lévy processes; First passage time

1. Introduction

Let L_t be Lévy process defined on the probability space $(\Omega, \mathcal{F}, \mathbb{P})$. Consider the smooth approximation of L_t [4, 10]

$$L_t^\epsilon := \frac{1}{\epsilon}\int_{t-\epsilon}^{t} L_s ds = \int_0^1 L_{t-\epsilon s} ds, \quad \epsilon > 0, \tag{1.1}$$

and the stationary approximation

$$L_t^\epsilon := \frac{1}{\epsilon}\int_0^t L(s) - L(s-\epsilon) ds = \frac{1}{\epsilon}\int_{t-\epsilon}^{t} L_s ds - \frac{1}{\epsilon}\int_{-\epsilon}^{0} L_s ds, \quad \epsilon > 0. \tag{1.2}$$

It can be deduced that, for each fixed t, L_t^ϵ converges to L_t almost surely. Since the limit of continuous approximants in either the uniform or Skorokhod J_1 topologies must be continuous, so L_t^ϵ does not converge to L_t uniformly or in Skorokhod J_1 topology.

For a real valued stochastic process $x = \{x_t\}_{t\geq 0}$ and $a > 0$, let $\tau_a(x)$ denote the first passage time

$$\tau_a(x) := \inf\{t \geq 0 : x_t > a\}.$$

Then a central question is whether $\tau_a(L^\epsilon)$ converges to $\tau_a(L)$. In this paper, we will show that L_t^ϵ converges to L_t in Skorokhod M_1 topology.

Theorem 10.1. *Let L_t be an arbitrary real valued Lévy process, L_t^ϵ be the smooth approximating process defined in* (1.1) *or stationary approximation defined in* (1.2)*, then*

$$L_t^\epsilon \to L_t \quad \textit{in the sense of probability in } D([0,\infty),\mathbb{R};M_1)$$

as ϵ tend to zero.

As a consequence, the first passage time of L_t^ϵ converges to that of L_t in distribution.

Theorem 10.2. *$\tau_a(L^\epsilon)$ converges to $\tau_a(L)$ in distribution.*

This paper is organized as follows. In next section, some basic concepts will be recalled. Then, we will review some results for Skorokhod spaces. In Section 3, we will prove Theorem 10.1. In Section 4, we will drive a simple "Wong–Zakai" type approximation in Skorokhod space.

2. On Skorokhod topology

In this section, we will review Skorokhold M_1 topology. Recall that for a function defined on a subset $\mathbb{T}$ of the real line $(-\infty,+\infty)$ to deserve the title cadlag, it must be right continuous and have left limits at each point of $\mathbb{T}$. Many of the well-studied stochastic processes have versions with cadlag sample paths. For $\mathbb{T}=[0,1]$, these processes can be analyzed as random elements of $D([0,1],\mathbb{R})$, the set of all real-valued cadlag functions on $[0,1]$.

Skorokhod define the M_1 metric using the **completed graphs** of the functions. For each $x \in D([0,1],\mathbb{R})$, the completed graph of x is the set

$$\Gamma_x := \{(t,z) \in [0,1]\times\mathbb{R} : z \in [[x(t-),x(t)]]\},$$

where $x(0-) := x(0)$ and $[[z_1,z_2]]$ is the segment of a straight line between z_1 and z_2:

$$[[z_1,z_2]] := \{z \in \mathbb{R} : z = az_1 + (1-a)z_2 \quad \text{for some} \quad a \in [0,1]\}.$$

Order on Γ_z is given by

$$(t_1,z_1) \le (t_2,z_2) \Leftrightarrow t_1 < t_2 \quad \text{or} \quad t_1 = t_2 \text{ and } |x(t_1-) - z_1| \le |x(t_2-) - z_2|.$$

A parametric representation of the completed graph Γ_x is a continuous nondecreasing function

$$(r,u) : [0,1] \to \Gamma_z, \quad (r,u)(0) = (0,z(0)), (r,u)(1) = (1,z(1)).$$

Let $\prod(x)$ be the set of all parametric representation of x.

For any $x, y \in D([0,1],\mathbb{R})$, Skorokhod [6] defines

$$d_{M_1}(x,y) = \inf_{(r_1,u_1)\in\prod(x),(r_2,u_2)\in\prod(y)} \{ \sup_{0\le t\le 1} |r_1(t) - r_2(t)| + \sup_{0\le t\le 1} |u_1(t) - u_2(t)|\}.$$

The topology generated by the metric d_{M_1} is called Skorokhod M_1 topology. Note that $D([0,1],\mathbb{R}; M_1)$ is separable but not complete (Page 274 of [5]). Moreover, its topologically equivalent metric $d_{M_1'}$ is complete (Theorem 12.8.1 of [7]). $D([0,T],\mathbb{R}; M_1)$ can be defined similarly. The metric of this space is denoted by $d_{M_1,T}$.

Example 10.1. Let $x(t) = I_{[\frac{1}{2},1]}(t)$, $x^n(t) = n(t - \frac{1}{2} + \frac{1}{n})I_{[\frac{1}{2}-\frac{1}{n},\frac{1}{2}]}(t) + I_{[\frac{1}{2},1]}(t)$, then $x^n(t) \to x(t)$ in $D([0,T],\mathbb{R}; M_1)$ but not $D([0,T],\mathbb{R}; J_1)$ as n tend to infinity.

Now, we turn to the Skorokhod topology on $D([0,\infty),\mathbb{R})$. The Skorokhod M_1 topology in $D([0,\infty),\mathbb{R}])$ is then induced by the metric

$$d_{M_1,\infty}(x,y) = \int_0^\infty e^{-t}(1 \wedge d_{M_1,t}(x,y))dt.$$

The Skorokhod J_1 and uniform topology in $D([0,\infty),\mathbb{R})$ can be defined similarly.

Let $Disc(x)$ be the discontinuous point of function x. The relationship between $D([0,\infty),\mathbb{R}; M_1)$ and $D([0,t],\mathbb{R}; M_1)$ has been exploited (Theorem 12.9.3 of [7]).

Proposition 10.1.

$$d_{M_1,\infty}(x^n, x) \to 0 \Leftrightarrow d_{M_1,t}(x^n, x) \to 0, \ \textit{for each } t \in \overline{Disc(x)}.$$

Let

$$w'(x,\delta) := \sup_{0\le t_1<t<t_2\le 1; t_2-t_1<\delta} \inf_{a\in[0,1]} |x(t) - (ax(t_1) + (1-a)x(t_2))|,$$

where $\inf_{a\in[0,1]} |x(t) - (ax(t_1) + (1-a)x(t_2))|$ is the distance between $x(t)$ and segment $[[x(t_1) \wedge x(t_2), x(t_1) \vee x(t_2)]]$.

The following tightness criterion is well known.

Proposition 10.2. *$\{X^n : n \ge 1\}$ is tight with respect to the M_1 topology iff*
1. $\lim_{k\to\infty} \limsup_n \mathbb{P}(\sup_{t\in[0,1]} |x^n(t)| > k) = 0$;
2. For any fixed $\Delta > 0$, $\lim_{\delta\to 0+} \limsup_n \mathbb{P}(w'(x^n,\delta) > \Delta) = 0$.

Above proposition has been enhanced as follows (Theorem 3.2 of [5]).

Proposition 10.3. *X^n converges to X in $D([0,\infty),\mathbb{R}; M_1)$ in probability iff $X^n(t)$ converges to $X(t)$ in probability for all t and $\{X^n : n \ge 1\}$ is tight with respect to the topology.*

3. Proof of Theorem 10.1

In this section, we will prove Theorem 10.1.

Proof. We prove the condition that L_t^ϵ is the smooth approximation defined in (1.1) and the proof of stationary approximation defined in (1.2) is analogous. By Lévy–Itô decomposition, we represent L_t as a sum of a continuous Wiener process σW_t and a Lévy process V_t without Gaussian part. Owing to the linearity of the smooth approximation, we describe L_t^ϵ as a sum

$$L_t^\epsilon = W_t^\epsilon + V_t^\epsilon := \frac{1}{\epsilon}\int_{t-\epsilon}^t \sigma W_s ds + \frac{1}{\epsilon}\int_{t-\epsilon}^t V_s ds.$$

For the continuous Wiener process, we prove that W_t^ϵ converges to σW_t in the sense of probability in $D([0,\infty),\mathbb{R};M_1)$ as ϵ tends to zero. In fact,

$$Y_t^\epsilon := W_t^\epsilon - \sigma W_t = \int_0^1 \sigma W_{t-\epsilon s} ds - \int_0^1 \sigma W_t ds = \sigma\int_0^1 (W_{t-\epsilon s} - W_t).$$

In the uniform topology, for an arbitrary $T>0, t\in[0,T]$, $W_{t-\epsilon s}$ almost surely converges to W_t as ϵ tends to zero. Skorokhod M_1 topology is weaker than uniform topology. These show that W_t^ϵ converges to σW_t in the sense of probability in $D([0,\infty),\mathbb{R};M_1)$ as ϵ tends to zero. For the Lévy process V_t without Gaussian part, we consider the convergency of V_t. We first prove that for any $T>0$ and $\mathbb{P}\kappa>0$, there exists

$$\lim_{\delta\downarrow 0}\limsup_{\epsilon\to 0}\mathbb{P}(\sup_{0\le t_1<t<t_2\le T, t_2-t_1\le\delta} M(V_{t_1}^\epsilon, V_t^\epsilon, V_{t_2}^\epsilon) > \kappa) = 0 \tag{3.1}$$

with

$$M(V_{t_1}^\epsilon, V_t^\epsilon, V_{t_2}^\epsilon) = \begin{cases} \min\{|V_t^\epsilon - V_{t_1}^\epsilon|, |V_t^\epsilon - V_{t_2}^\epsilon|\}, & V_t^\epsilon \notin [[V_{t_1}^\epsilon, V_{t_2}^\epsilon]] \\ 0, & V_t^\epsilon \in [[V_{t_1}^\epsilon, V_{t_2}^\epsilon]] \end{cases}.$$

Without loss of generality, let $T>0$ be fixed from now on. When $\kappa>0$ is fixed, the equivalent form of (3.1) is that for any $\theta>0$, there exists $\delta_0=\delta_0(\kappa,\theta,T)>0$ such that for any $\delta\in(0,\delta_0]$, there is $\epsilon_0=\epsilon_0(\delta,\kappa,\theta,T)>0$ such that for any $\epsilon<\epsilon_0$, the inequality

$$\mathbb{P}(\sup_{0\le t_1<t<t_2\le T, t_2-t_1\le\delta} M(V_{t_1}^\epsilon, V_t^\epsilon, V_{t_2}^\epsilon) > \kappa) \le \theta \tag{3.2}$$

holds true. We prove (3.2) by three steps.

- First, we reduce V_t to a compound Poisson process. In other words, we decompose V_t into a sum of a martingale with bounded jumps and small variance and a compound Poisson process with drift.

There exists $a=a(\kappa,\theta,T)$ such that $\nu(\{a\})=\nu(\{-a\})=0$ and

$$\frac{16T}{\kappa^2}\int_{\{|y|<a\}} y^2\nu(dy) \le \frac{\theta}{4}$$

For this a, denote $\mu_a := \mu - \int_{\{a\leq|y|\leq 1\}y\nu(dy)}$, and consider the independent Lévy processes

$$\eta_t := \Sigma_{s\leq t}\Delta V_s \mathbb{I}_{(\Delta V_s \geq a)} + \mu_a t,$$

$$\xi_t := V_t - \eta_t.$$

Furthermore, $V_t = \eta_t + \xi_t$, η_t is a compound Poisson process with the drift μ_a, and ξ_t is a zero mean martingale with the variance $\mathbb{E}\xi_t^2 = t\int_{\{|y|<a\}y^2\nu(dy)}$. With linearity, there is

$$V_t^\epsilon = V_t^{\epsilon,\eta} + V_t^{\epsilon,\xi},$$

with

$$V_t^{\epsilon,\eta} = \int_0^1 \eta_{t-\epsilon s} ds, \quad V_t^{\epsilon,\xi} = \int_0^1 \xi_{t-\epsilon s} ds.$$

Let D_ξ be the event $\{\omega : \sup_{t\in[0,T]} |V_t^{\epsilon,\xi}| < \frac{\kappa}{4}\}$. For any $\epsilon > 0$, we obtain that

$$\sup_{t\in[0,T]} |V_t^{\epsilon,\xi}| = \sup_{t\in[0,T]} \left| \int_0^1 \xi_{t-\epsilon s} ds \right| \leq \int_0^1 \sup_{t\in[0,T]} |\xi_t| ds = \sup_{t\in[0,T]} |\xi_t|.$$

Using the Doob martingale inequality, we acquire that

$$\mathbb{P}(D_\xi^c) \leq \mathbb{P}(\sup_{t\in[0,T]} |V_t^{\epsilon,\xi}| > \frac{\kappa}{4}) \leq \frac{16T\mathbb{E}|\xi_T|^2}{\kappa^2} \leq \frac{16T}{\kappa^2} \int_{\{|y|<a\}} y^2\nu(dy) \leq \frac{\theta}{4}.$$

This implies that for all $\omega \in D_\xi$ and $\epsilon > 0$, the inequality

$$\sup_{t\in[0,T]} |V_t^\epsilon - V_t^{\epsilon,\eta}| = \sup_{t\in[0,T]} |V_t^{\epsilon,\xi}| < \frac{\kappa}{4}$$

holds true. Thus, for all $\epsilon > 0$ and $0 \leq t_1 < t < t_2 \leq T$,

$$|M(V_{t_1}^\epsilon, V_t^\epsilon, V_{t_2}^\epsilon) - M(V_{t_1}^{\epsilon,\eta}, V_t^{\epsilon,\eta}, V_{t_2}^{\epsilon,\eta})| \leq \frac{\kappa}{2}.$$

• Second, consider the local extremum of $V_t^{\epsilon,\eta}$. Let $z = z(a, \theta, T) > 0$ be a level such that for the event

$$D_\eta := \left\{\omega : \sup_{t\in[0,T]} |\eta_t| \leq \frac{z}{2}\right\},$$

there exists $\mathbb{P}(D_\eta^c) \leq \frac{\theta}{4}$. Under this condition, for $\omega \in D_\eta$, the absolute value of jump sizes of η_t are no larger than z, that is $\sup_{t\in[0,T]} |\Delta\eta_t(\omega)| \leq z$. The finite intensity of η_t is

$$\lambda_a = \int_{|y|\geq a} \nu(dy) < \infty.$$

For the Lévy process η_t, consider its counting jump process Q_t. Q_t is a Poisson process with intensity λ_a. We use $\{\tau_k\}$, $\tau_0 = 0$ to represent the sequence of arrival times of η_t and denote the sequence of its jump sizes by $\{J_k\}$, $J_0 = 0$, that is $J_k = \eta_{\tau_k} - \eta_{\tau_{k-}}$. Hence, the process $V_t^{\epsilon,\eta}$ has the following path-wise representation:

$$V_t^{\epsilon,\eta} = \sum_{k=0}^{Q_t} J_k - \sum_{k=Q_{t-\epsilon}}^{Q_t} \frac{t-\tau_k-\epsilon}{\epsilon} J_k + \mu_a(t-\frac{\epsilon}{2}), \quad t \geq 0. \tag{3.3}$$

Let $m^* \geq 0$ be such that

$$\mathbb{P}(Q_T \leq m^*) > 1 - \frac{\theta}{4}.$$

Moreover, for $\beta > 0$, $m = 0, 1, ..., m^*$, consider the events

$$C_\beta^0 := C^0 = \{\omega : Q_T = 0\};$$

$$\begin{aligned} C_\beta^m : &= \{\omega : Q_T = m\} \cap \{\omega : \tau_k - \tau_{k-1} \geq 2\beta, k = 1, \ldots, m, \\ &\quad \text{and } \ T - \tau_m \geq 2\beta\}, \quad m = 1, ..., m^*; \end{aligned}$$

$$C_\beta^{m^*} := \bigcup_{m=0}^{m^*} C_\beta^m \subset \{Q_T \leq m^*\}.$$

On the condition $\{Q_T = m\}$, the jump times $\tau_1, ..., \tau_m$ are distributed on the interval $[0, T]$ obeying the probability law of the order statistics obtained from m samples of the uniform distribution on $[0, T]$. It is easy to see that there is $\beta = \beta(\theta, \kappa, T, m^*) > 0$ small enough such that

$$\mathbb{P}(C_\beta) > 1 - \frac{\theta}{3}.$$

For a fixed m, $m = 0, 1, ..., m^*$, consider $\epsilon_m < \beta$, $\omega \in D_\eta \cap C_\beta$. Under the condition $\epsilon < \epsilon_m$, by (3.3), we obtain the representation

$$V_t^{\epsilon,\eta}(\omega) = \begin{cases} \mu_a(t-\frac{\epsilon}{2}), & t \in [0, \tau_1(\omega)) \\ \sum_{j=0}^{k-1} J_j(\omega) - \frac{t-\tau_k}{\epsilon} J_k + \mu_a(t-\frac{\epsilon}{2}), & t \in [\tau_k(\omega), \tau_k(\omega)+\epsilon), \\ & k = 1, ..., m \\ \sum_{j=1}^{k-1} J_j(\omega) + \mu_a(t-\frac{\epsilon}{2}), & t \in [\tau_{k-1}(\omega)+\epsilon, \tau_k(\omega)), \\ & k = 2, ..., m \\ \sum_{j=1}^{m} J_j(\omega) + \mu_a(t-\frac{\epsilon}{2}), & t \in [\tau_m(\omega), T] \end{cases}.$$

The process $V_t^{\epsilon,\eta}$ has smooth paths on every interval. Next, we will show that when ϵ is small enough, the path of $V_t^{\epsilon,\eta}$ is either monotone or has at most one local extremum on each interval. In fact, $V_t^{\epsilon,\eta}$ is obviously monotone on $[0, \tau_1(\omega))$, $[\tau_{k-1}(\omega)+\epsilon, \tau_k(\omega)), k = 2, ..., m$, and $[\tau_m(\omega), T]$.

Now let $m = 1, ..., m^*$. For $t \in [\tau_k(\omega), \tau_k(\omega) + \epsilon), k = 1, ..., m$, consider the derivative of $V_t^{\epsilon,\eta}$ with regard to t

$$\frac{d}{dt} V_t^{\epsilon,\eta} = -\frac{J_k}{\epsilon} + \mu_a.$$

Due to $\omega \in D_\eta$, the jump sizes J_k are bounded, $|J_k(\omega)| \in [a, z]$. And $\omega \in C_\beta$ implies that the arrival times are separated by 2β. If $\mu_a = 0$ holds true, for $-\frac{J_k}{\epsilon}$, $k = 1, ..., m$, the sign is determined and $V_t^{\epsilon,\eta}$ is monotone. If $\mu_a \neq 0$ holds, we can choose a non-random $\epsilon_m = \epsilon_m(a, z, \beta, m)$ such that for $\epsilon < \epsilon_m$, the inequality $|\frac{J_k}{\epsilon}| < |\mu_a|$ is right. In other words, $\frac{d}{dt} V_t^{\epsilon,\eta}$ has the same sign as μ_a. So $V_t^{\epsilon,\eta}$ is monotone. Besides, it is known that when ϵ_m is small enough, for $\epsilon < \epsilon_m$, we obtain

$$\max_{t\in[0,T]} |\frac{d}{dt} V_t^{\epsilon,\eta} - \mu_a| \le \frac{\kappa}{4}.$$

Additionally for $\mu_a \neq 0$, we assume that for $\epsilon < \epsilon_m$

$$\max_{t\in[0,T]} |\frac{d}{dt} V_t^{\epsilon,\eta}| \le 2|\mu_a|.$$

Generally, for $\epsilon < \epsilon_m$ and for $\omega \in D_\eta \cap C_\beta$, the paths of $V_t^{\epsilon,\eta}$ have the following structure: $V_t^{\epsilon,\eta}$ are continuous on $[0, T]$, smooth and monotone on $[0, \tau_1(\omega))$, $[\tau_k(\omega), \tau_k(\omega) + \epsilon), k = 1, ..., m$, $[\tau_{k-1}(\omega) + \epsilon, \tau_k(\omega)), k = 2, ..., m$, $[\tau_m(\omega), T]$ and may have extrema at the boundaries of the intervals. The derivative of $V_t^{\epsilon,\eta}$ is bounded by a constant $2|\mu_a|$. Let $\epsilon^* = \wedge_{m=1}^{m^*} \epsilon_m$.

• Third, estimate the value of $M = M(V_{t_1}^{\epsilon,\eta}, V_t^{\epsilon,\eta}, V_{t_2}^{\epsilon,\eta})$. Let $\delta_0 \in (0, \beta \wedge \frac{\kappa}{8(|\mu_a|+1)})$, $\epsilon \in (0, \epsilon^*)$, $\omega \in D_\eta \cap C_\beta^m$, $m = 0, ..., m^*$. We consider two cases for $0 \le t_1 < t < t_2 \le T, t_2 - t_1 \le \delta \le \delta_0$.

(1) If the path of $t \mapsto V_t^{\epsilon,\eta}(\omega)$ is monotone on $[t_1, t_2]$, it is obvious that $M = 0$.

(2) If the path of $t \mapsto V_t^{\epsilon,\eta}(\omega)$ is not monotone on $[t_1, t_2]$, we assume that $\tau_n^* \in [t_1, t_2]$ is a local extremum and then

$$M \le \min\{|V_{\tau_n^*}^{\epsilon,\eta} - V_{t_1}^{\epsilon,\eta}|, |V_{\tau_n^*}^{\epsilon,\eta} - V_{t_2}^{\epsilon,\eta}|\} \le |V_{\tau_n^*}^{\epsilon,\eta} - V_{\tau_n^*+\delta_0}^{\epsilon,\eta}| \le 2|\mu_a|\delta_0 \le \frac{\kappa}{4}.$$

Overall, these imply that for all $\delta \le \delta_0$ and $\epsilon \in (0, \epsilon^*)$

$$\mathbb{P}(\sup_{0\le t_1<t<t_2\le T, t_2-t_1\le\delta} M(V_{t_1}^{\epsilon,\eta}, V_t^{\epsilon,\eta}, V_{t_2}^{\epsilon,\eta}) > \frac{\kappa}{2} | D_\eta \cap C_\beta) = 0,$$

and then

$$\begin{aligned}
&\mathbb{P}\left(\sup_{0\le t_1<t<t_2\le T, t_2-t_1\le\delta} M(V_{t_1}^{\epsilon}, V_t^{\epsilon}, V_{t_2}^{\epsilon}) > \kappa \right) \\
\le &\mathbb{P}\left(\sup_{0\le t_1<t<t_2\le T, t_2-t_1\le\delta} M(V_{t_1}^{\epsilon,\eta}, V_t^{\epsilon,\eta}, V_{t_2}^{\epsilon,\eta}) > \frac{\kappa}{2} | D_\eta \cap C_\beta \right) \\
&+ \mathbb{P}(D_\xi^c) + \mathbb{P}(D_\eta^c) + \mathbb{P}(C_\beta^c) \\
\le &0 + \frac{\theta}{4} + \frac{\theta}{4} + \frac{\theta}{3} < \theta.
\end{aligned}$$

(3.1) is proved.

For any $t \in [0, T]$, it is known that the process V_t^ϵ converges to V_t in the sense of probability as ϵ tends to zero. Combined with (3.1), for V_t^ϵ and V_t defined in the theorem, by the equivalent condition (Theorem 3.2 in [5]), we obtain that $V_t^\epsilon \to V_t$ in the sense of probability in $D([0,T), \mathbb{R}; M_1)$ as ϵ tends to zero. Furthermore, Whitt proved (Section 12.9 in [7]) that if for any $T > 0$, $V_t^\epsilon \xrightarrow{\mathbb{P}} V_t$ as $\epsilon \to 0$ in $D([0,T), \mathbb{R}; M_1)$ holds true, $V_t^\epsilon \xrightarrow{\mathbb{P}} V_t$ as $\epsilon \to 0$ in $D([0,\infty), \mathbb{R}; M_1)$ holds. □

4. Applications

Consider the following stochastic differential equation driven by additive Lévy process.

$$dX_t = f(X_t)dt + dL_t, \tag{4.1}$$

where f is Lipschitz continuous and satisfied growth condition.

For any Lévy process L_t, by standard Picard iteration, under local Lipschitz and growth conditions, Equation (4.1) admits a unique cadlag solution [1].

Let L_t^ϵ be a smooth approximation of L_t, such that L_t^ϵ converges to L_t in the sense of probability in $D([0,\infty), \mathbb{R}; M_1)$.

Consider the following smooth approximation equation

$$dX_t^\epsilon = f(_t^\epsilon)dt + dL_t^\epsilon. \tag{4.2}$$

Assume same initial data for (4.1) and (3.9), it can be shown that X_t^ϵ converges to X_t in the sense of probability in $D([0,\infty), \mathbb{R}; M_1)$. In fact, let $Y_t^\epsilon = X_t^\epsilon - L_t^\epsilon$ and $Y_t = X_t - L_t$, by the Grownwall inequality, we can show that Y_t^ϵ converges to Y_t in $C([0,\infty), \mathbb{R}; U)$. This is a "Wong–Zakai" type approximation result [8, 9].

Moreover, the first passage time $\tau_a(X^\epsilon)$ converges to $\tau_a(X)$ in distribution. It is clear that the distribution of the first passage time is determined by the running supremum of corresponding random process. The continuous mapping theorem implies the first passage time $\tau_a(X^\epsilon)$ converges to $\tau_a(X)$ in distribution. As a special case, Theorem 10.2 holds.

Acknowledgment

The authors acknowledge the support provided by NSFs of China (Nos. 11301197).

References

[1] Applebaum D. (2009). Lévy Processes and Stochastic Calculus, Cambridge University Press, 2nd edition, Cambridge, UK.

[2] Hintze R. and Pavlyukevich I. (2014). Small noise asymptotics and first passage times of integrated Ornstein–Uhlenbeck processes driven by α-stable Lévy processes, Bernoulli, 20, no. 1, pp. 265-281.

[3] Jacod J. and Shiryaev A. N. (2003). Limit Theorems for Stochastic Processes, Springer, Berlin, Heidelberg.

[4] Kurtz T. G., Pardoux E. and Protter P. (1985). Stratonovich stochastic differential equations driven by general semimartingales, Ann. Inst. Henri Poincaré Probab. Stat, 23, pp. 351-377.

[5] Pavlyukevich I. and Riedle M. (2015). Non-standard Skorokhod convergence of Lévy-driven convolution integrals in Hilbert spaces, Stochastic Analysis and Applications, 33(2), pp. 271-305.

[6] Skorokhod A. V. (1956). Limit theorems for stochastic processes, Theory of Probability and its Applications, 1(3), pp. 261-290.

[7] Whitt W. (2002). Stochastic-Process Limits: An Introduction to Stochastic-Process Limits and Their Application to Queues, Springer Science Business Media.

[8] Wong E. and Zakai M. (1965). On the relation between ordinary and stochastic differential equations, Internat. J. Engrg. Sci., 3, pp. 213-229.

[9] Wong E. and Zakai M. (1965). On the convergence of ordinary integrals to stochastic integrals, Ann. Math. Statist., 36, pp. 1560-1564.

[10] Zhang X. (2013). Derivative formulas and gradient estimates for SDEs driven by α-stable processes, Stochastic Processes and their Applications, 123(4), pp. 1213-1228.

Chapter 11

Error Estimation on Projective Integration of Expensive Multiscale Stochastic Simulation

Xiaopeng Chen[1,*] and A. J. Roberts[2]

[1]*Department of Mathematics, Shantou University, Shantou, Guangdong, 515063, P.R. China*
**xpchen@stu.edu.cn*

[2]*School of Mathematical Sciences, University of Adelaide, Adelaide, Australia*

Projective integration has the potential to be an effective method to compute the long time dynamic behaviour of multiscale systems. However, there is some intrinsic difficulty when the behaviour is described by a stochastic process. The improved projective integration of stochastic processes is explored for linear stochastic differential equations. The theory errors of projective integration are obtained by deriving the bias of the maximum likelihood estimation for the Ornstein–Uhlenbeck processes. The error estimation of parameters improves some results in either the statistic or econometrics literature. A number of examples are provided and Monte Carlo simulation demonstrates how the result is closed to the error formulas.

1. Introduction

Projective integration uses bursts of the microscale simulator, using microscale time steps, and computes an approximation to the system over a macroscale time step by extrapolation. We are concerned about the projective integration influenced by noise in the paper [4]. We estimate a linear stochastic differential equation from short bursts of data. The analytic solution of the linear stochastic differential equation then estimates the solution over a macroscale time step. We have explored how the noise affects the projective integration and presented two numerical algorithms in the empirical exploration. The simulation result shows that the two methods are stable and accurate. We consider a third method which is called Adams–Moulton inspired method in the present paper. It appears this version of an Adams–Moulton method is more robust to noise. Monte Carlo simulation result shows that the method is better in estimation of the errors of design parameters. We focus on the maximum likelihood estimation of the linear SDE model

$$dX = (a_0 + a_1 X)dt + (b_0 + b_1 X)dW\,, \tag{1.1}$$

for constants a_i and b_i.

*Corresponding author.

An early work of the projective integration simulate the errors of the design parameters [4], which depends on the bursts and macroscale time step. We give analysis errors of the parameters we estimate from the maximum likelihood in the present paper. First we analyse the errors of the estimation of the coefficients. There are some works about the estimation of the parameter from the stochastic differential equations [1, 2, 7, e.g.]. However, there are few works about the theory and technique for measuring the errors of parameters in the stochastic differential equations. The difficulty is to obtain a closed form of the parameter estimation. A discretisation approximation to the continuous time stochastic differential equations is used in estimation of the parameters by Kloeden et al. [5]. The method is to find an explicit expression for the likelihood function. Some weak convergence of the estimation parameters is obtained by Yoshida [9]. For approximation to the bias, a basic tool is by taking the stochastic Taylor expansions. Phillips and Tang et al. [6, 8] gave explicit expressions to the errors of some special stochastic differential equations, where the estimator is from the exact likelihood function. By using the Euler-like approximation, we present a bias and variance of the parameter estimation of Ornstein–Uhlenbeck processes. We compute the errors from a burst. Two examples are given and the Monte Carlo simulation result shows that the power law is quite closed to the analysis errors we compute. We also obtain a closed form expression of the parameter estimation from two bursts. A relationship of our results is the threshold autoregressive model. The threshold autoregressive model [3] is from the unequally spaced data in practice.

2. Adams–Moulton inspired method

In the former work we presented two versions of projective integration of the stochastic systems [4]. A variation on the theme of predictor corrector methods is analogous to deterministic Adams–Moulton methods. Here, instead of estimating the equation from just two bursts, the current and first predicted bursts, we also use the burst from the preceding time step. We fit the burst with the linear SDE equation (1.1).

(1) Execute the microscale simulator for a burst $[t_n, t'_n]$.
(2) For constant a_0 and b_0, take a projective integration time step of
$$\hat{X}_{n+1} = X'_n + a_0\Delta t + b_0\Delta W \quad \text{for some } \Delta W \sim N(0, \Delta t). \tag{2.1}$$
Then crudely predict to time t_{n+1} using (2.1).
(3) Compute a 'predictor' burst over $[t_{n+1}, t'_{n+1}]$.
(4) Fit the maximum likelihood estimation to the <u>three</u> bursts over $[t_{n-1}, t'_{n-1}]$, $[t_n, t'_n]$ and $[t_{n+1}, t'_{n+1}]$.
(5) The analytic solution is as follows: given $X(t'_n, \omega) = X'_n$,
$$\text{let } \varphi(t,\omega) = \exp\left[(a_1 - b_1^2/2)t + b_1 W(t)\right],$$
$$\text{then } X(t,\omega) = \varphi \times \left[\frac{X'_n}{\varphi(t'_n)} + (a_0 - b_0 b_1)\int_{t'_n}^{t} \frac{dt}{\varphi} + b_0 \int_{t'_n}^{t} \frac{dW}{\varphi}\right]. \tag{2.2}$$

The solution (2.2) then steps across a macroscale time to the numerical estimate

$$X_{n+1} = X(t_{n+1}, \omega) \quad \text{at time } t_{n+1} = t_n + \delta t + \Delta t\,. \tag{2.3}$$

(6) Correct the estimate of $X(t_{n+1}, \omega)$ via solution (2.2).

We apply the numerical algorithm to the stochastic differential equation

$$dX = -X\,dt + 0.03(1 + X)dW\,, \quad X(0, \omega) = 1\,. \tag{2.4}$$

Assessing errors of this scheme is a little more difficult as the scheme depends upon the previous step as well. Here we record a partial assessment of errors by taking two projective integration steps: the first using one long burst of the microscale process from time t_0, which is then split into two and treated as two regular bursts; and the second step, as in the previous paragraph, using a burst at time t_0 and a second burst from time t_1. Figure 1 shows the initial long burst from time $t = 0$ that is split into two for the initial step. We then assess the error in the second step which goes from $t \approx 0.5$ to $t \approx 1$. Figure 2 plots isosurfaces of the median absolute error when the algorithm, over a range of design parameters, was applied to the example (2.4). Roughly, the error fits

$$\text{MAE} \approx 0.046\rho^{0.02}(1 - r)^{2.1}(\Delta t)^{0.23}. \tag{2.5}$$

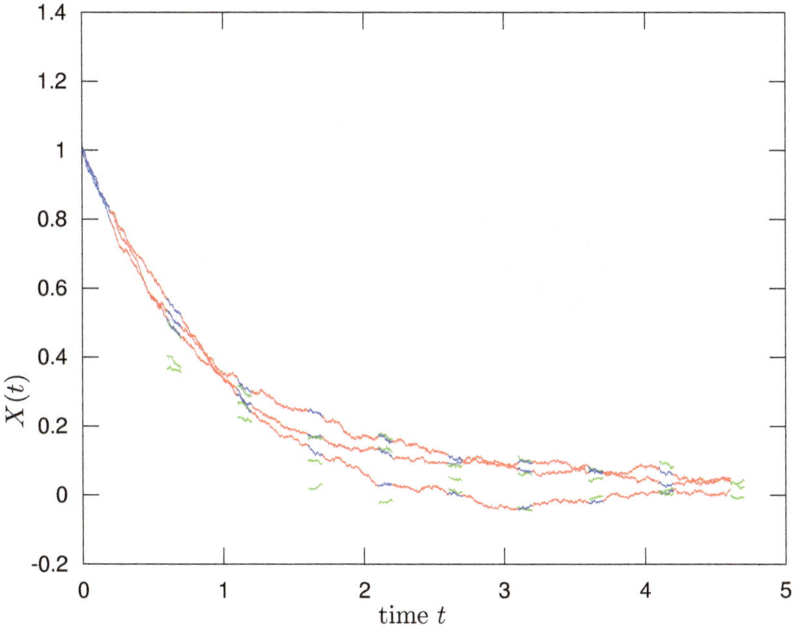

Fig. 1. Three realisations of the 'Adams–Moulton' projective integration applied to the stochastic differential equation (2.4) simulated with micro-time step $\eth t = 0.001$. In the projective integration: initial bursts (blue) lasted $\delta t = 0.1$; second bursts (green) started following the simple prediction (2.1); and lastly using (2.2) the stochastic projective time step (red) was $\Delta t = 0.4$.

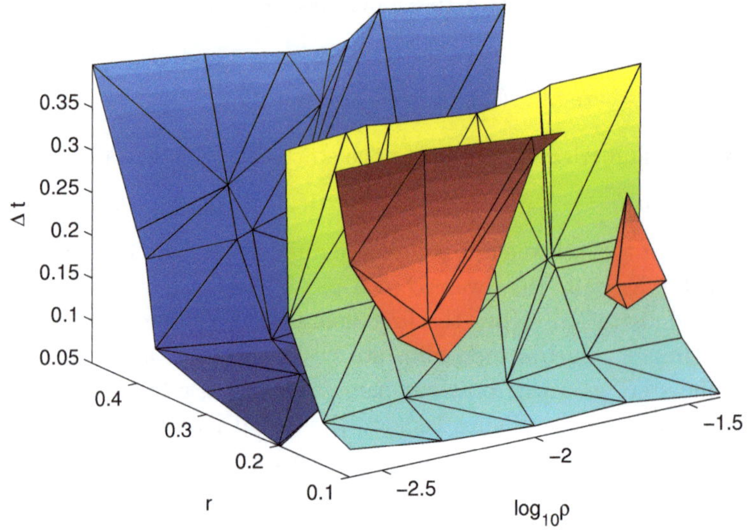

Fig. 2. Isosurfaces of an error in projective Adams–Moulton-like integration: the surfaces are for median absolute error MAE $= 0.01, 0.02, 0.03$ (in order of blue to red, obtained from 300 realisations). Roughly, the error fits MAE $\approx 0.046\rho^{0.02}(1-r)^{2.1}(\Delta t)^{0.23}$.

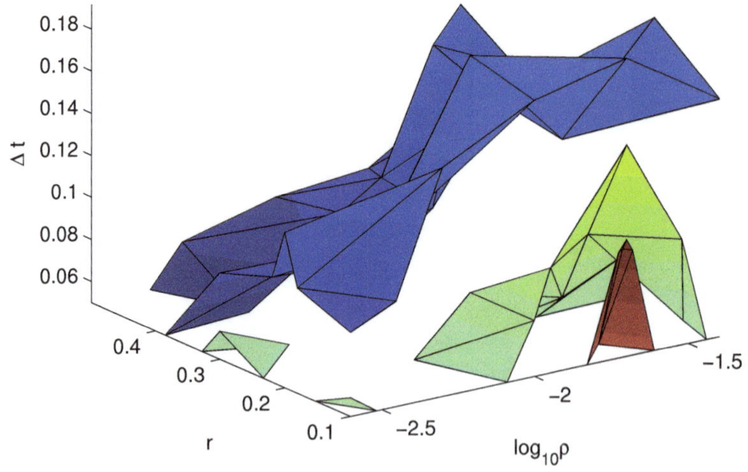

Fig. 3. Isosurfaces of the error ratio RMS / MAE in projective Adams–Moulton-like integration: the surfaces are for ratios $2, 6, 20$ (in order of blue to red). Note the change in the scale of the vertical axis.

Figure 2 plots the median absolute errors as these are robust to the outliers in any fat tailed error distribution. Figure 3 plots the ratio between the root mean square and the median absolute errors. Again the domain of design parameters above the blue surface indicates parameters for which there should be few, if any, outlier predictions.

3. The errors of the maximum likelihood estimation

Consider

$$dX = (aX + b)dt + c\,dW, \quad X(0) = x_0. \tag{3.1}$$

Then the solution is $X_t = e^{at}(x_0 + b\int_0^t e^{-as}ds + c\int_0^t e^{-as}dW_s)$. The solution X_t is a continuous Markov process with mean, variance function $E[X(t)] = \exp(at)E[X(0)]$, $\mathrm{Var}(X(t)) = -\frac{c^2}{2a} + (\mathrm{Var}(X(0)) + \frac{c^2}{2a})\exp(2at)$. If $x_0 \sim N(0, -\frac{c^2}{2a})$, then the process is a stationary Ornstein–Uhlenbeck process. If the distribution of initial condition is not $N(0, -\frac{c^2}{2a})$, we denote the process a non-stationary Ornstein–Uhlenbeck process. In this section we compute the errors of projective integration of the Ornstein–Uhlenbeck equation (3.1). We show the errors depend on the number of data, the burst and the projective integration time step.

The Euler scheme for the equation (3.1) is

$$X_{j+1} - X_j = (aX_j + b)\eth t_1 + c\eth W_j.$$

Then $X_{j+1} - X_j$ are independent Gaussian random variables with mean $(aX_j + b)\eth t_1$ and variance $c^2\eth t_1$. Therefore the transition density of the process is

$$p(y|x) = \frac{1}{\sqrt{2\pi c^2 \eth t_1}}\exp\Big(-\frac{1}{2}\frac{(y - x - (ax + b)\eth t_j)^2}{c^2\eth t_j}\Big).$$

The log-likelihood function for X_j, $j = 0, \dots, n$, is

$$\log L_n(a, b, c) = -\frac{1}{2}n\log 2\pi c^2 - \frac{1}{2}\sum \log \eth t_j - \frac{1}{2}\sum \frac{(X_{j+1} - X_j - (aX_j + b)\eth t_j)^2}{c^2\eth t_j}.$$

We find

$$\hat{a} = \frac{n}{\delta t}\frac{n\sum X_{j+1}X_j - n\sum X_j^2 - \sum(X_{j+1} - X_j)\sum X_j}{n\sum X_j^2 - (\sum X_j)^2};$$

$$\hat{b} = \frac{\sum X_{j+1} - \sum X_j - \hat{a}\eth t_1 \sum X_j}{\delta t};$$

$$\hat{c}^2 = \frac{1}{\delta t}\sum (X_{j+1} - X_j - (\hat{a}X_j + \hat{b})\eth t_j)^2,$$

where the data is from

$$X_{j+1} = (1 + a\eth t_1)X_j + b\eth t_1 + c\eth W_j. \tag{3.2}$$

The microscale increments $\eth X_j = X_{j+1} - X_j$, $j = 0, \dots, n-1$, from a burst of simulation of small time $\delta t = t_n - t_0$ and microscale time increments $\eth t_j = t_{j+1} - t_j$. We simply write $\sum_{j=0}^{n-1}$ as $\sum$ and consider the the same time span $\eth t_j$.

Theorem 11.1. *For the Ornstein–Uhlenbeck equation* (3.1)*, as $n \to \infty$, we have the following errors of the parameter estimation.*

$$E\hat{a} = a + C + O(n^{-1}).$$

$$\operatorname{Var}\hat{a} = D + O(n^{-1}).$$

$$E\hat{b} = b + C\frac{b}{a}(\frac{\exp(a\delta t) - 1}{a\delta t} - 1) + O(n^{-1}).$$

$$\operatorname{Var}\hat{b} = \frac{C^2c^2}{a^2\delta t^2}[\delta t + \frac{\exp(2a\delta t) - 1}{2a} - \frac{2(\exp(a\delta t) - 1)}{a}] + O(n^{-1}).$$

$$E\hat{c^2} = c^2 + F + O(n^{-1}).$$

$$\operatorname{Var}\hat{c^2} = \frac{3c^4}{\delta tn} + G + O(n^{-2}).$$

The constants C, D, F, G are defined in the following.

Proof. Let $\phi := 1 + a\eth t_1$. Without loss of generality, we assume $x_0 = 0$. From the discrete data, we compute

$$S(i,j) = EX_iX_j = \phi^{|i-j|}[\frac{c^2\eth t_1(1-\phi^{2j})}{1-\phi^2} + \frac{b^2}{a^2}(1-\phi^j)^2].$$

$$\sum EX_i = \frac{b}{a}(\frac{1-\phi^n}{1-\phi} - n).$$

$$\begin{aligned}\sum_{i>j} S(i,j) = &(\frac{c^2}{a}\frac{1}{1+\phi} + \frac{b^2}{a^2})(\frac{(n-1)\phi}{1-\phi} + \phi^n\frac{\phi-\phi^{-(n-2)}}{(1-\phi)^2})\\ &+(\frac{b^2}{a^2} - \frac{c^2}{a} - \frac{1}{1+\phi})(\frac{\phi-\phi^{2n-1}}{(1-\phi)^2(1+\phi)} - \frac{1-\phi^{n-1}}{(1-\phi)^2}\phi^n)\\ &-\frac{2b^2}{a^2}(\frac{\phi-\phi^n}{(1-\phi)^2} - \frac{(n-1)\phi}{1-\phi}).\end{aligned}$$

$$\begin{aligned}n^{-1}\sum EX_j^2 = &-\frac{c^2}{2a}[1 - \frac{\phi^{2n}-1}{2a\delta t}] + \frac{b^2}{a^2}[1 + \frac{2(1-\phi^n)}{a\delta t} + \frac{\phi^{2n}-1}{2a\delta t}]\\ &-(\frac{c^2}{8a\delta t} + \frac{b^2}{4a^3\delta t})\eth t_1(\phi^{2n}-1) + O(n^{-2}).\\ n^{-2}E(\sum X_j)^2 = &\frac{c^2}{a^2\delta t^2}[\delta t + \frac{\phi^{2n}-\phi}{2a} - \frac{2(\phi^n-\phi)}{a}]\\ &+\frac{b^2}{a^2}[1 + \frac{2(1-\phi^n)}{a\delta t} + \frac{(\phi^n-1)^2}{a^2\delta t^2}]\\ &-\frac{c^2}{\delta t^2}\eth t_1[\phi^{2n}-\phi] + O(n^{-2}).\end{aligned}$$

$$E(X_j\eth W_jX_i^2) = \begin{cases}0, & \text{if } i \le j;\\ 2c\eth t_1\phi^{i-j-1}(\phi^{2(i-j)-1}EX_j^2 & \\ \quad +\frac{b}{a}(\phi^j-1)EX_j), & \text{if } i > j.\end{cases}$$

Then

$$
\begin{aligned}
B_1 :&= \frac{1}{\eth t_1}\lim_{n\to\infty} n^{-2}\sum E\eth W_j X_i = c\lim_{n\to\infty} n^{-2}[\frac{n-1}{1-\phi} + \phi^n\frac{\phi-\phi^{-(n-1)}}{(1-\phi)^2}] \\
&= \frac{c(\exp(a\delta t)-1)}{a^2\delta t^2} - \frac{c}{a\delta t}. \\
B_2 :&= \frac{1}{\eth t_1}\lim_{n\to\infty} n^{-2}\sum_{i>j} E(X_j\eth W_j X_i^2) \\
&= \lim_{n\to\infty} 2cn^{-2}[\frac{c^2\phi^{-1}}{a(1-\phi^2)(1+\phi)}(\frac{\phi^2-\phi^{2n}}{1-\phi^2} - (n-1)\phi^{2n} - (n-1)\phi^2 \\
&\quad + \frac{\phi^2-\phi^{-2(n-2)}}{\phi^2-1}\phi^{2n}) + \phi^{-1}\frac{b^2}{a^2(1-\phi^2)}((n-1)\phi^2 \\
&\quad - \frac{\phi^2-\phi^{-2(n-2)}}{\phi^2-1}\phi^{2n} - 2\frac{\phi^2-\phi^{n+1}}{1-\phi} + 2\frac{\phi-\phi^{-n+2}}{\phi-1}\phi^{2n} \\
&\quad + \frac{\phi^2-\phi^{2n}}{1-\phi^2} - (n-1)\phi^{2n})].
\end{aligned}
$$

$$
\begin{aligned}
B_3 :&= \lim_{n\to\infty} n^{-2}E\sum X_j^2\sum X_i^2 \\
&= \frac{c^4}{a^2}(2\frac{\exp(2a\delta t)-1}{a^2\delta t^2} + 1) + \frac{2b^2c^2}{a^3}(2\frac{1-\exp(2a\delta t)}{a^2\delta t^2} - 1 - 2\frac{\exp(a\delta t)-1}{a^2\delta t^2} \\
&\quad - 2\frac{(1-\exp(2a\delta t))(1-\exp(a\delta t))}{a^2\delta t^2}) + \frac{b^4}{a^4}(1 + 2\frac{\exp(2a\delta t)-1}{a^2\delta t^2} \\
&\quad + 4\frac{(\exp(a\delta t)-1)^2}{a^2\delta t^2} - 4\frac{(1-\exp(2a\delta t))(1-\exp(a\delta t))}{a^2\delta t^2}) \\
&= \frac{2c^4}{a^2}(2\frac{1-\exp(2a\delta t)}{a^2\delta t^2} - \frac{1}{\delta t}) + \frac{4b^2c^2}{a^3}(\frac{1}{a\delta t} + 2\frac{\exp(a\delta t)+1}{a^2\delta t^2}) \\
&\quad + \frac{2b^4}{a^4}(-\frac{1}{a\delta t} + 6\frac{1-\exp(2a\delta t)}{a^2\delta t^2} - 8\frac{1-\exp(a\delta t)}{a^2\delta t^2}) \\
&= \frac{c^4}{a^2}(1 + 2\frac{1-\exp(2a\delta t)}{a\delta t} - \frac{1-\exp(4a\delta t)}{a\delta t}) + \frac{2b^2c^2}{a^3}(-\frac{1-\exp(4a\delta t)}{a\delta t} - 1 \\
&\quad + 2\frac{1-\exp(3a\delta t)}{a\delta t} - 2\frac{1-\exp(a\delta t)}{a\delta t}) + \frac{b^4}{a^4}(1 - \frac{1-\exp(4a\delta t)}{a\delta t} \\
&\quad - 6\frac{1-\exp(2a\delta t)}{a\delta t} - 4\frac{1-\exp(a\delta t)}{a\delta t} - 4\frac{1-\exp(3a\delta t)}{a\delta t}).
\end{aligned}
$$

$$
\begin{aligned}
B_4 :&= \lim_{n\to\infty} n^{-2}\sum S(i,j) = \lim_{n\to\infty} n^{-2}E(\sum X_j)^2 \\
&= \frac{c^2}{a^2\delta t^2}[\delta t + \frac{\exp(2a\delta t)-1}{2a} - \frac{2(\exp(a\delta t)-a)}{a}] \\
&\quad + \frac{b^2}{a^2}[1 + \frac{2(1-\exp(a\delta t))}{a\delta t} + \frac{(\exp(a\delta t)-1)^2}{a^2\delta t^2}]. \\
B_5 :&= \frac{1}{\eth t_1^2}\lim_{n\to\infty} n^{-2}\sum E(\eth W_i X_j)^2. \\
B_6 :&= \lim_{n\to\infty} n^{-2}(\sum EX_j)^2 = \lim_{n\to\infty} n^{-2}E(\sum X_j)^2
\end{aligned}
$$

$$= \frac{c^2}{a^2\delta t^2}[\delta t + \frac{\exp(2a\delta t)-1}{2a} - \frac{2(\exp(a\delta t)-1)}{a}]$$
$$+ \frac{b^2}{a^2}[1 + \frac{2(1-\exp(a\delta t))}{a\delta t} + \frac{(\exp(a\delta t)-1)^2}{a^2\delta t^2}].$$

Define

$$u_1 := \lim_{n\to+\infty} n^{-1}\sum EX_{j+1}X_j - \lim_{n\to+\infty} n^{-2}E\sum X_{j+1}\sum X_j$$
$$= \lim_{n\to+\infty} n^{-1}\sum EX_j^2 - \lim_{n\to+\infty} n^{-2}E(\sum X_j)^2$$
$$= -\frac{c^2}{2a}[\frac{2}{a\delta t} + 1 + \frac{\exp(2a\delta t)-1}{a^2\delta t^2} - \frac{\exp(2a\delta t)-1}{2a\delta t} - \frac{4(\exp(a\delta t)-1)}{a^2\delta t^2}]$$
$$+ \frac{b^2}{a^2}[\frac{\exp(2a\delta t)-1}{2a\delta t} - \frac{(\exp(a\delta t)-1)^2}{a^2\delta t^2}].$$
$$u_2 := \lim_{n\to+\infty} n^{-1}E\sum X_j^2 - \lim_{n\to+\infty} n^{-2}E(\sum X_j)^2 = u_1.$$

Note that

$$(1+\frac{x}{n})^n = \exp(x)(1 - \frac{x^2}{2n} + O(n^{-2})).$$

We have

$$E(n^{-1}\sum X_{j+1}X_j - n^{-2}\sum X_{j+1}\sum X_j - u_1) = O(\frac{1}{n}).$$
$$E(n^{-1}\sum X_j^2 - n^{-2}(\sum X_j)^2 - u_2) = O(\frac{1}{n}).$$

And

$$E(u_1 - n^{-1}\sum X_{j+1}X_j - n^{-2}\sum X_{j+1}\sum X_j)^2 = O(\frac{1}{n}).$$
$$E(u_2 - n^{-1}\sum X_j^2 - n^{-2}(\sum X_j)^2)^2 = O(\frac{1}{n}).$$

Now we analyse the errors, the errors are expressed by the data. Denote $t_1 = n^{-1}\sum X_{j+1}X_j - u_1$, $t_2 = n^{-1}\sum X_j^2 - u_2$, $t_3 = n^{-1}\sum X_{j+1}$, $t_4 = n^{-1}\sum X_j$. Let $f(x,y) = \frac{u_1+x}{u_2+y}$, then by the multivariate Taylor expansion

$$\frac{n^{-1}\sum X_{j+1}X_j - n^{-2}\sum X_{j+1}\sum X_j}{n^{-1}\sum X_j^2 - n^{-2}(\sum X_j)^2} = \frac{u_1+t_1-t_3t_4}{u_2+t_2-t_4^2} = \frac{u_1}{u_2} + A_1 + A_2 + O_p(n^{-\frac{3}{2}}),$$

where

$$A_1 = \frac{t_1}{u_2} - \frac{u_1t_2}{u_2^2},$$
$$A_2 = -\frac{t_3t_4}{u_2} + \frac{u_1t_4^2}{u_2^2} - \frac{t_1t_2}{u_2^2} + \frac{u_1t_2^2}{u_2^3}.$$

The computation gives

$$E(t_1 - t_2) = E(n^{-1}\sum X_{j+1}X_j - n^{-1}\sum X_j^2)$$

$$= a\eth t_1 n^{-1}\sum EX_j^2 + b\eth t_1 n^{-1}\sum EX_j.$$

$$Et_3t_4 = n^{-2}\sum S(i,j+1).$$

$$Et_4^2 = n^{-2}\sum S(i,j).$$

From equation (3.2), we have

$$EX_{j+1}X_i = \phi EX_jX_i + b\eth t_1 EX_j + cE\eth W_jX_i.$$

So

$$S(i,j+1) = \phi S(i,j) + b\eth t_1 EX_j + cE\eth W_jX_i.$$

$$E(t_4^2 - t_3t_4) = -a\eth t_1 n^{-2}\sum S(i,j) - b\eth t_1 n^{-1}\sum EX_j - cn^{-2}\sum E\eth W_jX_i.$$

$$\begin{aligned}E(t_2^2 - t_1t_2) &= u_2 a\eth t_1 n^{-1}\sum EX_j^2 + u_2 b\eth t_1 n^{-1}\sum EX_j\\ &\quad - (a\eth t_1 n^{-2}E\sum X_j^2\sum X_i^2\\ &\quad + b\eth t_1 n^{-2}E\sum X_j\sum X_i^2 + cn^{-2}E\sum \eth W_jX_j\sum X_i^2).\end{aligned}$$

Denote

$$C = \frac{b^2}{au_1}\left(\frac{\exp(a\delta t)-1}{a\delta t} - 1\right) - \frac{cB_1}{u_1} - \frac{cB_2}{u_1^2} - \frac{aB_3}{u_1^2}.$$

We have

$$E\hat{a} = a + C + O(n^{-1}). \tag{3.3}$$

$$B_3 - B_6 = \frac{1}{\delta t}n^{-1}\lim_{n\to\infty}\sum EX_j^2 - \lim_{n\to\infty} n^{-2}E(\sum X_j)^2.$$

$$B_5 - B_6 = \frac{1}{\delta t}n^{-1}\lim_{n\to\infty}\sum EX_j^2 - \lim_{n\to\infty} n^{-2}E(\sum X_j)^2.$$

$$D = \frac{a^2(B_3-B_6)}{u_1^2} + \frac{b^2(B_5-B_6)}{u_1^2} + \frac{c^2B_4}{u_1^2} + \frac{2acB_2}{u_1^2}.$$

Then

$$\mathrm{Var}\,\hat{a} = D + O(n^{-1}). \tag{3.4}$$

$$\begin{aligned}\hat{b} &= \frac{\sum X_{j+1} - \sum X_j - \hat{a}\eth t_1\sum X_j}{\delta t}\\ &= \frac{\sum X_{j+1} - \sum X_j - a\eth t_1\sum X_j}{\delta t} - Cn^{-1}\sum X_j + O_p(n^{-1}).\end{aligned}$$

So

$$E\hat{b} = b + C\frac{b}{a}\left(\frac{\exp(a\delta t)-1}{a\delta t} - 1\right) + O(n^{-1}).$$

$$\operatorname{Var}\hat{b} = \frac{C^2c^2}{a^2\delta t^2}[\delta t + \frac{\exp(2a\delta t)-1}{2a} - \frac{2(\exp(a\delta t)-1)}{a}] + O(n^{-1}).$$

Since

$$\hat{c}^2 = \frac{1}{\delta t}\sum(X_{j+1} - X_j - (aX_j + b)\eth t_j)^2 + Cn^{-1}\sum X_j^2 + O(n^{-1}).$$

Denote

$$F = C\lim_{n\to\infty} n^{-1}\sum EX_j^2;$$

$$G = \frac{1}{\eth t_1^2}\lim_{n\to\infty} n^{-2}\sum E(\eth W_i X_j)^2.$$

We have

$$E\hat{c^2} = c^2 + F + O(n^{-1}).$$

$$\operatorname{Var}\hat{c^2} = \frac{3c^4}{\delta tn} + G + O(n^{-2}).$$

Remark 11.1. Phillips et al. [6] used the estimator

$$\hat{c}^2 = \frac{1}{n}\sum(X_{j+1} - X_j)^2.$$

Example 11.1. Consider the constant coefficient SDE

$$dX = a\,dt + c\,dW, \quad X(0) = x_0. \tag{3.5}$$

The solution of (3.5) is $X(t) = x_0 + at + cW$. The microscale increments $\eth X_j = X_{j+1} - X_j$, $j = 0, \ldots, n-1$, from a burst of simulation of small time $\delta t = t_n - t_0$ are distributed normally with mean $a\eth t_j$ and variance $b^2\eth t_j$, $\eth t_j = t_{j+1} - t_j$.

Solve $\hat{a}$ and $\hat{b}$.

$$\hat{a} = \frac{\delta X}{\delta t}, \tag{3.6}$$

where $\delta X = X_{t_n} - X_{t_0}$.

$$\hat{c}^2 = \sum\frac{1}{\delta t}(\eth X_j - \hat{a}\eth t_j)^2. \tag{3.7}$$

The errors of parameter estimates depend on the number of data values and microscale time step.

$$E\hat{a} = a;$$

$$\operatorname{Var}\hat{a} = \frac{c^2}{\delta t};$$

$$E\hat{c}^2 = c^2(1 - \frac{1}{n});$$

$$\begin{aligned}\operatorname{Var}\hat{c}^2 &= \operatorname{Var}\left[\sum\frac{1}{\delta t}(\eth X_j - (a + b\frac{\delta W}{\delta t})\eth t_j)^2\right] \\ &= \frac{c^4}{\delta t^2}\operatorname{Var}\sum\left[(\eth W_j)^2 - 2\eth W_j \times \frac{\delta W}{\delta t}\eth t_j + \frac{\delta W^2}{\delta t^2}\eth t_j^2\right]\end{aligned}$$

$$
\begin{aligned}
&= \frac{c^4}{\delta t^2} \operatorname{Var}\left[\sum(\eth W_j)^2 - \frac{(\delta W)^2}{n}\right] \\
&= \frac{c^4}{\delta t^2}\left[\operatorname{Var}(\sum(\eth W_j)^2) + \operatorname{Var}\frac{(\delta W)^2}{n} - 2\operatorname{Cov}(\sum(\eth W_j)^2, \frac{(\delta W)^2}{n})\right] \\
&= \frac{c^4}{\delta t^2}\left[\operatorname{Var}(\sum(\eth W_j)^2) + \operatorname{Var}\frac{(\delta W)^2}{n} - \frac{2}{n}\operatorname{Cov}(\sum(\eth W_j)^2, \sum(\eth W_j)^2)\right] \\
&= \frac{c^4}{\delta t^2}\left[(1-\frac{2}{n})\operatorname{Var}(\sum(\eth W_j)^2) + \frac{1}{n^2}\operatorname{Var}(\delta W)^2\right] \\
&= \frac{2}{n}c^4 - \frac{2}{n^2}b^4.
\end{aligned}
$$

Example 11.2.

$$dX = aX dt + c\,dW, \quad X(0) = x_0. \tag{3.8}$$

The solution is $X_t = e^{at}(x_0 + c\int_0^t e^{-as} dW_s)$. We use the Euler scheme for the equation (3.8)

$$X_{j+1} - X_j = aX_j \eth t_j + c\eth W_j \tag{3.9}$$

From the maximum likelihood estimation, we find

$$\hat{a} = \frac{\sum X_{j+1}X_j - \sum X_j^2}{\sum X_j^2 \eth t_j}. \tag{3.10}$$

$$\hat{c}^2 = \frac{1}{n}\sum \frac{(X_{j+1} - X_j - \hat{a}X_j\eth t_j)^2}{\eth t_j}. \tag{3.11}$$

Let $\phi := 1 + a\eth t_1$. From the data, we compute

$$EX_i^2 = c^2\eth t_1 \frac{1-\phi^{2i}}{1-\phi^2}; \quad EX_iX_{i-1} = c^2\eth t_1 \phi \frac{1-\phi^{2i}}{1-\phi^2}.$$

Define

$$
\begin{aligned}
u_1 &:= \lim_{n\to+\infty} n^{-1}E\sum X_j^2 = -\frac{c^2}{2a}(1 - \frac{\exp(2a\delta t) - 1}{2a\delta t}). \\
u_2 &:= \lim_{n\to+\infty} n^{-1}\sum EX_{j+1}X_j = u_1.
\end{aligned}
$$

Now we analyse the errors, the errors are expressed by the data. Let $f(x,y) = \frac{x}{y}$, then by the multivariate Taylor expansion

$$
\begin{aligned}
\frac{n^{-1}\sum X_{j+1}X_j}{n^{-1}\sum X_j^2} &= \frac{u_2}{u_1} + \frac{1}{u_1}(n^{-1}\sum X_{j+1}X_j - u_2) - \frac{u_2}{u_1^2}(n^{-1}\sum X_j^2 - u_1) - \frac{1}{u_1^2} \\
&(n^{-1}\sum X_{j+1}X_j - u_2)(n^{-1}\sum X_j^2 - u_1) + \frac{u_2}{u_1^3}(n^{-1}\sum X_j^2 - u_1)^2 + O_p(n^{-\frac{3}{2}}).
\end{aligned}
$$

Compute

$$n^{-2}\sum(EX_iX_j)^2 = \frac{1}{u_1^2}\frac{2c^4}{a^2(1+\phi)^2}n^{-2}[\frac{(n-1)\phi^2}{1-\phi^2} - \frac{\phi^4 - \phi^{2n+2}}{(1-\phi^2)^2} + \frac{\phi^2 - \phi^{4n-2}}{(1-\phi^2)^2(1+\phi^2)}$$

$$-\frac{\phi^{2n}-\phi^{4n-2}}{(1-\phi^2)^2}] - \frac{4c^4}{a^2(1+\phi)^2}n^{-1}\frac{1-\phi^{2n}}{1-\phi^2}$$
$$+\frac{c^4}{a^2(1+\phi)^2}n^{-1}[n-\frac{2(1-\phi^{2n})}{1-\phi^2}+\frac{1-\phi^{4n}}{1-\phi^4}].$$

$$E\frac{n^{-1}\sum X_jX_{j-1}}{n^{-1}\sum X_j^2} = \frac{2}{u_1}a\eth t_1 n^{-1}\sum EX_j^2 - \frac{1}{u_1^2}a\eth t_1 n^{-2}\sum(EX_j^2)^2$$
$$-\frac{2a\eth t_1}{u_1^2}n^{-2}\sum(EX_iX_j)^2 + O(n^{-2}).$$

We show that

$$E\hat{a} = a - C + O(n^{-1}),$$

where

$$C = \frac{c^4}{u_1^2 a}\left[\frac{1}{2}+\frac{7}{8a\delta t}-\frac{1}{8a^2\delta t^2}-\frac{3\exp(2a\delta t)}{2a\delta t}+\frac{\exp(4a\delta t)}{8a\delta t}+\frac{\exp(4a\delta t)}{8a^2\delta t^2}\right].$$

$$\frac{n^{-1}\sum X_jX_{j-1}}{n^{-1}\sum X_j^2} = \frac{1}{u_1^2}\Big[\operatorname{Cov}(n^{-1}\sum X_{j+1}X_j, n^{-1}\sum X_{i+1}X_i)$$
$$+\operatorname{Cov}(n^{-1}\sum X_i^2, n^{-1}\sum X_j^2)$$
$$-2\operatorname{Cov}(n^{-1}\sum X_{j+1}X_j, n^{-1}\sum X_i^2)\Big].$$

$$\operatorname{Var}\frac{n^{-1}\sum X_jX_{j-1}}{n^{-1}\sum X_j^2} = \frac{n^{-2}}{u_1^2}\Big[a^2\eth t_1^2\operatorname{Cov}(\sum X_i^2, \sum X_j^2)$$
$$+c^2\operatorname{Cov}(\sum X_j\eth W_j, \sum X_j\eth W_j)\Big] + O(n^{-3})$$
$$= \frac{c^4\eth t_1^2}{u_1^2}(-\frac{1}{2a\delta t}-\frac{1-\exp(2a\delta t)}{4a^2\delta t^2}) + O(n^{-3})$$
$$= \frac{1}{n}(1-\exp(2a\eth t_1)) + O(n^{-3}).$$

We have

$$\operatorname{Var}\hat{a} = -\frac{2a}{\delta t} + O(n^{-1}).$$

From the result of $\hat{a}$, we have

$$\hat{c}^2 = \frac{1}{\delta t}\sum(X_{j+1}-X_j-\hat{a}X_j\eth t_j)^2 = \frac{1}{\delta t}\sum(X_{j+1}-X_j-aX_j\eth t_j)^2 + O_P(\frac{1}{n}).$$

Then

$$E\hat{c}^2 = c^2 + O(\frac{1}{n});$$

$$\operatorname{Var}\hat{c}^2 = \frac{2c^4}{n} + O(\frac{1}{n^2}).$$

4. Monte Carlo simulation results

Since the errors of the expression of Theorem 11.1 is hard to apply. For the stationary case [8], the expressions are also complicated. We give a simulation order of the parameters from the Monte Carlo simulation in this section. Firstly, we simulate the parameters from the maximum likelihood estimation for some SDEs. Secondly, we set varied parameters to simulate the power law of the errors. Then we have the simulated errors. The power law we simulate equation (3.5) is quite closed to the analytic expression errors we compute.

$$
\begin{aligned}
&E\hat{a} \propto a;\\
&\operatorname{Var}\hat{a} \propto \frac{c^2}{\delta t};\\
&E\hat{c}^2 \propto c^2;\\
&\operatorname{Var}\hat{c}^2 \propto \frac{1}{n}c^4.
\end{aligned}
$$

Table 1. We simulate the SDE $dX = a\,dt + c\,dW$, $X(0) = 1$. The burst $\delta t = 2$. The number of data $n = 1000$. The maximum likelihood expressions are equations (3.6) and (3.7). The errors are expressed as the mean and variance of a and c^2.

Parameters	a	c	a	c	a	c
True values	1	0.2	2	0.5	3	0.8
Mean	0.9991	0.0400	1.9896	0.2493	3.0144	0.6386
Variance	0.0197	3.0320e-06	0.1139	1.2376e-04	0.2992	7.8635e-04

Table 2. We simulate the SDE $dX = a\,dt + c\,dW$, $X(0) = 2$. The burst $\delta t = 5$. The number of data $n = 1500$. The maximum likelihood expressions are equations (3.6) and (3.7). The errors are expressed as the mean and variance of a and c^2.

Parameters	a	c	a	c	a	c
True values	1	0.2	2	0.5	3	0.8
Mean	0.9969	0.0399	2.0016	0.2494	2.9945	0.6387
Variance	0.0078	2.0527e-06	0.0488	7.9425e-05	0.1183	5.4581e-04

Table 3. We simulate the SDE $dX = aX\,dt + c\,dW$, $X(0) \sim N(0, -\frac{c^2}{2a})$. The burst $\delta t = 2$. The number of data $n = 500$. The maximum likelihood expressions are equations (3.10) and (3.11). The errors are expressed as the mean and variance of a and c^2.

Parameters	a	c	a	c	a	c
True values	-1	0.2	- 2	0.5	-3	0.8
Mean	-1.7954	0.0398	-2.8753	0.2493	-3.8746	0.6389
Variance	2.8198	6.5877e-06	3.4996	2.4565e-04	4.7498	0.0016

Table 4. We simulate the SDE $dX = aX\,dt + c\,dW$, $X(0) \sim N(0, -\frac{c^2}{2a})$. The burst $\delta t = 4$. The number of data $n = 1000$. The maximum likelihood expressions are equations (3.10) and (3.11). The errors are expressed as the mean and variance of a and c^2.

Parameters	a	c	a	c	a	c
True values	-2.1	0.03	- 3.5	0.04	-4.5	0.05
Mean	-2.6047	8.9910e-04	-3.9577	0.0016	-4.8337	0.0025
Variance	1.5889	1.6351e-09	2.3352	4.9424e-09	2.4358	1.2806e-08

Table 5. We simulate the SDE $dX = (aX + b)dt + c\,dW$. $X(0) = 2$. The burst $\delta t = 1$. The number of data $n = 1500$. The maximum likelihood expressions are equations (3.10) and (3.11). The errors are expressed as the mean and variance of a, b and c^2.

Parameters	a	b	c	a	b	c
True values	3.5	1	0.3	4	2	0.4
Mean	3.5003	0.9832	0.08990	4.0000	2.0172	0.1597
Variance	2.4085e-04	0.1939	1.0845e-05	1.3803e-04	0.2961	3.2811e-05

Table 6. We simulate the SDE $dX = (aX + b)dt + c\,dW$. $X(0) = 2$. The burst $\delta t = 1.7$. The number of data $n = 5500$. The maximum likelihood expressions are equations (3.10) and (3.11). The errors are expressed as the mean and variance of a, b and c^2.

Parameters	a	b	c	a	b	c
True values	1.2	0.1	0.2	2.5	1	0.8
Mean	1.2007	0.1041	0.0400	2.5006	1.0060	0.6395
Variance	0.0016	0.0944	5.3574e-007	2.2978e-004	0.6926	1.4544e-004

The simulation power law of (3.2) is a little different from the analysis expression. This is because of the errors of simulation from the maximum likelihood estimation. The errors depend on the burst time δt.

$$E\hat{a} \propto a;$$
$$\operatorname{Var}\hat{a} \propto \frac{ac}{\delta t};$$
$$E\hat{c}^2 \propto c^2;$$
$$\operatorname{Var}\hat{c}^2 \propto \frac{1}{n}c^4.$$

The simulation equation (1.7) of the power law is as follows.

$$E\hat{a} \propto a;$$
$$\operatorname{Var}\hat{a} \propto \frac{c^2}{a^{3.7}b^{0.5}\delta t^{5.2}};$$
$$E\hat{b} \propto b;$$
$$\operatorname{Var}\hat{b} \propto \frac{c^2}{a\delta t^2};$$
$$E\widehat{c^2} \propto c^2;$$
$$\operatorname{Var}\widehat{c^2} \propto \frac{c^4}{n}.$$

4.1. *Two bursts*

$$dX = (aX + b)dt + c\,dW, \quad X(0) = 0. \tag{4.1}$$

Suppose the data is from two same length bursts, denoted as δt. Denote the data by $X_0, X_1, \ldots, X_n$, $X_{1,0}, \ldots, X_{1,n}$. The Euler scheme for the equation (4.1) are

$$X_{j+1} - X_j = (aX_j + b)\eth t_1 + c\eth W_j,$$

$j = 0, \ldots, n-1$, and

$$X_{1,j+1} - X_{1,j} = (aX_{1,j} + b)\eth t_1 + c\eth W_{1,j},$$

$j = 0, \ldots, n-1$.
$X_{1,0} = \exp a(\delta t + \Delta t)(x_0 + b\int_0^{\delta t+\Delta t}\exp(-as)ds + c\int_0^{\delta t+\Delta t} e^{-as}dW_s)$. Denote $dX_j = X_{j+1} - X_j$, $dX_{1,j} = X_{1,j+1} - X_{1,j}$. Ignoring the $X_{1,0}$ component, we find

$$\hat{a} = \frac{n}{\delta t}\frac{2n\sum X_j dX_j - 2n\sum X_{1,j}dX_{1,j} - (\sum dX_j + \sum dX_{1,j})(\sum X_j + \sum X_{1,j})}{2n\sum X_j^2 + 2n\sum X_{1,j}^2 - (\sum X_j + \sum X_{1,j})^2}; \tag{4.2}$$

$$\begin{aligned}\hat{b} &= \frac{(\sum X_{j+1} - \sum X_j - \hat{a}\eth t_1 \sum X_j) + (\sum X_{1,j+1} - \sum X_{1,j} - \hat{a}\eth t_1 \sum X_{1,j})}{2\delta t}\\ &= \frac{n}{\delta t}\Big[\frac{(\sum dX_j + \sum dX_{1,j})(\sum X_j^2 + \sum X_{1,j}^2)}{2n\sum X_j^2 + 2n\sum X_{1,j}^2 - (\sum X_j + \sum X_{1,j})^2}\\ &\quad - \frac{(\sum X_j dX_j + \sum X_{1,j}dX_{1,j})(\sum X_j + \sum X_{1,j})^2}{2n\sum X_j^2 + 2n\sum X_{1,j}^2 - (\sum X_j + \sum X_{1,j})^2}\Big];\end{aligned} \tag{4.3}$$

$$\begin{aligned}\hat{c}^2 = \frac{1}{2\delta t}\Big\{&\Big[\sum (dX_j - (\hat{a}X_j + \hat{b})\eth t_j)^2\Big]\\ &+ \Big[\sum (dX_{1,j} - (\hat{a}X_{1,j} + \hat{b})\eth t_{1,j})^2\Big]\Big\}.\end{aligned} \tag{4.4}$$

The simulation equation (4.1) of the power law is as follows.

$$\begin{aligned}&E\hat{a} \propto a;\\ &\operatorname{Var}\hat{a} \propto \frac{c^2}{a^4 b \delta t^{3.4}};\\ &E\hat{b} \propto b;\\ &\operatorname{Var}\hat{b} \propto \frac{c^2}{a^{0.5}\delta t^{1.2}};\\ &E\hat{c} \propto c;\\ &\operatorname{Var}\hat{c} \propto \frac{c^2}{n}.\end{aligned}$$

Table 7. We simulate the SDE $dX = (aX + b)dt + c\,dW$. $X(0) = 2$. The two bursts $\delta t = 0.5$, $\Delta t = 1.2$. The number of data $n = 500$. The errors are expressed as the mean and variance of a, b and c.

Parameters	a	b	c	a	b	c
True values	5.2	2.1	0.2	3.2	1.3	0.1
One burst mean	5.214	2.110	0.199	3.207	1.290	0.099
One burst variance	0.035	0.474	0.006	0.053	0.329	0.003
Two bursts mean	5.213	2.101	0.200	3.205	1.299	0.100
Two bursts variance	0.0000	0.2441	0.004	0.000	0.130	0.002

5. Errors of projective integration

The maximum likelihood estimation equation is

$$d\hat{X} = (\hat{a}\hat{X} + \hat{b})dt + \hat{c}\,dW, \quad \hat{X}(0) = 0. \tag{5.1}$$

From equation (4.1) and equation (5.1) we have

$$d(\hat{X} - X) = \hat{a}(\hat{X} - X)dt + (\hat{a} - a)X dt + (\hat{b} - b)dt + (\hat{c} - c)\,dW. \tag{5.2}$$

From the analytic solution of equation (5.2),

$$\begin{aligned} E(\hat{X}_n - X_n) &= E(\hat{a} - a)\int_0^{\delta t} X(s)\exp(\hat{a}(\delta t - s))ds \\ &\quad + E(\hat{b} - b)\int_0^{\delta t} \exp(\hat{a}(\delta t - s))ds \\ &\leq \delta t^{1/2} M(C^{1/2} + D^{1/2}) + \mathrm{O}(n^{-1}), \end{aligned}$$

where M is a constant. Consider the equations of the time step Δt.

$$d\hat{Y} = (\hat{a}\hat{Y} + \hat{b})dt + \hat{c}\,dW, \quad \hat{Y}(0) = \hat{X}_n. \tag{5.3}$$

$$d\hat{X} = (\hat{a}\hat{X} + \hat{b})dt + \hat{c}\,dW, \quad \hat{X}(0) = X_n. \tag{5.4}$$

The errors on projective integration time step Δt is

$$\begin{aligned} E(Y_{\delta t+\Delta t} - X_{\delta t+\Delta t}) &= E\exp(\hat{a}\Delta t)(\hat{X}_n - X_n) \\ &\leq \delta t^{1/2} M(C^{1/2} + D^{1/2}) + \mathrm{O}(\Delta t) + \mathrm{O}(n^{-1}). \end{aligned}$$

Acknowledgement

The first author is partially supported by National Natural Science Foundation of China (NSFC) [11501344], STU Scientific Research Foundation for Talents (SRFSU) [NTF15004], Natural Science Foundation of Guangdong (NSFG) [2017A030313005], National Natural Science Foundation of China (NSFC) [11571122].

Table 8. We simulate the SDE $dX = (aX + b)dt + c\,dW$. $X(0) = 2$. The two bursts $\delta t = 0.7$, $\Delta t = 1.8$. The number of data $n = 550$. The maximum likelihood expressions are equations (4.2), (4.3) and (4.4). The errors are expressed as the mean and variance of a, b and c^2.

Parameters	a	b	c
True values	5.2	2.1	0.2
Mean	5.2000	2.0270	0.0469
Variance	8.6288e-016	6.6933e-006	1.2685e-010
Parameters	a	b	c
True values	3.2	1.3	0.1
Mean	3.1999	1.1347	0.0210
Variance	2.2978e-004	0.0719	2.5659e-006

Table 9. We simulate the SDE $dX = (aX + b)dt + c\,dW$. $X(0) = 2$. The maximum likelihood expressions are equations (4.2), (4.3) and (4.4). The errors are expressed as the burst, the projective time step, the number of data.

Parameters	n	δt	Δt	n	δt	Δt
True values	550	0.1	0.6	650	0.2	1.1
Mean		-0.9980			-3.0684	
Variance		0.1958			1.3594	

References

[1] Y. Ait-Sahalia, Maximum likelihood estimation of discretely sampled diffusions: a closed-form approximation approach. Econometrica, **70**, 223–262, 2002.

[2] Jaya P. N. Bishwal, Parameter estimation in stochastic differential equations. Lecture Notes in Mathematics, 2008.

[3] K. S. Chan, Consistency and limiting distribution of the least squares estimator of a Threshold Autoregressive model. Ann. Statist., **21**, 520–533, 1993.

[4] X. P. Chen, A. J. Roberts and I. G. Kevrekidis, Projective integration of expensive stochastic processes. ANZIAM J., **52**(E), C661–C677, 2011.

[5] P. E. Kloeden, E. Platen, H. Schurz and M. Sorensen, On effects of discretization on estimators of drift parameters for diffusion processes. J. Appl. Prob., **33**, 1061–1076, 1996.

[6] P. C. B. Phillips and J. Yu, A two-stage realized volatility approach to estimation of diffusion processes with discrete data. Journal of Econometrics, **150**, 139–150, 2009.

[7] M. I. Stefano, Simulation and Inference for Stochastic Differential Equations: With R Examples. Springer, New York, 2008.

[8] C. Y. Tang and S. X. Chen, Parameter estimation and bias correction for diffusion processes. Journal of Econometrics, **149**, 65-81, 2009.

[9] N. Yoshida, Estimation for diffusion processes from discrete observation. Journal of Multivariate Analysis, **41**, 220–242, 1992.

Chapter 12

Approximation for Stochastic Wave Equations with Singular Perturbation

Yan Lv[1,*] and Wei Wang[2,†]

[1]*School of Science, Nanjing University of Science and Technology, Nanjing, P.R. China*

[2]*Department of Mathematics, Nanjing University, Nanjing, P.R. China*

[*]*lvyan@njust.edu.cn;* [†]*wangweinju@nju.edu.cn.*

Some results on approximation for stochastic wave equations with singular perturbation and white noise are reviewed, which include the so-called Smolukowski–Kramers approximation on both finite time interval and infinite time interval.

1. Introduction

Due to the presence of turbulence, the media such as the atmosphere or the ocean has random fluctuation, the wave propagation in such random media is usually described by stochastic wave equations [8, 9].

Let $D \subset \mathbb{R}^n$, $1 \leq n \leq 3$, be a regular bounded domain with smooth boundary ∂D. One typical stochastic wave equation model is in the following form

$$\nu u_{tt}^{\nu} + u_t^{\nu} = \Delta u^{\nu} + f(u^{\nu}) + \epsilon\sigma(x, u^{\nu})\dot{W} \tag{1.1}$$

with some boundary condition on ∂D and positive parameters ν and ϵ. Δ is the Laplace operator on D and $\{W(t)\}_{t\geq 0}$ is an infinite dimensional Wiener process defined on a canonical complete probability space $(\Omega,\ \mathcal{F},\ \{\mathcal{F}_t\}_{t\in\mathbb{R}},\ \mathbb{P})$ (which is determined later) and nonlinear term f satisfies some increasing properties. Here ν describes the mass of the particle and $\epsilon\sigma(x, u^{\nu})$ describes the strength of the noise, which is detailed later. There are a few works on the stochastic wave equation under different assumptions on f and σ [26, 27]. As the mass of one particle is small, an interesting problem is the limit or the approximation of u^{ν} as $\nu \to 0$. This paper reviews some results on the approximation of u^{ν} in different sense and under different assumptions.

Cerrai & Freidlin [6] first derived the approximation of u^{ν} with $\epsilon\sigma(x, u)\dot{W} = \dot{W}^Q$ an Q-Wiener process [17] on finite time interval. In this paper we first (section 3–4) review some results on the approximation on infinite time interval in the sense of both almost sure [21] and distribution [24] with $\epsilon = \sqrt{\nu}$. More precisely, in next section, we consider the following stochastic wave equation

$$\nu u_{tt}^{\nu} + u_t^{\nu} = \Delta u^{\nu} + f(u^{\nu}) + \sqrt{\nu}\dot{W} \tag{1.2}$$

$$u^\nu(x,0) = u_0\,, \quad u_t^\nu(x,0) = u_1\,, \quad u^\nu|_{\partial D} = 0\,. \tag{1.3}$$

By some increasing assumptions on nonlinearity f a global pull-back random attractor $\{\mathbf{A}^\nu(\omega)\}_\omega$ is constructed in space $L^2(D) \times H_0^1(D)$. Then we show that the random attractor $\mathbf{A}^\nu(\omega)$ is upper semicontinuous almost surely, that is

$$\lim_{\nu \to 0} \operatorname{dist}(\mathbf{A}^\nu(\omega), \mathbf{A}^0) = 0,$$

where $\operatorname{dist}(\cdot,\cdot)$ is the Hausdorff semidistance and $\mathbf{A}^0$ is the extension of the global attractor of the limit system

$$u_t = \Delta u + f(u)\,, \quad u|_{\partial D} = 0\,. \tag{1.4}$$

This continuous result implies the convergence of stationary solution of the stochastic wave equations to that of the nonlinear heat equation (1.4).

In section 4 we review the approximation of u^ν in the sense of distribution. Typically we construct a global measure attractor $\mathcal{A}^\nu$ for (1.2)–(1.3) and show the continuity property of $\mathcal{A}^\nu_{u_1}$, the part of $\mathcal{A}^\nu$ supported on $H_0^1(D)$, in ν at $\nu = 0$.

Furthermore (section 5), we review an approximation result in the sense of almost sure [22] with $\epsilon\sigma(x,u)\dot{W} = \dot{W}^Q$. We construct a random inertial manifold $\mathcal{M}^\nu$ for the damped nonlinear stochastic wave equation

$$\nu u_{tt}^\nu + u_t^\nu = \Delta u^\nu + f(u^\nu) + \dot{W} \tag{1.5}$$

$$u^\nu(x,0) = u_0\,, \quad u_t^\nu(x,0) = u_1\,, \quad u^\nu|_{\partial D} = 0\,. \tag{1.6}$$

By a Rohlin's result on the classification of homomorphisms of Lebesgue space, the random set $\mathcal{M}^\nu(\omega)$ is shown to be approximated almost surely by a random set $\mathcal{M}(\psi^\nu\omega)$ with $\mathcal{M}(\omega)$ the random inertial manifold of the following stochastic nonlinear heat equation

$$u_t = \Delta u + f(u) + \dot{W}\,, \quad u|_{\partial D} = 0\,. \tag{1.7}$$

Here $\psi : \Omega \to \Omega$ is a $\mathbb{P}$-measure preserved mapping.

In the last section we show a large deviation result for (1.2)–(1.3) by the weak convergence method. The result shows that the rate function for the $\{u^\nu\}_\nu$ coincides with that for the following stochastic heat equation

$$\tilde{u}_t^\nu = \Delta \tilde{u}^\nu + f(\tilde{u}^\nu) + \sqrt{\nu}\dot{W}\,, \tag{1.8}$$

which implies the singular perturbation νu_{tt}^ν is smaller perturbation than the small random perturbation $\sqrt{\nu}\dot{W}$.

2. Preliminary

This section gives some notations and some preliminary results. Let $D \subset \mathbb{R}^n$ with $1 \le n \le 3$ be a bounded domain with smooth boundary ∂D. We just consider the case of additive noise, that is $\epsilon\sigma(x,u^\nu) = \epsilon$ and W is an $L^2(D)$ valued Q-Wiener process. The covariance operator Q is trace class, and there is a complete orthonormal system $\{e_k(x)\}_{k=1}^\infty$ in $L^2(D)$ and real numbers $\{b_k\}_{k=1}^\infty$ such that

$$W(t,x) = \sum_{k=1}^\infty \sqrt{b_k} e_k \beta_k(t), \tag{2.1}$$

where $\{\beta(t)\}_{k=1}^{\infty}$ are real mutually independent Brownian motions. Then

$$B_0 := \mathrm{Tr}Q = \sum_{k=1}^{\infty} b_k < \infty\,. \tag{2.2}$$

We also need the following assumption on Q

$$B_1 = \sum_{k=1}^{\infty} \lambda_k b_k < \infty. \tag{2.3}$$

Furthermore, we assume Q commutes with Δ on D with zero Dirichlet boundary, that is $Q\Delta = \Delta Q$. Then $\{e_k(x)\}_{k=1}^{\infty}$ are eigenfunctions of Δ on D with zero boundary and denote by $-\lambda_k < 0$, the eigenvalue with respect to e_k for $k = 1, 2, \cdots$.

For our purpose we assume the probability space $(\Omega, \mathcal{F}, \{\mathcal{F}_t\}_t, \mathbb{P})$ to be the canonical probability space with Wiener measure $\mathbb{P}$ [1]. To be more precise, W is the identity on Ω, with

$$\Omega = \big\{w \in C(\mathbb{R}, L^2(D)) : w(0) = 0\big\}$$

and the filtration is defined as

$$\mathcal{F}_t := \bigvee_{s\le t} \mathcal{F}_s^t, \quad t \in \mathbb{R},$$

where

$$\mathcal{F}_s^t = \sigma\{W(t_2) - W(t_1) : s \le t_1 \le t_2 \le t\}$$

is the smallest σ-algebra generated by random variable $W(t_2) - W(t_1)$ for all $s \le t_1 \le t_2 \le t$. Let $\theta_t : (\Omega, \mathcal{F}, \{\mathcal{F}_t\}_{t\in\mathbb{R}}, \mathbb{P}) \to (\Omega, \mathcal{F}, \{\mathcal{F}_t\}_{t\in\mathbb{R}}, \mathbb{P})$ be a metric dynamical system (driven system), that is,

- $\theta_0 = \mathrm{id}$,
- $\theta_t\theta_s = \theta_{t+s}$ for all s, $t \in \mathbb{R}$,
- the map $(t, \omega) \mapsto \theta_t\omega$ is measurable and $\theta_t\mathbb{P} = \mathbb{P}$ for all $t \in \mathbb{R}$.

On this canonical probability space Ω, we choose θ_t to be the Wiener shift [1]

$$\theta_t\omega(\cdot) = \omega(\cdot + t) - \omega(t), \quad t \in \mathbb{R}\,, \quad \omega \in \Omega_0\,, \tag{2.4}$$

which preserves the Wiener measure $\mathbb{P}$ on Ω. Furthermore, θ_t is ergodic under Wiener measure $\mathbb{P}$. Write $W(t, x)$ as $W(t, x, \omega)$ to show the dependence on $\omega \in \Omega$, then

$$W(\cdot, x, \theta_t\omega) = W(\cdot + t, x, \omega) - W(t, x, \omega).$$

In this view, the stochastic wave equation (1.1) is driven by θ_t.

Denote by $\langle\cdot,\cdot\rangle$ the usual inner product in $L^2(D)$. In the following part for any $m \in \mathbb{R}$ we denote by $H^m(D)$ the completion of $C_0^\infty(D)$, the smooth function defined on D with compact support, with respect to the norm

$$\|u\|_m^2 := \|u\|_{H^m(D)}^2 = \sum_{k=1}^{\infty} \lambda_k^m \langle u, e_k\rangle^2\,.$$

For $m = 0$ we denote by $H = H^0 = L^2(D)$ and $H_0^1(D)$ the space of functions in $H^1(D)$ which vanishes on boundary ∂D. Next we present some assumptions on nonlinearity f.

($\mathbf{H_{cub}}$) The nonlinear function f is twice continuously differentiable and there exist positive constants C_i, $i = 1, \dots, 4$ such that for any $s \in \mathbb{R}$

(1) $|f(s)| \leq C_1(1 + |s|^3)$
(2) $|f'(s)| \leq C_2(1 + |s|^2)$
(3) $F(s) \geq C_3(|s|^4 - 1)$
(4) $sf(s) \geq C_4(F(s) - 1)$

with $F(s) = \int_0^s f(r)dr$.

($\mathbf{H_{lip}}$) The nonlinear function f is Lipschitz continuous with Lipschitz constant L_f, and there is a constant $C > 0$ such that

$$|f(s)| \leq L_f(1 + |s|) \quad \text{for any} \quad s \in \mathbb{R}\,. \tag{2.5}$$

Furthermore,

$$L_f \leq \sqrt{\lambda_1}\,. \tag{2.6}$$

Now let $A = -\Delta$ with zero boundary condition on ∂D and define the following linear operator

$$\mathbb{A}_\nu = \begin{pmatrix} 0 & 1 \\ \frac{-1}{\nu}A & \frac{-1}{\nu} \end{pmatrix}$$

with $\mathcal{D}(\mathbb{A}_\nu) = (H_0^1(D) \cap H^2(D)) \times L^2(D)$. Then equation (1.1) can be rewritten as

$$\frac{d}{dt}U(t) = \mathbb{A}_\nu U(t) + \mathbb{F}(U(t)) + \Sigma\dot{\mathbb{W}}(t) \tag{2.7}$$

with $U = (u, u_t) \in H_0^1(D) \times L^2(D)$ and

$$\mathbb{F}(U) = (0, f(u))^*\,, \quad \Sigma\dot{\mathbb{W}}(t) = (0, \epsilon\dot{W})^*\,.$$

Here * denotes the transpose of a vector. Sometimes we write $\mathcal{H} = H_0^1(D) \times L^2(D)$ and $\mathcal{H}^1 = (H^2(D) \cap H_0^1(D)) \times H^1(D)$. Then we have the following existence result for equation (1.1) [8, 17, e.g.].

Theorem 12.1. *Assume (2.2) and* ($\mathbf{H_{cub}}$) or ($\mathbf{H_{lip}}$) *hold. Then equation (1.1) has a unique solution* $(u^\nu, u_t^\nu) \in L^2(\Omega, C(0, T; \mathcal{H}))$ *for any initial data* $u(0, x) = u_0 \in H_0^1(D)$ *and* $u_t(0, x) = u_1 \in L^2(D)$.

We intend to construct random attractor for system (1.2)–(1.3). We give a brief introduction of the random dynamical system (RDS). For detail we refer to [1].

Definition 12.1. Let $(H, \|\cdot\|_H)$ be a Polish space. A measurable mapping

$$\phi : \mathbb{R}^+ \times \Omega \times H \to H$$

having the cocycle property

$$\phi(0, \omega)x = x, \quad \phi(t, \theta_\tau\omega) \circ \phi(\tau, \omega)x = \phi(t + \tau, \omega)x$$

for $t, \tau \in \mathbb{R}^+$, $x \in H$ and $\omega \in \Omega$ is called RDS with respect to $(\theta_t)_{t\in\mathbb{R}}$. If ϕ is continuous with respect to x for $t \geq 0$ and $\omega \in \Omega$, ϕ is called a continuous RDS.

A set valued map $B : \Omega \to 2^H$ is called a random closed set if $B(\omega)$ is closed, nonempty and

$$\omega \mapsto \inf_{x \in B(\omega)} \|x - y\|_H$$

is a random variable for any $y \in H$. A random set B is said to be tempered if for a.e. $\omega \in \Omega$ and all $\gamma > 0$

$$\lim_{t \to \infty} e^{-\gamma t} d\big(B(\theta_{-t}\omega)\big) = 0,$$

where $d(B) = \sup_{x \in B} \|x\|_H$. Let $\mathcal{D}$ be the collection of all tempered random sets in H. And for such set we write as $\mathcal{D}$-set. Then we give the following definition of random attractor, see [12, 13, 29].

Definition 12.2. A random set $\mathcal{A}(\omega) \in \mathcal{D}$ with $\mathcal{A}(\omega)$ compact is called a random (pullback) attractor for RDS ϕ in $\mathcal{D}$ if

$$\phi(t, \omega)\mathcal{A}(\omega) = \mathcal{A}(\theta_t \omega) \ \forall t \geq 0, \ \ a.s. \ \omega \in \Omega,$$

and for all $B \in \mathcal{D}$ and a.e. $\omega \in \Omega$ we have

$$\lim_{t \to \infty} d\big(\phi(t, \theta_{-t}\omega)B(\theta_{-t}\omega), \mathcal{A}(\omega)\big) = 0,$$

where $d(X, Y) = \sup_{x \in X} \inf_{y \in Y} \|x - y\|_H, \ \ X, Y \subseteq H.$

For a detailed description of random attractor, we need the following random invariant manifold.

Definition 12.3. A random set $M(\omega)$ is called invariant for RDS φ if

$$\varphi(t, \omega, M(\omega)) \subset M(\theta_t \omega), \quad \text{for any } t \geq 0\,.$$

If an invariant set $M(\omega)$ is represented by a Lipschitz or C^k mapping $h(\cdot, \omega)$: $H_1 \to H_2$ with $H = H_1 \oplus H_2$ such that $M(\omega) = \{\xi + h(\xi, \omega) \mid \xi \in H_1\}$, then we call $M(\omega)$ a Lipschitz or C^k invariant manifold. Furthermore, if H_1 is finite dimensional and $M(\omega)$ attracts exponentially all the orbits of φ, then we call $M(\omega)$ a random stochastic inertial manifold of φ.

However, direct application to the random dynamical theory to stochastic differential equations is difficult. For this we introduce the following linear system

$$\nu \eta_{tt} + \eta_t + A\eta = \sqrt{\nu}\dot{W}, \tag{2.8}$$

$$\eta(0) = 0\,, \eta_t(0) = 0\,. \tag{2.9}$$

Then we have the following result [21].

Lemma 12.1. *Assume (2.2) holds.*

(1) Equations (2.8)–(2.9) have a unique solution (η, η_t) which is a Gaussian process in $\mathcal{H}$. Further (η, η_t) has continuous paths on $\mathcal{H}^1$.

(2) $t \mapsto \|(\eta(t,\omega),\eta_t(t,\omega))\|_{\mathcal{H}}$ *grows sublinearly, that is, a.s.*

$$\lim_{t\to\pm\infty} \frac{\|(\eta(t,\omega),\eta_t(t,\omega))\|_{\mathcal{H}}}{t} = 0.$$

(3) For any $t \geq 0$ *and positive integer* k*,*

$$\mathbb{E}\Big[\|\eta(t)\|_1^k + \|\eta_t(t)\|_0^k\Big] \leq C$$

for some constant $C > 0$.

(4) The process $\eta(t,\omega)$ *converges to zero a.s..*

Now for fixed $\omega \in \Omega$ consider the following system

$$\nu v_{tt} + v_t + Av + f(v+\eta) = 0, \tag{2.10}$$

$$v(0) = u_0, \quad v_t(0) = u_1, \tag{2.11}$$

where η solves (2.8)–(2.9). By the classical analysis [32], for any $(u_0, u_1) \in \mathcal{H}$, (2.10)–(2.11) have a unique solution $(v, v_t) \in C(0,T;\mathcal{H})$ for any $T > 0$ and a.s. $\omega \in \Omega$ (The global existence of solution follows the boundedness of solution in section 3). Then system (2.10)–(2.11) defines a continuous random dynamical system $\Phi^\nu(t,\omega)$ on $\mathcal{H}$ as

$$\Phi^\nu(t,\omega)(u_0,u_1) = (v(t,\omega), v_t(t,\omega)). \tag{2.12}$$

Define an isomorphism transformation on $\mathcal{H}$ as

$$T(\omega, x) = x + \big(\eta(\omega), \eta_t(\omega)\big), \quad x \in \mathcal{H}, \tag{2.13}$$

then the system (1.2)–(1.3) generates a random dynamical system $\Psi^\nu(t,\omega)$ on $\mathcal{H}$ as

$$\Psi^\nu(t,\omega) = T \circ \Phi^\nu \circ T^{-1}(t,\omega) = \big(u(t,\omega), u_t(t,\omega)\big).$$

For more about the dynamics between two conjugate RDSs, see [14].

At the end of this section we introduce a Rohlin's result on the classification of homomorphisms of Lebesgue space.

Definition 12.4. Two random variables X and Y defined on a same probability space $(\Omega, \mathcal{F}, \mathbb{P})$ are called equivalent if and only if there is a measurable preserving map $\psi : \Omega \to \Omega$ such that $X(\psi\omega) = Y(\omega)$ for almost all $\omega \in \Omega$.

If X is equivalent to Y, then X has same distribution as that of Y. Rohlin's result on the classification of the homomorphisms of a Lebesgue space [28] gives an inverse result.

Recall that a homomorphism ψ from probability space $(\Omega_1, \mathcal{F}_1, \mathbb{P}_1)$ to probability space $(\Omega_2, \mathcal{F}_2, \mathbb{P}_2)$ is a measurable mapping such that $\psi\mathbb{P}_1 = \mathbb{P}_2$. If ψ is measurably invertible, then ψ is an isomorphism. A probability space $(\Omega, \mathcal{F}, \mathbb{P})$ is a Lebesgue space [1, Appendix A] if this probability space is isomorphic to a probability space which is the disjoint union of an at most countable (possibly empty) set $\{\omega_1, \omega_2, \ldots\}$ of points each of positive measure and the space $([0,s), \mathcal{B}, \lambda)$, where $\mathcal{B}$ is the σ-algebra of Lebesgue measurable subsets of the interval $[0,s)$ and λ is the Lebesgue measure. Here $s = 1 - \sum p_n$ where p_n is the measure of the point ω_n. For a measure space, the signature is the mass of its non-atomic part plus the non-increasing sequence of the weights of its atoms.

Rohlin's classification theorem on the homomorphisms of Lebesgue space states the following [28].

Theorem 12.2. *A homomorphisms of Lebesgue space is determined by the signature of the quotient measure space and the signatures of the condition measure spaces associated with the homomorphism.*

Then we have the following corollary.

Corollary 12.1. *Random variables X and Y are equivalent if and only if for almost all values taken by these variables, the condition measure spaces are isomorphic; that is, they have the same signature.*

One special case for the above condition measure spaces having the same signature is when almost all conditional measures are purely non-atomic.

The canonical probability space $(\Omega_0, \mathcal{F}_0, \mathbb{P}_0)$ is a Lebesgue space [1]. Now we consider $\eta^{*\nu}(\omega)$ and $\eta^*(\omega)$ which have the same distribution on the same probability space $(\Omega_0, \mathcal{F}_0, \mathbb{P}_0)$. Moreover, the distribution is Gaussian, so almost all conditional measures are purely non-atomic. Then by the above corollary, there is a measure preserving mapping $\psi^\nu : \Omega_0 \to \Omega_0$ such that

$$\eta^*(\psi^\nu \omega) = \eta^{*\nu}(\omega). \tag{2.14}$$

3. Approximation of random attractor

This section is concerned with the approximation of random dynamics for (1.2)–(1.3) for almost all $\omega \in \Omega$. In this section we assume $(\mathbf{H}_{cub})$ and (2.2) hold. Then a global random attractor is constructed for the RDS $\Psi^\nu(t, \omega)$ defined by (1.2)–(1.3), which is shown to be upper semicontiuous at $\nu = 0$.

3.1. *Existence result on random attractor*

We first present a sufficient condition on the existence of random attractor. Assume $(E, \|\cdot\|_E)$ be a Polish space and let C_E be the collection of all the bounded subsets of E. We review briefly the basic concept of the Kuratowskii measure of non-compactness and some basic properties [15, 31, e.g.].

For $A \in C_E$, define the diameter of A as

$$\operatorname{diam}(A) = \sup\{\|x - y\|_E : x, y \in A\}.$$

Definition 12.5. For a bounded set $A \subset E$, the Kuratowskii measure of non-compactness $\alpha(A)$ of A is defined as

$$\alpha(A) = \inf \Big\{ \varepsilon > 0 : A \text{ has finite cover sets } \{X_i\}$$

$$\text{with diam}(X_i) \le \varepsilon, X_i \in C_E\Big\}.$$

If A is an unbounded set in E, define $\alpha(A) = \infty$. The following properties will be useful in our approach.

Lemma 12.2. *The Kuratowskii measure of non-compactness $\alpha(A)$ of A satisfies the following properties*

(1) $\alpha(A) = 0$ if and only if the closure $Cl_E(A)$ of A in E is compact in E.
(2) If $B_1 \subset B_2 \subset E$, $\alpha(B_1) \le \alpha(B_2)$.
(3) $\alpha(A) = \alpha(Cl_E A)$.
(4) If A_τ is a family of nonempty, closed bounded sets for $\tau > \tau_0$ such that $A_\tau \subset A_s$, $\tau \ge s$ and $\alpha(A_\tau) \to 0$ as $\tau \to \infty$, then $\cap_{\tau > \tau_0} A_\tau$ is a nonempty compact set in E.
(5) If $(E, \|\cdot\|_E)$ is a Banach space, then $\alpha(B_1 + B_2) \le \alpha(B_1) + \alpha(B_2)$ for B_1, $B_2 \in C_E$.

Remark 12.1. If the set $A \in \mathcal{D}$ is a random set, then $\alpha(A)$ is a random variable.

Now for a random dynamical system $\phi(t,\omega)$ on E we introduce the following definition, which can be generalized to a general cocycle [7, e.g.].

Definition 12.6. A continuous random dynamical system ϕ is called almost surely $\mathcal{D}$-α-contraction if for any set $A \in \mathcal{D}$

$$\alpha(\phi(t, \theta_{-t}\omega)A(\theta_{-t}\omega)) \to 0 \;\; a.s., \;\; t \to \infty.$$

We give the following result about the sufficient condition for a random dynamical system is almost surely $\mathcal{D}$-α-contraction.

Theorem 12.3. *For a continuous random dynamical system $\phi(t,\omega)$ on E if almost surely the following holds*

(1) $\phi(t,\omega) = \phi_1(t,\omega) + \phi_2(t,\omega)$.
(2) For any tempered random variable $a \ge 0$, there exists $r = r(a) < \infty$ a.s. such that for the closed ball B_a with radius a in E, $Cl_E\big(\phi_1(t, \theta_{-t}\omega)B_a(\theta_{-t}\omega)\big)$ is compact in E for $t > r(a)$.
(3) $\|\phi_2(t, \theta_{-t}\omega)u\|_E \le k(t, a, \theta_{-t}\omega)$, $t > 0$, $u \in B_a(\theta_{-t}\omega)$ with $k(t, a, \omega)$ is a measurable function with respect to (t, x, ω) which satisfies $k(t, a, \theta_{-t}\omega) \to 0$ as $t \to \infty$.

Then $\phi(t,\omega)$ is almost surely $\mathcal{D}$-α-contraction.

Proof. Let $A \in \mathcal{D}$ and a tempered random variable $a > 0$ such that $A \subset B_a$. Then for any $\varepsilon > 0$ with $t > r(a)$

$$\alpha(\phi(t, \theta_{-t}\omega)A(\theta_{-t}\omega)) \le \alpha(\phi(t, \theta_{-t}\omega)B_a(\theta_{-t}\omega))$$

$$
\begin{aligned}
&= \alpha(\phi_1(t,\theta_{-t}\omega)B_a(\theta_{-t}\omega) + \phi_2(t,\theta_{-t}\omega)B_a(\theta_{-t}\omega)) \\
&\leq \alpha(\phi_2(t,\theta_{-t}\omega)B_a(\theta_{-t}\omega)) + \varepsilon \\
&\leq \operatorname{diam}(\phi_2(t,\theta_{-t}\omega)B_a(\theta_{-t}\omega)) + \varepsilon \\
&\leq 2k(t,a,\theta_{-t}\omega) + \varepsilon.
\end{aligned}
$$

Then the arbitrary ε yields the result. □

For the almost surely $\mathcal{D}$-α-contraction random dynamical system ϕ on E, we give the following result about a sufficient condition of the existence of the global random attractor in E. The argument is just a small adaption of that for deterministic system [15, 31, e.g.].

Theorem 12.4. *Random dynamical system $\phi(t,\omega)$ is continuous in E and*

(1) $\phi(t,\omega)$ has an absorbing set $B_0(\omega) \in \mathcal{D}$.
(2) $\phi(t,\omega)$ is almost surely $\mathcal{D}$-α-contraction.

Then $\phi(t,\omega)$ possesses a global random attractor in E.

Proof. Set

$$
\mathcal{A}(\omega) = \bigcap_{\tau>0} Cl_E\Big\{ \bigcup_{t\geq\tau} \phi(t,\theta_{-t}\omega)B_0(\theta_{-t}\omega)\Big\}.
$$

We claim that $\mathcal{A}(\omega)$ is the unique global random attractor of $\phi(t,\omega)$ in $\mathcal{D}$. In fact, it is known that ([12, 13, 29, e.g.]) $\mathcal{A}(\omega)$ is nonempty, invariant and attracting all the sets in $\mathcal{D}$. So we just prove the compactness of $\mathcal{A}(\omega)$. Let

$$
A_\tau(\omega) = Cl_E\Big\{ \bigcup_{t\geq\tau} \phi(t,\theta_{-t}\omega)B_0(\theta_{-t}\omega)\Big\}.
$$

Then $A_\tau(\omega) \subset B_0(\omega) \in \mathcal{D}$ and $\alpha(A_\tau(\omega)) \to 0$ a.s. as $\tau \to \infty$. By the property (4) in Lemma 12.2, $\mathcal{A}(\omega)$ is compact. The proof is completed. □

To apply the above result we need some bounded estimates of solution in spaces $\mathcal{H}$ and $\mathcal{H}^1$.

3.2. *Boundedness of solutions in $\mathcal{H}$ and $\mathcal{H}^1$*

We just give brief estimates on the solutions, for details we refer to [21].

Let $\rho = v_t + \delta v$, where $\delta \in (0,\delta_0)$ with $\delta_0 = \min(\frac{1}{4}, \frac{\lambda_1}{2})$, then

$$
\nu\rho_t + (1-\delta\nu)\rho + (A - \delta(1-\delta\nu))v + f(u) = 0. \tag{3.1}
$$

Multiplying ρ in H on both sides of (3.1), we have

$$
\begin{aligned}
&\frac{1}{2}\frac{d}{dt}(\|v\|_1^2 + \nu\|\rho\|_0^2) + \delta\|v\|_1^2 + (1-\delta\nu)\|\rho\|_0^2 \\
&\quad - \delta(1-\delta\nu)\langle v,\rho\rangle + \langle f(u),\rho\rangle = 0.
\end{aligned}
$$

By the assumption ($\mathbf{H}_{\text{cub}}$) and a similar discussion for deterministic system [32, e.g.] we have the following estimates for some constant $C > 0$

$$\begin{aligned}&\|v(t,\omega)\|_1^2 + \|v_t(t,\omega)\|_0^2\\ &\le Q(t,\omega)e^{-\alpha t} + C\int_0^t e^{-\alpha(t-s)}g(s,\omega)\,ds\end{aligned}$$

with $Q(t,\omega)$, $g(t,\omega)$, which grow at most polynomial, just depend on (η,η_t) and the domain D. By the property of $(\eta\,,\eta_t)$, let

$$r_1^2(\omega) = 2C\lim_{t\to\infty}\int_0^t e^{-\alpha(t-s)}g(s,\omega)\,ds$$

which is a $\mathbb{P}$-a.s. finite random variable. Then for any $\mathcal{D}$-set B in $\mathcal{H}$, there is $T_0(B) > 0$ such that for any $(u_0,u_1) \in B$

$$\|v(t,\theta_{-t}\omega)\|_1^2 + \|v_t(t,\theta_{-t}\omega)\|_0^2 \le r_1^2(\omega), \quad \text{for } t \ge T_0(B) \ \ \mathbb{P}\text{-}a.s.. \tag{3.2}$$

We further need the estimates in $\mathcal{H}^1$ for (v,v_t) with $(u_0,u_1) \in \mathcal{H}^1$. For some $\beta > 0$ small enough, multiplying equation (2.10) by $\Delta(v_t + \beta v)$ in H, we have

$$\begin{aligned}&\frac{1}{2}\frac{d}{dt}\left[\nu\|\nabla v_t\|_0^2 + \|\Delta v\|_0^2 - 2\beta\nu\langle v_t,\Delta v\rangle\right]\\ &\quad + (1-\beta\nu)\|\nabla v_t\|_0^2 + \beta\|\Delta v\|_0^2 + \beta\langle\nabla v_t,\nabla v\rangle\\ &= \beta\langle f(u),\Delta v\rangle + \langle f(u),\Delta v_t\rangle.\end{aligned}$$

We have, with TrQ small enough, a $\mathbb{P}$-a.s. finite random variable $r_2(\omega)$ such that for any $\mathcal{D}$-set B in $\mathcal{H}^1$, there is $T_1(B) > 0$ such that for any $(u_0,u_1) \in \mathcal{H}^1$

$$\nu\|v_t(t,\theta_{-t}\omega)\|_1^2 + \|v(t,\theta_{-t}\omega)\|_2^2 \le r_2^2(\omega), \ \ t \ge T_1(B), \ \ \mathbb{P}\text{-}a.s.. \tag{3.3}$$

By the above analysis and Theorem 12.4, to construct a random attractor we just prove that $\Phi^\nu(t,\theta_{-t}\omega)$ is almost surely $\mathcal{D}$-α-contraction in $\mathcal{H}^1$. Write

$$\begin{aligned}v(t,\theta_{-t}\omega) &= S_\nu(t)v(0) + \int_0^t S_\nu(t-s)G(s,\theta_{s-t}\omega)ds\\ &= v_1(t,\theta_{-t}\omega) + v_2(t,\theta_{-t}\omega),\end{aligned}$$

with

$$G(t,\theta_{-t}\omega) = \big(0, f(v(t,\theta_{-t}\omega) + \eta(\omega))\big).$$

Then v_1 and v_2 satisfy

$$\nu v_{1tt} + v_{1t} + Av_1 = 0, \ \ v_1(0) = u_0, \ \ v_{1t}(0) = u_1 \tag{3.4}$$

and

$$\nu v_{2tt} + v_{2t} + Av_2 + f(u) = 0, \ \ v_2(0) = 0, \ \ v_{2t}(0) = 0. \tag{3.5}$$

By a standard calculation for (3.4), we have for some positive constant $\gamma > 0$ independent of ν

$$\|\tilde{v}_1(t)\|_0^2 + \|v_1(t)\|_1^2 \le e^{-\gamma t}(\|v(0)\|_1^2 + \|v_t(0)\|_0^2) \tag{3.6}$$

where $\tilde{v}_1 = v_{1t} + \delta v_1$. Then the condition (3) of Theorem 12.3 holds. Notice that $(v_2(0), v_{2t}(0)) = (0,0) \in \mathcal{H}^1$, by the estimates in $\mathcal{H}^1$ we have

$$\|v_{2t}(t,\theta_{-t}\omega)\|_1^2 + \|v_2(t,\theta_{-t}\omega)\|_2^2 \le r_2^2(\omega), \quad t > T_2, \quad \mathbb{P}\text{-}a.s. \tag{3.7}$$

for some $T_2 > 0$. For any tempered random ball $B_a \subset H^2(D) \cap H_0^1(D)$, by the assumptions of f, the estimates of (3.6), (3.7) and the embedding $H^2(D) \subset L^\infty(D)$, $\{f(v(s,\theta_{-s}\omega) + \eta(\omega)) : 0 \le s \le t, v(0) \in B_a(\omega)\}$ is bounded in $H^2(D) \cap H_0^1(D)$ and therefore a compact set of $H_0^1(D)$ a.s.. Then $\{v_2(s,\theta_{-s}\omega) : 0 \le s \le t, v(0) \in B_a(\omega)\}$ is precompact in $\mathcal{H}^1$. So condition (2) of Theorem 12.3 holds. By Theorem 12.3, $\Phi^\nu(t,\theta_{-t}\omega)$ is almost surely $\mathcal{D}$-α-contraction in $\mathcal{H}^1$. That means $\Phi^\nu(t,\theta_{-t}\omega)$ possesses a compact global random attractor $\mathcal{A}^\nu(\omega) \subset \mathcal{H}^1$. We summarize the main result of this section as the following theorem.

Theorem 12.5. *Suppose the assumptions on f hold and* $\mathrm{Tr}Q$ *is small, system (1.2)–(1.3) possesses a global random attractor $\mathcal{A}^\nu(\omega)$ in $\mathcal{H}^1$.*

3.3. *Upper semicontinuity of $\mathcal{A}^\nu(\omega)$*

We next show the upper semicontinuous in probability with respect to ν for the random attractor $\mathcal{A}^\nu(\omega)$. First we present an abstract result [5].

Let $\phi(t)$ be a dynamical system on a Polish space $(E, \|\cdot\|_E)$. And there exists a global compact attractor $\mathcal{A}^0$ in E for the dynamical system ϕ. We perturb the dynamical system $\phi(t)$ by a random element depending on a parameter $\nu \in (0,1]$, so that we obtain an RDS

$$\phi^\nu : \mathbb{R}^+ \times \Omega \times E \to E.$$

If $\mathbb{P}$-a.s. $\omega \in \Omega$ and all $t \in \mathbb{R}^+$

(H1) $$\phi^\nu(t,\theta_{-t}\omega)x \to \phi(t)x, \quad x \in E \;\; as \;\; \nu \searrow 0$$

uniformly on bounded sets of E. Then we can have the following result (see [5]).

Theorem 12.6. *Assume that for all $\nu \in (0,1]$ there is a random attractor $\mathcal{A}^\nu(\omega)$ for RDS ϕ^ν which satisfies* (**H1**), *and there exists a random compact set $K(\omega)$ such that, $\mathbb{P}$-a.s.*

(H2) $$\lim_{\nu\searrow 0} \mathrm{dist}_E(\mathcal{A}^\nu(\omega), K(\omega)) = 0,$$

where $\mathrm{dist}_E(M,N) = \sup_{m\in M} \inf_{n\in N} \|m-n\|_E$ *for M, $N \subset E$. Then*

$$\lim_{\nu\searrow 0} \mathrm{dist}_E(\mathcal{A}^\nu(\omega), \mathcal{A}^0) = 0 \;\; \textit{with probablity one.}$$

Consider the following heat equation

$$z_t + Az + f(z) = 0, \quad z(0) = z_0. \tag{3.8}$$

Under the assumption ($\mathbf{H}_{\text{cub}}$) on f, the system (3.8) possesses a global attractor $\widetilde{\mathcal{A}} \subset H_0^1(D) \cap H^2(D)$ [15, 32, e.g.]. Associated with the attractor $\widetilde{\mathcal{A}}$ we define the set

$$\mathcal{A}^0 = \{(x, y) : x \in \widetilde{\mathcal{A}}, \ \ y = -Ax - f(x)\}, \tag{3.9}$$

which is a natural embedding of the attractor $\widetilde{\mathcal{A}}$ into $\mathcal{H}^1$.

Then we have the following result.

Theorem 12.7.

$$\lim_{\nu \to 0} \text{dist}_{\mathcal{H}}(\mathcal{A}^\nu, \mathcal{A}^0) = 0 \ \ \textit{with probability one.}$$

We justify the assumptions **(H1)**–**(H2)**. The assumption **(H2)** is in fact the property of the random absorbing sets which is followed from the construction of the random attractors. Next we justify **(H1)**. Since we consider upper semicontinuity of the attractor, we just verify the condition **(H1)** for the orbits in the attractor. For this we need some further estimates.

Consider the following linear stochastic differential equation restricted on $\mathcal{A}^\nu(\omega)$

$$\nu y_{tt}^\nu + y_t^\nu + Ay^\nu + F^\nu(t, \theta_{-t}\omega) = 0 \tag{3.10}$$

with

$$F^\nu(t, \omega) = f'(u^\nu)(v_t^\nu + \eta_t^\nu).$$

Applying the Hölder inequality and noticing the assumption on f' with the embedding $H_1 \subset L^6$ we have for some positive constant C

$$\|F^\nu(t, \omega)\|_0 \leq C(1 + \|u^\nu\|_1^2)\|v_t^\nu + \eta_t^\nu\|_0. \tag{3.11}$$

Notice that (3.10) is a linear equation. Let $y_t^\nu(t) = z^\nu(t)$, we rewrite (3.10) as

$$\begin{aligned} y_t^\nu(t) &= z^\nu(t) \\ \nu z_t^\nu(t) &= -Ay^\nu(t) - z^\nu(t) + F^\nu(t, \theta_{-t}\omega). \end{aligned}$$

Multiplying $(y^\nu(t), z^\nu(t))$ in $\mathcal{H}$ on both sides of the above equations,

$$\begin{aligned} &\frac{1}{2}\frac{d}{dt}\Big[\|y^\nu(t)\|_1^2 + \nu\|z^\nu(t)\|_0^2\Big] \\ &\leq -\min\{\lambda_1, \tfrac{1}{2}\}\Big[\|y^\nu(t)\|_1^2 + \|z^\nu(t)\|_0^2\Big] + \|F^\nu(t, \theta_{-t}\omega)\|_0\|z^\nu(t)\|_0. \end{aligned}$$

Noticing that we just consider the system (3.10) on $\mathcal{A}^\nu(\omega)$, by (3.11) and applying the Cauchy inequality, Gronwall lemma, there is a random variable $r_3(\omega)$, which is a.s. finite, such that

$$\nu\|y_t^\nu\|_0^2 \leq r_3^2(\omega).$$

By (2.10)–(2.11), v_t^ν satisfies the equation (3.10). Then we have

$$\nu\|v_{tt}^\nu(t, \theta_{-t}\omega)\|_0^2 \leq r_3^2(\omega). \tag{3.12}$$

Proof of Theorem 12.7

Let $\phi(t)$ be the dynamical system generated by (3.8) and $\phi^\nu(t,\omega) = \Phi^\nu(t,\omega)$. It is enough to just verify the condition **(H1)** for the orbits in the random attractor $\mathcal{A}^\nu(\omega)$. That is for a fixed $\omega \in \Omega$, we prove for any $\nu_n \to 0$ as $n \to \infty$ there is a subsequence, which we still write as ν_n, such that $(v^{\nu_n}(t,\theta_{-t}\omega), v_t^{\nu_n}(t,\theta_{-t}\omega)) \to (v(t), v_t(t))$ with $(v^{\nu_n}(0), v_t^{\nu_n}(0)) = (v(0), v_t(0))$, with (v, v_t) satisfying system (3.8).

Let $V^{\nu_n} = v^{\nu_n} - v$, then V^{ν_n} satisfies the following equation

$$V_t^{\nu_n} = \Delta V^{\nu_n} - f(v^{\nu_n} + \eta^{\nu_n}) + f(v) - \nu_n v_{tt}^{\nu_n} \tag{3.13}$$

with $V^{\nu_n}(0) = 0$. Since we restrict the system on $\mathcal{A}^\nu(\omega)$, by the assumptions of f', f is Lipschtiz with Lipschtiz constant $L(\omega)$ which is an a.s. finite random variable. Taking scalar product in $L^2(D)$ with $-\Delta V^{\nu_n}$, we have

$$\begin{aligned}\frac{1}{2}\frac{d}{dt}\|V^{\nu_n}\|_1^2 &\le \langle f(v^{\nu_n} + \eta^{\nu_n}) - f(v), \Delta V^{\nu_n}\rangle + \nu_n\langle v_{tt}^{\nu_n}, \Delta V^{\nu_n}\rangle \\ &\le L(\omega)\|V^{\nu_n}\|_1^2 + L(\omega)\|\Delta V^{\nu_n}\|_0\|\eta^{\nu_n}\|_0 + \nu_n\|v_{tt}^{\nu_n}\|_0\|\Delta V^{\nu_n}\|_0.\end{aligned}$$

Then

$$\|V^{\nu_n}(t)\|_1 \to 0 \ \ a.s. \ \text{as} \ n \to \infty, \ \ for \ \ t \in \mathbb{R} \tag{3.14}$$

by (3.12) and the fact that $\eta^{\nu_n} \to 0$ as $n \to \infty$ a.s..

Multiplying both sides of (4.12) with $V_t^{\nu_n}$ in $L^2(D)$, we have

$$\|V_t^{\nu_n}\|_0^2 \le \|V^{\nu_n}\|_1\|V_t^{\nu_n}\|_1 + \|f(v^{\nu_n} + \eta^{\nu_n}) - f(v)\|_0\|V_t^{\nu_n}\|_0 + \nu\|v_{tt}^{\nu_n}\|_0\|V_t^{\nu_n}\|_0.$$

Then also by $\eta^{\nu_n} \to 0$ as $n \to \infty$ a.s., (3.12) and (3.14),

$$\|V_t^{\nu_n}(t)\|_0 \to 0 \ \ a.s. \ \text{as} \ n \to \infty, \ \ for \ \ t \in \mathbb{R}. \tag{3.15}$$

Thus the condition of **(H1)** in Theorem 12.6 is satisfied which ends the proof.

4. Approximation of measure attractor

This section is concerned with the approximation of the dynamics of (1.2)–(1.3) in the sense of distribution.

In the following we emphasize the superscript ν and denote by (u^ν, u_t^ν) the solution to equations (1.2)–(1.3). Let $\mathcal{M}(\mathcal{H})$ be the set of all the probability measures on $\mathcal{H}$. For $\mathfrak{u} \in \mathcal{M}(\mathcal{H})$ and a sequence $\mathfrak{u}_n \subset \mathcal{M}(\mathbb{H})$ we say

$$\mathfrak{u}_n \to \mathfrak{u} \iff \int_{\mathcal{H}} \vartheta(h)\mathfrak{u}_n(dh) \to \int_{\mathcal{H}} \vartheta(h)\mathfrak{u}(dh)$$

for any bounded continuous function $\vartheta \in C_b(\mathcal{H}, \mathbb{R}) := C_b(\mathcal{H})$. This is called the weak convergence between measures and the topology of such weak convergence can be metrizable [3]. Denoted by d_{P} the associate metric to weak convergence, which is called the *Prokhorov* metric [3].

For our system, define

$$\mathbb{X}_2 = \{\mathfrak{u} \in \mathcal{M}(\mathcal{H}) : \int_{\mathcal{H}} \|z\|_{\mathcal{H}}^2 \mathfrak{u}(dz) < \infty\}, \tag{4.1}$$

then $(\mathbb{X}_2, d_{\mathrm{P}})$ is a complete metric space. We also need the following space for integer $m \geq 3$,

$$\mathbb{X}_m = \{\mathfrak{u} \in \mathcal{M}(\mathcal{H}) : \int_{\mathcal{H}} \|z\|^m_{\mathcal{H}} \mathfrak{u}(dz) < \infty\}, \tag{4.2}$$

with complete metric d_{P}.

Denote by ϕ^ν the solution operator for system (1.2)–(1.3), that is,

$$\phi^\nu(t, (u_0, u_1)) = (u^\nu(t, x, \omega), u^\nu_t(t, x, \omega)), \quad (u_0, u_1) \in \mathcal{H}. \tag{4.3}$$

Then define a family of maps $\{\mathbb{S}^\nu(t)\}_\nu$ on $\mathcal{M}(\mathcal{H})$ as

$$\int_{\mathcal{H}} \vartheta(h)[\mathbb{S}^\nu(t)\mathfrak{u}](dh) = \int_{\mathcal{H}} \mathbb{E}\vartheta(\phi^\nu(t, h))\mathfrak{u}(dh) \tag{4.4}$$

for any $\vartheta \in C_b(\mathcal{H})$. Then $\mathbb{S}^\nu(t) : \mathcal{M}(\mathcal{H}) \to \mathcal{M}(\mathcal{H})$ defines a continuous semigroup on $\mathbb{X}_m$ with

(1) $\mathbb{S}^\nu(0) = Id_{\mathbb{X}_m}$,
(2) $\mathbb{S}^\nu(t+s) = \mathbb{S}^\nu(t)\mathbb{S}^\nu(s), \quad t, s \geq 0.$

Remark 12.2. The semigroup $\{\mathbb{S}^\nu(t)\}_{t\geq 0}$ is also called a Markov semigroup [18, e.g.] defined by system (1.2)–(1.3).

Next we construct an attractor for semigroup $\mathbb{S}^\nu(t)$. For this we need some estimates in $\mathbb{X}_m$ with appropriate m.

4.1. *Some estimates on solutions*

Let (u_0, u_1) be random variables with values in $H^1_0(D) \times L^2(D)$ and $\rho^\nu = u^\nu_t + \delta u^\nu$ for some $0 < \delta < \min\{\frac{1}{4}, \frac{\lambda_1}{2}\}$, then

$$\nu\rho^\nu_t = -(1 - \nu\delta)\rho^\nu + \Delta u^\nu + \delta(1 - \nu\delta)u^\nu + f(u^\nu) + \sqrt{\nu}\dot{W}(t).$$

By the Itô formula,

$$\begin{aligned}
&\frac{1}{2}\frac{d}{dt}\Big[\|u^\nu\|_1^2 + \nu\|\rho^\nu\|_0^2\Big] \\
&= -\delta\|u^\nu\|_1^2 + (\nu\delta - 1)\|\rho^\nu\|_0^2 + \delta(1 - \nu\delta)\langle u^\nu, \rho\rangle + \\
&\quad \langle f(u^\nu), \rho\rangle + \epsilon\langle \dot{W}(t), \rho\rangle + \frac{1}{2}B_0,
\end{aligned}$$

and further by the definition of F and ρ,

$$\begin{aligned}
&\frac{1}{2}\frac{d}{dt}\Big[\|u^\nu\|_1^2 + \nu\|\rho^\nu\|_0^2 - 2\langle F(u^\nu), 1\rangle\Big] \\
&= -\delta\|u^\nu\|_1^2 + (\nu\delta - 1)\|\rho^\nu\|_0^2 + \delta(1 - \nu\delta)\langle u^\nu, \rho^\nu\rangle + \\
&\quad \delta\langle f(u^\nu), u^\nu\rangle + \epsilon\langle \dot{W}(t), \rho^\nu\rangle + \frac{1}{2}B_0.
\end{aligned} \tag{4.5}$$

Then with the assumption $(\mathbf{H}_{\mathrm{cub}})$ and standard estimates [24, e.g.], we have for any $t \geq 0$,

$$\mathbb{E}\Big[\|u^\nu\|_1^2 + \nu\|\rho^\nu\|_0^2 - 2\langle F(u^\nu), 1\rangle\Big] \leq C\mathbb{E}\left(1 + C'_\alpha + \|u_0\|_1^2 + \|u_1\|_0^2\right) \tag{4.6}$$

for some positive constant C.

For a later purpose we need estimate $\mathbb{E}\|u^\nu(t)\|_1^m$, for positive integer m. Let

$$U(t) = \|u^\nu\|_1^2 + \nu\|\rho^\nu\|_0^2 - 2\langle F(u^\nu), 1\rangle.$$

By (4.5) and noticing that $\nu\|\rho^\nu(t)\|_0^2 \le U(t)$, applying Itô formula to $U^m(t)$ yields a positive constant C'_m,

$$\begin{aligned}\frac{d}{dt}U^m(t) &\le mU^{m-1}(t)\frac{dU(t)}{dt} + \frac{1}{2}m(m-1)U^{m-1}(t)B_0\\ &\le -C'_m U^m(t) + \frac{C'_m}{2} + \epsilon m U^m(t)\langle\rho^\nu, \dot{W}\rangle\,.\end{aligned} \tag{4.7}$$

Then by Gronwall lemma, for some positive constant C_m,

$$\mathbb{E}U^m(t) \le C_m\mathbb{E}\left(1 + \|u_0\|_1^{2m} + \|u_1\|_0^{2m}\right), \quad t > 0. \tag{4.8}$$

We also need an estimate on $\int_0^t \|u_t^\nu(s)\|_0^2 ds$. Applying the Itô formula to $\|u_t^\nu\|_0^2$ yields

$$\frac{d}{dt}\left[\frac{\nu}{2}\|u_t^\nu\|_0^2 + \frac{1}{2}\|u^\nu\|_1^2 - \langle F(u^\nu), 1\rangle\right] = -\|u_t^\nu\|_0^2 + \frac{1}{2}B_0 + \sqrt{\nu}\langle\dot{W}, u_t^\nu\rangle.$$

Then

$$\int_0^t \|u_t^\nu(s)\|_0^2 ds = -H(t) + H(0) + \sqrt{\nu}\int_0^t \langle u_t^\nu(s), dW(s)\rangle\,,$$

where $H(t) = \frac{1}{2}\Big[\nu\|u_t^\nu\|_0^2 + \|u^\nu\|_1^2 - 2\langle F(u^\nu), 1\rangle - \frac{1}{2}B_0 t\Big]$. Then for any $T > 0$, by (4.6), there is a positive constant C_T such that

$$\mathbb{E}\int_0^T \|u_t^\nu(s)\|_0^2\, ds \le C_T\mathbb{E}\left(1 + \|u_0\|_1^2 + \|u_1\|_1^2\right). \tag{4.9}$$

Due to the nonlinearity, in order to have some information of the long time behavior of solutions, we further estimate (u, u_t) in space $\mathcal{H}^1$. Assume (u_0, u_1) to be random variables taking values in $\mathcal{H}^1$. By the Itô formula we have

$$\begin{aligned}&\frac{d}{dt}\left[\frac{\nu}{2}\|\rho^\nu\|_1^2 + \frac{1}{2}\|u^\nu\|_2^2 - \frac{1}{2}\delta(1-\nu\delta)\|u^\nu\|_1^2\right]\\ &= (\nu\delta - 1)\|\rho^\nu\|_1^2 - \delta\|u^\nu\|_2^2 + \delta^2(1-\nu\delta)\|u^\nu\|_1^2 - \langle f(u^\nu), \Delta\rho^\nu\rangle\\ &\quad -\sqrt{\nu}\langle\Delta\rho^\nu, \dot{W}\rangle + \frac{1}{2}B_1.\end{aligned} \tag{4.10}$$

Then by some standard estimates [24], we have for some positive constant C, for any $t \ge 0$,

$$\mathbb{E}\left[\nu\|\rho^\nu\|_1^2 + \|u^\nu\|_2^2\right] \le C\mathbb{E}\left(1 + \|u_0\|_2^2 + \|u_1\|_1^2 + \|u_0\|_1^6 + \|u_1\|_0^6\right). \tag{4.11}$$

We also need an estimate on $\mathbb{E}\|u^\nu(t)\|_2^{2m}$ for any positive integer m. Let

$$V(t) = \frac{\nu}{2}\|\rho^\nu\|_1^2 + \frac{1}{2}\|u^\nu\|_2^2 - \frac{1}{2}\delta(1-\nu\delta)\|u^\nu\|_1^2.$$

Applying Itô formula to $V^m(t)$, we have, by a similar method to estimate $U^m(t)$,

$$\mathbb{E}\|u^\nu(t)\|_2^{2m} \le C\mathbb{E}\left(1 + \|u_0\|_2^{6m} + \|u_1\|_1^{6m}\right), \quad t > 0 \tag{4.12}$$

for some positive constant C which is independent of ν. Now we give a uniform estimate for u_t^ν with respect to ν. In fact, let $v^\nu = u_t^\nu$, then v^ν solves

$$v_t^\nu = \frac{1}{\nu}\Delta u^\nu - \frac{1}{\nu}v^\nu + \frac{1}{\nu}f(u^\nu) + \frac{1}{\sqrt{\nu}}\dot{W}(t), \quad v(0) = v_0. \tag{4.13}$$

Applying the Itô formula we have for some positive constant C

$$\begin{aligned} &\mathbb{E}\|u_t^\nu(t)\|_0^2 = \mathbb{E}\|v^\nu(t)\|_0^2 \\ &\le C\left(1 + \mathbb{E}\|u_0\|_2^2 + \mathbb{E}\|u_1\|_1^2 + \mathbb{E}\|u_0\|_1^6 + \mathbb{E}\|u_1\|_0^6\right). \end{aligned} \tag{4.14}$$

4.2. *Measure attractor*

This section construct a measure attractor for the semigroup $\{\mathbb{S}^\nu(t)\}_{t\ge 0}$ in some appropriate space $\mathbb{X}_m$.

We apply a splitting technique of semigroup. Let $u^\nu = p^\nu + q^\nu$ with

$$\nu p_{tt}^\nu + p_t^\nu = \Delta p^\nu\,, \quad p^\nu(0) = u_0\,, \quad p_t^\nu(0) = u_1\,, \tag{4.15}$$

$$\nu q_{tt}^\nu + q_t^\nu = \Delta q^\nu + f(u^\nu) + \sqrt{\nu}\dot{W}\,, \quad q^\nu(0) = 0\,, \quad q_t^\nu(0) = 0\,, \tag{4.16}$$

with Dirichlet boundary condition. Let

$$\begin{aligned} \phi_1^\nu(t,(u_0,u_1)) &= (p^\nu(t,x), p_t^\nu(t,x))\,, \\ \phi_2^\nu(t,(u_0,u_1)) &= (q^\nu(t,x,\omega), q_t^\nu(t,x,\omega))\,. \end{aligned}$$

Then $\phi^\nu(t,(u_0,u_1)) = \phi_1^\nu(t,(u_0,u_1)) + \phi_2^\nu(t,(u_0,u_1))$. By a standard analysis, there is positive constant $\gamma > 0$ such that for any $(u_0,u_1) \in \mathcal{H}$,

$$\mathbb{E}\|\phi_1^\nu(t,(u_0,u_1))\|_{\mathcal{H}} \le e^{-\gamma t}\mathbb{E}\|(u_0,u_1)\|_{\mathcal{H}}\,, \quad t \ge 0\,. \tag{4.17}$$

And by a similar analysis for (u^ν, v^ν) in last subsection,

$$\mathbb{E}\|\phi_2^\nu(t,(u_0,u_1))\|_{\mathcal{H}_1}^2 \le C_\nu \mathbb{E}\|(u_0,u_1)\|_{\mathcal{H}}^6\,, \quad t \ge 0 \tag{4.18}$$

for some positive constant $C_\nu > 0$ which may depend on ν. Then we define maps $\mathbb{S}_1^\nu(t)$ and $\mathbb{S}_2^\nu(t)$ on $\mathbb{X}_2$ as

$$\int_{\mathcal{H}} \vartheta(h)[\mathbb{S}_1^\nu(t)\mathfrak{u}](dh) = \int_{\mathcal{H}} \mathbb{E}\vartheta(\phi_1^\nu(t,h))\mathfrak{u}(dh)\,, \tag{4.19}$$

$$\int_{\mathcal{H}} \vartheta(h)[\mathbb{S}_2^\nu(t)\mathfrak{u}](dh) = \int_{\mathcal{H}} \mathbb{E}\vartheta(\phi_2^\nu(t,h))\mathfrak{u}(dh)\,. \tag{4.20}$$

Here we note that $\mathbb{S}_2^\nu(t)$ may not be a semigroup. It is natural to think that for any $\mathfrak{u} \in \mathbb{X}_2$, $\mathbb{S}^\nu(t)\mathfrak{u} = [\mathbb{S}_1^\nu(t)\mathfrak{u}] * [\mathbb{S}_2^\nu(t)\mathfrak{u}]$. Here * denotes the convolution between two measures. Unfortunately, it is untrue. But we have the following asymptotic property

$$\mathbb{S}^\nu(t)\mathfrak{u} - \mathbb{S}_1^\nu(t)\mathfrak{u} * \mathbb{S}_2^\nu(t)\mathfrak{u} \to 0\,, \quad t \to \infty\,. \tag{4.21}$$

In fact for any $\vartheta \in C_b(\mathcal{H})$, we have, by Fatou lemma and (4.17),

$$\begin{aligned}
&\limsup_{t\to\infty}\Big|\int_{\mathcal{H}}\vartheta(h)\mathbb{S}^\nu(t)\mathfrak{u}(dh) - \int_{\mathcal{H}}\vartheta(h)[\mathbb{S}_1^\nu(t)\mathfrak{u} * \mathbb{S}_2^\nu(t)\mathfrak{u}](dh)\Big| \\
&= \limsup_{t\to\infty}\Big|\int_{\mathcal{H}}\mathbb{E}\vartheta(\phi^\nu(t,h_2))\mathfrak{u}(dh_2) \\
&\qquad - \int_{\mathcal{H}}\int_{\mathcal{H}}\mathbb{E}\vartheta(\phi_1^\nu(t,h_1)+\phi_2^\nu(t,h_2))\mathfrak{u}(dh_1)\mathfrak{u}(dh_2)\Big| \\
&\le \int_{\mathcal{H}}\int_{\mathcal{H}}\mathbb{E}\limsup_{t\to\infty}\Big|\vartheta(\phi^\nu(t,h_2)) - \vartheta(\phi_1^\nu(t,h_1)+\phi_2^\nu(t,h_2))\Big|\mathfrak{u}(dh_1)\mathfrak{u}(dh_2) \\
&= 0.
\end{aligned}$$

Now define $\widetilde{\mathbb{S}}^\nu(t) = \mathbb{S}_1^\nu(t) * \mathbb{S}_2^\nu(t)$ by

$$\widetilde{\mathbb{S}}^\nu(t)\mathfrak{u} = \mathbb{S}_1^\nu(t)\mathfrak{u} * \mathbb{S}_2^\nu(t)\mathfrak{u}$$

for any $\mathfrak{u} \in \mathbb{X}_2$. In the following we prove that for any $\mathfrak{u} \in \mathbb{X}_2$, if $\mathfrak{u} \in \mathbb{X}_6$, $\{\widetilde{\mathbb{S}}^\nu(t)\mathfrak{u}\}_{t\ge 0}$ is tight. In fact by (4.18), for any $t > 0$,

$$\int_{\mathcal{H}}\|h\|^2_{\mathcal{H}_1}\mathbb{S}_2^\nu(t)\mathfrak{u}(dh) = \int_{\mathcal{H}}\mathbb{E}\|\phi_2^\nu(t,h)\|^2_{\mathcal{H}^1}\mathfrak{u}(dh) \le C_\nu\int_{\mathcal{H}}\|h\|^6_{\mathcal{H}}\mathfrak{u}(dh)\,. \tag{4.22}$$

Then by the Chebyshev inequality, for any $\epsilon > 0$, there is set $\mathbb{K}_\epsilon \subset \mathcal{H}^1$ such that for any $t \ge 0$,

$$\mathbb{S}_2^\nu(t)\mathfrak{u}(\mathbb{K}_\epsilon) \ge 1 - \epsilon\,. \tag{4.23}$$

Now for integer $m \ge 6$ and $r > 0$, we define the r-ball in $\mathbb{X}_m$ as

$$\mathbb{B}_m^r = \{\mathfrak{u} \in \mathbb{X}_2 : \int_{\mathcal{H}}\|h\|^m_{\mathcal{H}}\mathfrak{u}(dh) < r\}. \tag{4.24}$$

For any $\mathfrak{u} \in \mathbb{B}_m^r$ with $r > 0$, by estimates (4.8),

$$\int_{\mathcal{H}}\|h\|^m_{\mathcal{H}}\mathbb{S}^\nu(t)\mathfrak{u}(dh) = \int_{\mathcal{H}}\mathbb{E}\|\phi(t,h)\|^m_{\mathcal{H}}\mathfrak{u}(dh) \le C_{m,\nu}(1 + e^{-\gamma t}r),$$

with $C_{m,\nu}$ is one positive constant depending on m and ν. Then take $R = 2C_{m,\nu}$, for any $r > 0$, there is $t_r > 0$ such that $t > t_r$, $\mathbb{S}^\nu(t)\mathbb{B}_m^r \subset \mathbb{B}_m^R$. That means semigroup $\mathbb{S}^\nu(t)$ has an absorbing set $\mathbb{B}_m^R$ in $\mathbb{X}_m$. Moreover, as $m \ge 6$, $\mathbb{B}_m^R \subset \mathbb{X}_6$, by the estimate (4.23) $\{\mathbb{S}^\nu(t)\mathbb{B}_m^R\}_t$ is precompact. Then $\{\mathbb{S}^\nu(t)\}_{t\ge 0}$ has a global compact attractor. Write

$$\mathbb{A}^\nu = \bigcap_{t>0}\overline{\bigcup_{s\ge t}\mathbb{S}^\nu(s)\mathbb{B}_m^R}^{d_{\mathrm{P}}}\,.$$

Then $\mathbb{A}^\nu$ is a global attractor for semigroup $\{\mathbb{S}^\nu(t)\}_{t\ge 0}$, which is compact and attracts all bounded sets in $\mathbb{X}_m$. Moreover, $\mathbb{A}^\nu$ is supported on the space $\mathbb{H}_1$. We call the global attractor $\mathbb{A}^\nu$ the measure attractor for semigroup $\{\mathbb{S}^\nu(t)\}_{t\ge 0}$. Then we draw the following result.

Theorem 12.8. *Assume (2.3) holds. The semigroup $\{\mathbb{S}^\nu(t)\}_{t\ge 0}$ has a compact global attractor $\mathbb{A}^\nu$, supported on space $\mathcal{H}^1$, which attracts all bounded sets in $\mathbb{X}_m$.*

Remark 12.3. The attractor $\mathbb{A}^\nu$ is also called Schmalfuß attractor [11, 30] for the Markov semigroup defined by (1.2)–(1.3).

4.3. *Continuity of attractor* $\mathbb{A}^\nu$

By Theorem 12.8, $\mathbb{A}^\nu \subset \mathcal{M}(\mathcal{H}^1)$. So we consider the part of $\mathbb{A}^\nu$ in space $\mathcal{M}(H_0^1(D)\cap H^2(D))$. In order to define a semigroup, we fix the initial value u_1 and define a semigroup $\{S_{u_1}^\nu(t)\}_{t\geq 0}$ on space $\mathcal{M}(H_0^1(D)\cap H^2(D))$. Then we construct a global attractor $\mathcal{A}_{u_1}^\nu$ for $\{S_{u_1}^\nu(t)\}_{t\geq 0}$ and prove the continuity property of $\mathcal{A}_{u_1}^\nu$ at $\nu=0$.

We fix the initial value of u_1. Then $u^\nu(t;u_0,u_1)$ just depends on u_0 and indeed define a semigroup $\{S_{u_1}^\nu(t)\}_{t\geq 0}$ on space $\mathcal{M}(H_0^1(D)\cap H^2(D))$ by

$$\int_{H_0^1(D)} \vartheta(h)S_{u_1}^\nu(t)\mu(dh) = \int_{H_0^1(D)} \mathbb{E}\vartheta(u^\nu(t,h;u_1))\mu(dh) \tag{4.25}$$

for any $\mu \in \mathcal{M}(H_0^1(D)\cap H^2(D))$ and $\vartheta \in C_b(H_0^1(D)\cap H^2(D))$.

Define space X_2 by

$$X_2 = \Big\{\mu \in \mathcal{M}(H_0^1(D)\cap H^2(D)) : \int_{H_0^1(D)} \|h\|_1^2\mu(dh) < \infty\Big\},$$

and subspace $X_m \subset X_2$ by

$$X_m = \Big\{\mu \in \mathcal{M}(H_0^1(D)\cap H^2(D)) : \int_{H_0^1(D)} \|h\|_1^m\mu(dh) < \infty\Big\}.$$

Then (X_m, d_{P}), $m\geq 2$, is a complete metric space. For integer $m\geq 6$ and $r>0$, define r-ball in X_m as

$$B_m^r = \left\{\mu \in X_2 : \int_{H_0^1(D)} \|h\|_{H_0^1(D)}^m\mu(dh) < \infty\right\}.$$

For any $\mu \in B_m^r$ with $r>0$, by the estimate (4.8),

$$\begin{aligned}
&\int_{H_0^1(D)} \|h\|_{H_0^1(D)}^m S_{u_1}^\nu(t)\mu(dh)\\
&= \int_{H_0^1(D)} \mathbb{E}\|u^\nu(t;h,u_1)\|_{H_0^1(D)}^m\mu(dh)\\
&\leq C_m(1+e^{-\gamma t})
\end{aligned}$$

with positive constant C_m independent of ν. Then let $R=2C_m$, B_m^R is an absorbing set for $\{S_{u_1}^\nu(t)\}_t$ in X_m. Then by the same analysis of the previous section, semigroup $\{S_{u_1}^\nu(t)\}_t$ has a compact attractor $\mathcal{A}_{u_1}^\nu \subset X_m$ which attracts all bounded sets of X_m with $m\geq 6$ and $\mathcal{A}_{u_1}^\nu$ is supported on space $H_0^1(D)\cap H^2(D)$. Moreover, for any $\epsilon>0$, there is $T(\nu,\epsilon)>0$ such that

$$\mathrm{dist}(S_{u_1}^\nu(t)B_m^R, \mathcal{A}_{u_1}^\nu) \leq \epsilon,\ \ t > T(\nu,\epsilon,R). \tag{4.26}$$

And $\bigcup_{0<\nu<1}\mathcal{A}_{u_1}^\nu$ is contained in a compact set $\mathbf{B}\subset X_m$.

Now we construct a measure attractor for the deterministic equations

$$u_t = \Delta u + f(u),\ \ u(0)=u_0 \ \text{ in } D, \tag{4.27}$$

$$u(t) = 0 \ \text{ on } \Gamma. \tag{4.28}$$

For any $u_0 \in H_0^1(D) \cap H^2(D)$, (4.27)–(4.28) have a unique solution $u(t,x) \in C(\mathbb{R}^+; H^2(D))$. Let $S(t)$ be the semigroup on $\mathcal{M}(H_0^1(D) \cap H^2(D))$ defined as

$$\int_{H_0^1(D)} \vartheta(h) S(t)\mu(dh) = \int_{H_0^1(D)} \vartheta(u(t,h))\mu(dh) \tag{4.29}$$

for any $\mu \in \mathcal{M}(H_0^1(D) \cap H^2(D))$ and $\vartheta \in C_b(H_0^1(D) \cap H^2(D))$. By some classical estimates, semigroup $\{S(t)\}_t$ has one global attractor $\mathcal{A}$ which is compact and attracts all bounded sets in X_m, $m \geq 6$, and by a proper choice of $R > 0$, B_m^R is also an absorbing set of $S(t)$.

Now we recall some results about the continuity of semigroup depending on a parameter. Let X be a complete separable metric space with metric d_X. For any A, $B \subset X$, define

$$\mathrm{dist}(A, B) = \sup_{x \in A} \inf_{y \in B} d_X(x,y)\,.$$

Consider a semigroup $S_\lambda(t) : X \to X$, $t \in \mathbb{T}$, which depends on a parameter $\lambda \in \Lambda$. Suppose for any $\lambda \in \Lambda$, $S_\lambda(t)$ has a compact attractor $\mathcal{M}_\lambda$. We assume the following conditions (**A**) hold:

(1) Λ is compact with metric d_Λ and λ_0 is a nonisolated point of Λ.
(2) For each $\lambda \in \Lambda$, the semigroup $\{S_\lambda(t)\}_{t\in\mathbb{T}}$ possesses an absorbing set B_λ.
(3) There exists a bounded absorbing set B_0 and for each $\lambda \in \Lambda$, $B_\lambda \subset B_0$.

Denote by $\mathcal{O}_\epsilon(\mathcal{M}_\lambda)$ the ϵ neighborhood of $\mathcal{M}_\lambda$. Then $\mathcal{O}_\epsilon(\mathcal{M}_\lambda)$ is an absorbing set. Assume that we know a function $T(\lambda, \epsilon, B_0)$ such that

$$\mathrm{dist}(S_\lambda(t, B_0), \mathcal{M}_\lambda) \leq \epsilon, \quad t \geq T(\lambda, \epsilon, B_0)\,.$$

Then we give the following result on the continuity of $\mathcal{M}_\lambda$ [16].

Theorem 12.9. *Under assumptions (**A**). If for any $\epsilon > 0$ there exists $\delta > 0$ such that for any λ with $d_\Lambda(\lambda, \lambda_0) < \delta$ there exists a point $T_\lambda \geq T(\lambda, \epsilon, B_0)$ which satisfies*

$$d_X(S_\lambda(T_\lambda)x, S_{\lambda_0}(T_\lambda)x) < \epsilon\,, \quad \textit{for all } x \in B_0\,. \tag{4.30}$$

Then $\mathcal{M}_\lambda$ is continuous at the point λ_0.

Next we verify the conditions in Theorem 12.9. First by the construction of attractors for $S_{u_1}^\nu(t)$ and $S(t)$, condition (**A**) holds. Hence it remains to verify (4.30). By the estimates on linear system [24, section 7.2], we have

$$\begin{aligned} & d_{\mathrm{P}}^2(S_{u_1}^\nu(t)\mu, S(t)\mu) \leq d_{\mathrm{W}}(S_{u_1}^\nu(t)\mu, S(t)\mu) \qquad (4.31)\\ = & \sup_{|\vartheta|_L \leq 1} \left\{ \left| \int_{H_0^1(D)} \vartheta(h)\, S_{u_1}^\nu(t)\mu(dh) - \int_{H_0^1(D)} \vartheta(h)\, S(t)\mu(dh) \right| \right\} \\ \leq & \sup_{|\vartheta|_L \leq 1} \left\{ \left| \int_{H_0^1(D)} \mathbb{E}\Big[\vartheta(u^\nu(t,h)) - \vartheta(u(t,h))\Big] \mu(dh) \right| \right\} \end{aligned}$$

$$\leq \int_{H_0^1(D)} \mathbb{E}\|u^\nu(t,h) - u(t,h)\|_1 \mu(dh)$$

for any $\mu \in X_m$. Here d_P is the Prokhorov metric on measure space and d_W is the Wasserstein metric on measure space. Now it remains to estimate $\mathbb{E}\|u^\nu(t,h) - u(t,h)\|_1$ for $h \in H_0^1(D) \cap H^2(D)$. In the following we give a uniform convergence of $u^\nu(t, u_0, u_1)$ with respect to initial value on bounded set.

We still need some splitting of solution. Let $u = u^1 + u^2$ with u^1 and u^2 solving

$$u_t^1 = \Delta u^1\,, \quad u^1(0) = u_0\,, \quad \text{in } D\,, \quad u^1(t)|_\Gamma = 0\,, \quad t > 0 \tag{4.32}$$

and

$$u_t^2 = \Delta u^2 + f(u^1 + u^2)\,, \quad u^2(0) = 0\,, \quad \text{in } D\,, \quad u^2(t)|_\Gamma = 0\,, \quad t > 0 \tag{4.33}$$

respectively. And also let $u^\nu = p^\nu + q^\nu$ with p^ν and q^ν solving (4.15) and (4.16) respectively.

Then, by the estimates for linear system [24, section 7.2], for any $t > 0$ there is some positive constant C_t such that

$$\mathbb{E}\|p^\nu - u^1\|_1^2 \leq C_t \nu^2 \mathbb{E}\big[\|u_0\|_1^2 + \|u_1\|_0^2\big]\,. \tag{4.34}$$

Next we prove $\mathbb{E}\|q^\nu(t) - u^2(t)\|_1^2 \to 0$ as $\nu \to 0$ for any fixed $t > 0$ and this convergence is uniform in B_0. For this we introduce random process $\eta^\nu(t)$

$$\nu \eta_{tt}^\nu + \eta_t^\nu = \Delta \eta^\nu + \sqrt{\nu}\dot{W}(t)\,, \; \eta^\nu(0) = 0\,, \; \eta_t^\nu(0) = 0$$

and consider $z^\nu = q^\nu - \eta^\nu$, then

$$\nu z_{tt}^\nu + z_t^\nu = \Delta z^\nu + f(u^\nu)\,, \quad z^\nu(0) = 0\,, z_t^\nu(0) = 0.$$

Noticing that $\mathbb{E}\|f(u^\nu)\|_1 \leq C_1 \mathbb{E}\|u^\nu\|_1^3 + C_2$, we can prove that $(z^\nu, z_t^\nu) \in C\big(0,T; \big(H^2(D) \cap H_0^1(D)\big) \times H^1(D)\big)$ for any $T > 0$. Moreover,

$$\mathbb{E}\big[\|z^\nu(t)\|_2^4 + \|z_t^\nu(t)\|_0^4\big] \leq C_t \mathbb{E}(\|u_0\|_1^{12} + \|u_1\|_0^{12})$$

for some positive constant C_t which is independent of ν.

Now let $Z^\nu = z_t^\nu$, then

$$\nu Z_{tt}^\nu + Z_t^\nu = \Delta Z^\nu + f'(u^\nu)u_t^\nu\,, \; Z^\nu(0) = 0\,, \; Z_t^\nu(0) = z_{tt}^\nu(0).$$

By the estimates on (z^ν, z_t^ν), $\nu\mathbb{E}\|z_{tt}^\nu(0))\|_0^2$ is bounded by a constant which is independent of ν. Still by the linear estimates, for any $t > 0$, $\mathbb{E}\|p_t^\nu(t)\|_0^2$ and $\mathbb{E}\|\eta_t^\nu(t)\|_0^2$ are also bounded by a constant which is independent of ν. Then we have

$$\nu\mathbb{E}\|z_{tt}^\nu(t)\|_0^2 = \nu\mathbb{E}\|Z_t^\nu(t)\|_0^2 \leq C_t \mathbb{E}(\|u_0\|_1^{12} + \|u_1\|_0^{12}) \tag{4.35}$$

and

$$\nu^2\mathbb{E}\|z_{tt}^\nu(t)\|_0^4 = \nu^2\mathbb{E}\|Z_t^\nu(t)\|_0^4 \leq C_t \mathbb{E}(\|u_0\|_1^{24} + \|u_1\|_0^{24}) \tag{4.36}$$

for some positive constant C_t which is independent of ν.

Now let $U^\nu = z^\nu - u^2$, then

$$U_t^\nu = \Delta U^\nu + f(u^\nu) - f(u) - \nu z_{tt}^\nu, \quad U^\nu(0) = 0\,, \tag{4.37}$$

and multiplying U^ν on both sides of the above equation in $L^2(D)$ yields

$$\frac{1}{2}\frac{d}{dt}\|U^\nu\|_0^2 \leq -\lambda_1 \|U^\nu\|_0^2 + \langle f(u^\nu) - f(u), U^\nu\rangle - \nu\langle z_{tt}^\nu, U^\nu\rangle\,. \tag{4.38}$$

Notice that

$$\begin{aligned}
&\langle f(u^\nu) - f(u), U^\nu\rangle \\
&= \langle f(u^\nu) - f(u), u^\nu - u - \eta^\nu - p^\nu + u^1\rangle \\
&\leq \alpha\|U^\nu\|_0^2 + |\langle f(u^\nu) - f(u), \eta^\nu\rangle| + |\langle f(u^\nu) - f(u), p^\nu - u^1\rangle| \\
&\leq \alpha\|U^\nu\|_0^2 + \alpha\|p^\nu - u^1\|_0^2 + \alpha\|\eta^\nu\|_0^2 \\
&\quad + \|f(u^\nu) - f(u)\|_0(\|p^\nu - u^1\|_0 + \|\eta^\nu\|_0)\,.
\end{aligned}$$

By the assumption on f and the embedding $H_0^1(D) \hookrightarrow L^6(D)$

$$\mathbb{E}\|f(u^\nu) - f(u)\|_0^2 \leq \left(1 + \mathbb{E}\|u^\nu\|_1^6 + \mathbb{E}\|u\|_1^6\right)\,.$$

Then by (4.34), (4.35) and the linear estimate, for any $t > 0$, Gronwall lemma yields a positive constant C_t such that

$$\mathbb{E}\|U^\nu\|_0^2 \leq C_t \nu[\mathbb{E}\|u_0\|_1^{12} + \mathbb{E}\|u_1\|_0^{12}]\,. \tag{4.39}$$

Furthermore, we need an estimate on $\mathbb{E}\|U^\nu\|_0^4$. In fact, by (4.38) we have for some positive constants C such that

$$\begin{aligned}
\frac{1}{2}\frac{d}{dt}\|U^\nu\|_0^4 \leq -\alpha\|U^\nu\|_0^4 &+ C\|f(u^\nu) - f(u)\|_0^2(\|p^\nu - u^1\|_0^2 + \|\eta^\nu\|_0^2) \\
&+ C(\|p^\nu - u^1\|_0^4 + \|\eta^\nu\|_0^4) + C\nu^4\|z_{tt}^\nu\|_0^4\,.
\end{aligned}$$

Then by (4.34), (4.36) for any $t > 0$, Gronwall lemma yields a positive constant C_t such that

$$\mathbb{E}\|U^\nu\|_0^4 \leq C_t \nu^2[\mathbb{E}\|u_0\|_1^{24} + \mathbb{E}\|u_1\|_0^{24}]\,. \tag{4.40}$$

Multiplying $-\Delta U^\nu$ on both sides of (4.37) and integrating on D, we have

$$\begin{aligned}
\frac{d}{dt}\|U^\nu\|_1^2 &= -2\|\Delta U^\nu\|_0^2 - 2\langle f(u^\nu) - f(u)\,, \Delta U^\nu\rangle + 2\nu\langle z_{tt}^\nu\,, \Delta U^\nu\rangle \\
&\leq -\|\Delta U^\nu\|_0^2 - 2\langle f(u^\nu) - f(u)\,, \Delta U^\nu\rangle + \nu^2\|z_{tt}^\nu\|_0^2\,.
\end{aligned}$$

Notice that, by the embedding of $H^2(D) \hookrightarrow L^\infty(D)$,

$$\begin{aligned}
\langle f(u^\nu) - f(u), \Delta U^\nu\rangle &\leq 2\|f(u^\nu) - f(u)\|_0^2 + \frac{1}{2}\|\Delta U^\nu\|_0^2 \\
&\leq 16\|u^\nu - u\|_0^2\left(\|(u^\nu)\|_2^4 + \|u\|_2^4\right) + \frac{1}{2}\|\Delta U^\nu\|_0^2
\end{aligned}$$

and $u^\nu - u = p^\nu - u^1 + U^\nu + \eta^\nu$, then by estimates (4.12), (4.40) and Gronwall lemma yields that for any $t > 0$, there is a positive constant C_t such that

$$\mathbb{E}\|U^\nu\|_1^2 \leq C_t \nu(\mathbb{E}\|u_0\|_1^{24} + \|u_1\|_0^{24})\,.$$

Then for any $t > 0$ there is a positive constant C_t which is independent of ν such that

$$\mathbb{E}\|u^\nu(t) - u(t)\|_1^2 \le C_t\nu\big[\mathbb{E}\|u_0\|_1^{24} + \mathbb{E}\|u_1\|_0^{24}\big].$$

Thus for any $t > 0$ and any $\mu \in B_{24}^R$, we have

$$\begin{aligned}
&\int_{H_0^1(D)} \mathbb{E}\|u^\nu(t,h) - u(t,h)\|_1 \mu(dh)\\
&\le \int_{H_0^1(D)} \mathbb{E}\|u^\nu(t,h) - u(t,h)\|_1^2 \mu(dh)\\
&\le \nu C_t R.
\end{aligned}$$

Thus we proved (4.30) for $S_{u_1}^\nu(t)$ in space X_{24}. Then by Theorem 12.9, we draw the following result.

Theorem 12.10. *Assume (2.3) holds. For any $u_1 \in L^2(D)$, $\mathcal{A}_{u_1}^\nu$, attractor for semigroup $\{S_{u_1}^\nu(t)\}_{t\ge 0}$ defined by (4.25) in X_{24}, is continuous at $\nu = 0$, that is*

$$\lim_{\nu\to 0} \max\big\{dist(\mathcal{A}_{u_1}^\nu, \mathcal{A}), dist(\mathcal{A}, \mathcal{A}_{u_1}^\nu)\big\} = 0\,.$$

Here $\mathcal{A}$ is the attractor for semigroup $\{S(t)\}_{t\ge 0}$ defined by (4.29) in X_{24}.

5. Approximation of the random inertial manifold

This section considers the approximation in the almost sure sense, but on a long time scale. We consider equations (1.5)–(1.6). Here, one big difficulty in approximating the random invariant manifolds of system (1.5)–(1.6) by that of the limit system (1.7), is that second order derivatives in time of u^ν and u cannot be treated pathwise in the usual phase space. The difficulty for u^ν can be overcome by introducing the following stationary process $z^{*\nu}$ solving the linear system

$$\nu z_{tt}^\nu + z_t^\nu = -Az^\nu + \dot{W} \tag{5.1}$$

and consider $u^\nu - z^{*\nu}$. However, for the limit equation (1.7), if we introduce stationary process z^* solving the linear system

$$z_t = -Az + \dot{W}\,, \tag{5.2}$$

and it is difficult to consider $u - z^*$ with discontinuous process z_t^*. Fortunately, by Rohlin's classification of homomorphisms on Lebesgue space, as the distribution of $z^*(\theta_t\omega)$ is the same as that of $z^{*\nu}(\theta_t\omega)$, there is a measure preserving mapping ψ^ν on the probability space $(\Omega, \mathcal{F}, \mathbb{P})$ such that

$$z^*(\psi^\nu\theta_t\omega) = z^{*\nu}(\theta_t\omega).$$

So we can consider $z^*(\psi^\nu\theta_t\omega)$ instead of $z^*(\theta_t\omega)$.

The approximation result (Theorem 12.14) on random invariant manifolds implies that the approximate system is

$$\tilde{u}_t^\nu(t,x) = -A\tilde{u}^\nu(t,x) + f(\tilde{u}^\nu(t,x)) + W_t^\nu(t,x), \tag{5.3}$$

where the Wiener process $W^\nu(t,x) = \psi^\nu W(t,x)$. This approximating result also shows that, different from the Smolukowski–Kramers approximation on finite time interval, as $t \to \infty$, for small $\nu > 0$ and almost all $\omega \in \Omega$, the solution $u^\nu(t,x,\omega)$ to SWE (1.5) is approximated by $u(t,x,\psi^\nu\omega)$, the solution to SHE (5.3) on the $\psi^\nu\omega$ path.

5.1. *Random invariant manifold*

Next we construct a random invariant manifold for SWE (1.5). For this we assume ($\mathbf{H}_{\mathrm{lip}}$) and (2.2) hold which implies the existence of a unique stationary solution to SWE (1.5) [2]. We denote by A the Laplacian on the domain D with zero boundary condition.

We intend to construct a global random invariant manifold to the unique stationary solution of system (1.5) [2]. However, directly constructing an explicit expression to a stationary solution for a nonlinear SPDEs is very difficult. We use the transformation $\bar{u}^\nu = u^\nu - z^{*\nu}$, and we construct a global random invariant manifold $\widetilde{M}(\omega)$ for $\bar{u}^\nu$, then $M(\omega) = \widetilde{M}(\omega) + z^*(\omega)$ is the global random invariant manifold of system (1.5) [22].

Next, for technical reasons we make the change of variables

$$\bar{u}_t^\nu = -\frac{1}{2\nu}\bar{u}^\nu + \frac{1}{\nu}\bar{v}^\nu \quad \text{and} \quad \bar{U}^\nu = (\bar{u}^\nu, \bar{v}^\nu). \tag{5.4}$$

The above change of variables is similar to that for the deterministic wave equation [10] and to that in previous research on stochastic wave equations [19, 20]. For convenience, we give a simple description of the construction of the random inertial manifold. By the definition of $\bar{U}^\nu$ we have a random differential equation

$$\bar{U}_t^\nu(t,\omega) = C\bar{U}^\nu(t,\omega) + F(\bar{U}^\nu(t,\omega), \theta_t\omega) \tag{5.5}$$

where

$$C = \begin{bmatrix} -\frac{1}{2\nu} & \frac{1}{\nu} \\ \frac{1}{4\nu} - A & -\frac{1}{2\nu} \end{bmatrix}, \quad F(\bar{U}^\nu, \omega) = \begin{bmatrix} 0 \\ f(\bar{u}^\nu + z^{*\nu}) \end{bmatrix}.$$

We apply the Lyapunov–Perron method to construct a random invariant manifold for equation (5.5) based upon a stationary solution $(u^{*\nu}, u_t^{*\nu})$.

Let $E = H_0^1(D) \times L^2(D)$, we introduce the following Banach space

$$C_{\eta,E}^- = \left\{ U \in C((-\infty, 0]; E) : \sup_{t\le 0} e^{-\eta t}\|U(t)\|_E < \infty \right\} \tag{5.6}$$

with norm

$$|U|_{C_{\eta,E}^-} = \sup_{t\le 0} e^{-\eta t}\|U(t)\|_E\,.$$

Let $N > 0$ be an integer. Set

$$E_{11} = \mathrm{span}\left\{ \begin{bmatrix} e_k \\ 0 \end{bmatrix}, \begin{bmatrix} 0 \\ e_k \end{bmatrix} : k = 1, \ldots, N \right\}$$

$$E_{22} = \mathrm{span}\left\{ \begin{bmatrix} e_k \\ 0 \end{bmatrix}, \begin{bmatrix} 0 \\ e_k \end{bmatrix} : k = N+1, N+2, \ldots \right\}.$$

It is evident that $E = E_{11} \oplus E_{22}$, that E_{11} is orthogonal to E_{22} by the orthogonality of $\{e_k\}$, and that $\dim E_{11} = 2N$. Moreover, both E_{11} and E_{22} are invariant subspaces of the operator C.

Since the eigenvalues of A are $-k^2$ with corresponding eigenvectors e_k, $k = 1, 2, \ldots$, by restricting C to E_{11}, the eigenvalues of $C|_{E_{11}}$ are

$$\lambda_k^{\pm} = \frac{-1 \pm \sqrt{1-4\nu k^2}}{2\nu}, \quad k = 1, 2, \ldots, N,$$

with corresponding eigenvectors

$$e_k^{\pm} = \begin{bmatrix} e_k \\ \pm\frac{\sqrt{1-4\nu k^2}}{2} e_k \end{bmatrix}, \quad k = 1, 2, \ldots, N.$$

Let

$$E_1 = \text{span}\{e_k^+ : k = 1, \ldots, N\}, \quad E_{-1} = \text{span}\{e_k^- : k = 1, \ldots, N\}.$$

By this definition $E_{11} = E_1 \oplus E_{-1}$, and E_1 and E_{-1} are invariant subspaces of the operator C. Let P_1 and P_{-1} be the corresponding spectral projections [25] and P_{22} be the unique orthogonal projection onto E_{22}. Then there exists a decomposition $E = E_1 \oplus E_{-1} \oplus E_{22}$ with projections P_1, P_{-1}, P_{22} respectively. Note that E_1 is not orthogonal to E_{-1}. To overcome this we invoke an equivalent inner product on E, as defined for the deterministic wave equations [23], to ensure E_1 is orthogonal to E_{-1}.

Let $U_i = (u_i, v_i)$, $i = 1, 2$ be two elements of E or E_{11}, E_{22}. Assume $\frac{1}{2\sqrt{\nu}} > N+1$, and define the new inner products on E_{11} and E_{22} as

$$\langle U_1, U_2 \rangle_{E_{11}} = \langle \nu A u_1, u_2 \rangle + \frac{1}{4}\langle u_1, u_2 \rangle + \langle v_1, v_2 \rangle,$$

$$\langle U_1, U_2 \rangle_{E_{22}} = \langle -A u_1, u_2 \rangle + \left(\frac{1}{4\nu} - 2(N+1)^2\right) \langle u_1, u_2 \rangle + \langle v_1, v_2 \rangle,$$

where $\langle \cdot, \cdot \rangle$ is the usual inner product of $L^2(D)$. Define the new inner product on E by

$$\langle U, V \rangle_E = \langle U_{11}, V_{11} \rangle_{E_{11}} + \langle U_{22}, V_{22} \rangle_{E_{22}}$$

where $U = U_{11} + U_{22}$ and $V = V_{11} + V_{22}$ with $U_{ii}, V_{ii} \in E_{ii}$, $i = 1, 2$. The corresponding norm is denoted by $\|\cdot\|_E$.

Since $\frac{1}{2\sqrt{\nu}} > N + 1$, $\langle \cdot, \cdot \rangle_{E_{11}}$ is equivalent to the usual inner product on E_{11}, and $\langle \cdot, \cdot \rangle_{E_{22}}$ is equivalent to the usual inner product on E_{22}. Hence the new inner product $\langle \cdot, \cdot \rangle_E$ is equivalent to the usual inner product on E [23].

In terms of this new inner product, by the orthogonality of $\sin kx$, direct methods verify that $E_{-1} \perp E_{22}$ and $E_1 \perp E_{22}$. Moreover, $E_1 \perp E_{-1}$. Let $E_2 = E_{-1} \oplus E_{22}$, then $E_1 \perp E_2$.

By the definition of the new inner product, for $U = (0, v) \in E$,

$$\|U\|_E = \|v\|_{L^2(D)}, \tag{5.7}$$

and for any $U = (u, v) \in E$,

$$\|U\|_E \geq \sqrt{\frac{1}{4} - \nu(N+1)^2} \|u\|_{L^2(D)}. \tag{5.8}$$

Let C_1, C_2, C_{-1}, C_{22} denote $C|_{E_1}, C|_{E_2}, C|_{E_{-1}}, C|_{E_{22}}$ respectively. Then similar to Mora's bounds [23],

$$\|e^{C_1 t}\| \le e^{\lambda_N^+ t} \quad \text{for } t \le 0\,, \tag{5.9}$$

$$\|e^{C_{-1} t}\| \le e^{\lambda_N^- t} \quad \text{for } t \ge 0\,, \tag{5.10}$$

$$\|e^{C_{22} t}\| \le e^{\lambda_{N+1}^+ t} \quad \text{for } t \ge 0\,. \tag{5.11}$$

By the bounds (5.10) and (5.11), we have

$$\|e^{C_2 t}\| \le e^{\lambda_{N+1}^+ t} \quad \text{for } t \ge 0\,.$$

For the nonlinearity F, in terms of the new norm, by (5.7) and (5.8),

$$\begin{aligned}\|F(\bar{U}_1,\omega) - F(\bar{U}_2,\omega)\|_E &= \|f(\bar{u}_1 + z^{*\nu}) - f(\bar{u}_2 + z^{*\nu})\|_{L^2(D)} \\ &\le L_f\|\bar{u}_1 - \bar{u}_2\|_{L^2(D)} \\ &\le \frac{L_f}{\sqrt{\frac{1}{4} - \nu(N+1)^2}}\|\bar{U}_1 - \bar{U}_2\|_E \\ &\le 3L_f\|\bar{U}_1 - \bar{U}_2\|_E.\end{aligned}$$

So F is Lipschitz with respect to $\bar{U}$ and the Lipschitz constant is independent of ν provided the parameter ν is small.

Notice that by choosing $\alpha = \lambda_N^+$, $\beta = \lambda_{N+1}^+$ and $\eta = (\lambda_N^+ + \lambda_{N+1}^+)/2$, for $\nu > 0$ small enough, we have the following gap condition hold for some $K > 0$

$$KL_f\left(\frac{1}{\alpha - \eta} + \frac{1}{\eta - \beta}\right) < 1\,.$$

Then a similar discussion to that by Liu [19] and Lu & Schmalfuß [20] leads to the following theorem.

Theorem 12.11. *There exists $\nu_0 > 0$ such that for any $\nu \in (0, \nu_0)$, there is an N-dimensional inertial manifold $\overline{\mathcal{M}}_E^\nu(\omega)$ for equation (5.5), which is represented by*

$$\overline{\mathcal{M}}_E^\nu(\omega) = \{(\xi, h^\nu(\xi,\omega)) : \xi \in E_1\}$$

with

$$h^\nu(\cdot,\omega) : E_1 \to E_2$$

being Lipschitz continuous. Moreover, if $f \in C^1(L^2(D), L^2(D))$, then the random invariant manifold is C^1, that is $h^\nu \in C^1(E_1, E_2)$.

If further the following gap condition holds

$$KL_f\left(\frac{1}{\alpha - \eta} + \frac{1}{\eta - \beta}\right) + K^2 L_h L_f \frac{1}{\alpha - \eta} < 1\,, \tag{5.12}$$

then $\overline{\mathcal{M}}_E^\nu(\omega)$ is a random inertial manifold for (5.5).

For our purposes we need some estimates of the solution on the random invariant manifold $\overline{\mathcal{M}}_E^\nu(\omega)$. For $\bar{U}_0^\nu = (\xi, h^\nu(\xi,\omega)) \in \overline{\mathcal{M}}_E^\nu(\omega)$, by the invariance of $\overline{\mathcal{M}}_E^\nu(\omega)$,

$\bar{U}^\nu(t,\omega)$, the solution of (5.5) with $\bar{U}^\nu(0) = \bar{U}_0^\nu$ lies on $\overline{\mathcal{M}}_E^\nu(\theta_t\omega)$. Then, by the construction of the random invariant manifold and the uniqueness of solutions, for $t \le 0$ [22],

$$\begin{aligned}\bar{U}^\nu(t,\omega) = e^{Ct}\xi + \int_0^t P_1 e^{C(t-s)} F(\bar{U}^\nu(s,\omega),\omega)\, ds \\ + \int_{-\infty}^t (P_{-1} + P_{22}) e^{C(t-s)} F(\bar{U}^\nu(s,\omega),\omega)\, ds\,.\end{aligned}$$

Notice that

$$\|F(\bar{U}^\nu)\|_E \le \|f(\bar{u}^\nu + z^{*\nu})\|_{L^2(D)} \le 3L_f(\|\bar{U}^\nu\|_E + \|z^{*\nu}\| + 1).$$

Then, by the gap condition (5.12), a direct calculation yields

$$|\bar{U}^\nu|_{C^-_{\eta,E}} \le R_1(\omega) \tag{5.13}$$

for some tempered random variable R_1. Here the Banach space $C^-_{\eta,E}$ is defined by (5.6) through replacing H by E. Next we need the same estimate on $\bar{U}_t^\nu$. Since $\bar{U}^\nu(t,\omega)$ lies on $\overline{\mathcal{M}}^\nu(\omega)$, we have

$$\bar{U}^\nu(t,\omega) = \bar{U}_N^\nu(t,\omega) + h^\nu(\bar{U}_N^\nu(t,\omega),\omega)$$

with $\bar{U}_N^\nu(0,\omega) = \xi$ and $\bar{U}_N^\nu(t,\omega) \in P_1E$ for $t \in \mathbb{R}$. Moreover,

$$\dot{\bar{U}}_N^\nu(t,\omega) = C_1\bar{U}_N^\nu(t,\omega) + P_1F(\bar{U}_N^\nu(t,\omega) + h^\nu(\bar{U}_N^\nu(t,\omega),\omega)).$$

Then by (5.13) and (5.9), the spectrum gap condition (5.12) and the Lipschitz property of h^ν,

$$|\dot{\bar{U}}_N^\nu(t,\omega)|_{C^-_{\eta,E}} \le R_2'(\omega). \tag{5.14}$$

Notice that

$$\dot{\bar{U}}^\nu(t,\omega) = \dot{\bar{U}}_N^\nu(t,\omega) + Dh^\nu(\bar{U}_N^\nu(t,\omega),\omega)\dot{\bar{U}}_N^\nu(t,\omega),$$

then by the bound (5.14), for some tempered random variable R_2,

$$|\dot{\bar{U}}^\nu(t,\omega)|_{C^-_{\eta,E}} \le R_2(\omega). \tag{5.15}$$

5.2. *Approximation of random inertial manifold*

We next derive the approximation of $\overline{\mathcal{M}}_E^\nu$ for small $\nu > 0$. First we consider the stochastic heat equation (1.7). Recall the stationary process z^* that solves (5.2). We make the transformation $\tilde{u} = u - z^*$, then $\tilde{u}$ satisfies

$$\tilde{u}_t(t,\omega) = \Delta\tilde{u}(t,\omega) + f(\tilde{u}(t,\omega) + z^*(\theta_t\omega)). \tag{5.16}$$

Notice that under our assumptions, the stochastic nonlinear heat equation (1.7) has a unique stationary solution. Then we have the following theorem.

Theorem 12.12. *Assume $f \in C^1(L^2(D), L^2(D))$ and $N > 0$ large enough, then the random equation (5.16) has an N-dimensional C^1 random inertial manifold $\widetilde{\mathcal{M}}_{L^2(D)}(\omega)$ with*

$$\widetilde{\mathcal{M}}_{L^2(D)}(\omega) = \{(\zeta, h(\zeta,\omega)) : \zeta \in P_N L^2(D)\}$$

where P_N is the orthogonal projection from $L^2(D)$ to the N-dimensional space span$\{e_1, e_2, \ldots, e_N\}$.

Now we define the following random set in E

$$\widetilde{\mathcal{M}}_E(\omega) = \left\{(\tilde{u}_0, \tilde{u}_t(0, \tilde{u}_0)) : \tilde{u}_0 \in \widetilde{\mathcal{M}}_{L^2(D)}(\omega)\right\}, \tag{5.17}$$

and bounded random set

$$\widetilde{\mathcal{M}}_{ER}(\omega) = \left\{(\tilde{u}_0, \tilde{u}_t(0, \tilde{u}_0)) : \tilde{u}_0 = \zeta + h(\zeta, \omega) \in \widetilde{\mathcal{M}}_{L^2(D)}(\omega), \|\zeta\|_{L^2(D)} < R\right\}, \tag{5.18}$$

where $\tilde{u}(t, \tilde{u}_0)$ is the unique solution of equation (5.16) with initial value $\tilde{u}(0, \tilde{u}_0) = \tilde{u}_0$, and where $R > 0$ is an arbitrary constant.

As mentioned at the begin of this section, we have to consider (5.16) on $\psi^\nu\omega$. For this we define $\tilde{u}^\nu(t, \omega)$ on the fiber $\psi^\nu\omega$ solving

$$\tilde{u}^\nu_t(t, \omega) = \Delta\tilde{u}^\nu(t, \omega) + f(\tilde{u}^\nu(t, \omega) + z^*(\psi^\nu\theta_t\omega)). \tag{5.19}$$

Then by the Lyapunov–Perron method we have an N-dimensional random invariant manifold which is exactly $\widetilde{\mathcal{M}}_{L^2(D)}(\psi^\nu\omega)$. We give a relation between $\overline{M}^\nu_E(\omega)$ and $\widetilde{\mathcal{M}}_{ER}(\psi^\nu\omega)$.

We need some estimates of the solution of equation (5.19) on $\widetilde{\mathcal{M}}_{L^2(D)}(\psi^\nu\omega)$. For $\tilde{u}_0 = (\zeta, h(\zeta, \psi^\nu\omega)) \in \widetilde{\mathcal{M}}_{L^2(D)}(\psi^\nu\omega)$, then $\tilde{u}^\nu(t, \omega)$, the solution to equation (5.19) with $\tilde{u}^\nu(0, \omega) = \tilde{u}_0$, by the invariance of $\widetilde{\mathcal{M}}_{L^2(D)}(\psi^\nu\omega)$, lies on $\widetilde{\mathcal{M}}_{L^2(D)}(\psi^\nu\theta_t\omega)$. Then, by a similar discussion for $\bar{U}^\nu$, for $t \le 0$,

$$\begin{aligned}\tilde{u}^\nu(t, \omega) = e^{At}\zeta &+ \int_0^t P_N e^{A(t-s)} f(\tilde{u}^\nu(s, \omega) + z^{*\nu}(s, \omega))\, ds \\ &+ \int_{-\infty}^t (\mathrm{Id} - P_N) e^{A(t-s)} f(\tilde{u}^\nu(s, \omega) + z^{*\nu}(s, \omega))\, ds\,.\end{aligned}$$

By the gap condition and the Lipschitz property of f, for some tempered random variable R_3,

$$|\tilde{u}^\nu|_{C^-_{\eta, L^2(D)}} \le R_3(\omega).$$

Furthermore, we need the following estimate of $\tilde{u}^\nu_{tt}$ with $\tilde{u}^\nu(t, \omega)$ lying on the random invariant manifold $\widetilde{\mathcal{M}}_{ER}(\psi^\nu\theta_t\omega)$.

Lemma 12.3. *Assume the conditions of Theorem 12.12 hold. There is $R > 0$ such that for $\|\zeta\|_{L^2(D)} \le R$ and $\zeta \in P_N L^2(D)$, then almost surely*

$$\nu\|e^{-\eta t}\tilde{u}^\nu_{tt}(t, \zeta + h(\zeta, \psi^\nu\omega), \omega)\|_{L^2(D)} \to 0\,, \quad t \le 0\,,$$

where $\tilde{u}^\nu$ is the unique solution of the equation (5.16) with $\tilde{u}^\nu(0) = \zeta + h(\zeta, \psi^\nu\omega)$.

Proof. By the invariance of $\widetilde{\mathcal{M}}_{L^2(D)}(\psi^\nu\omega)$ for $\tilde{u}_0 = \zeta + h(\zeta, \psi^\nu\omega)$,

$$\tilde{u}^\nu(t, \tilde{u}_0, \omega) = \tilde{u}^\nu_N(t, \zeta, \omega) + h(\tilde{u}^\nu_N(t, \zeta, \omega), \psi^\nu\omega)$$

with

$$\tilde{u}^\nu_{N,t} = \Delta\tilde{u}^\nu_N + P_N f(\tilde{u}^\nu_N + h(\tilde{u}^\nu_N, \psi^\nu\omega) + z^{*\nu}(t, \omega)), \quad \tilde{u}^\nu_N(0) = \zeta\,. \tag{5.20}$$

Here we use the equality $z^*(\psi^\nu\theta_t\omega) = z^{*\nu}(\theta_t\omega) = z^{*\nu}(t,\omega)$. By a similar discussion to that for the Lyapunov–Perron method to construct a random invariant manifold, we can construct a unique solution $\tilde{u}_N$ to (5.20) which is in the space $C^-_{\eta,P_N L^2(D)}$. Then

$$\|e^{-\eta t}\tilde{u}^\nu_N(t,\omega)\|_{L^2(D)} \le C(\omega), \quad t \le 0\,,$$

for some random constant $C(\omega)$ which is independent of parameter ν. By Theorem 12.12, $h \in C^1$ and Lipschitz, we have

$$\nu\|\tilde{u}^\nu_t\|_{L^2(D)} = \nu\|\tilde{u}^\nu_{N,t}\|_{L^2(D)} + \nu L_h\|\tilde{u}^\nu_{N,t}\|_{L^2(D)}\,. \tag{5.21}$$

Then by (5.20) and above estimate for $e^{-\eta t}\tilde{u}^\nu_N$,

$$\nu\|e^{-\eta t}\tilde{u}^\nu_t(t,\omega)\|_{L^2(D)} \to 0 \quad \text{almost surely for any } t \le 0\,, \quad \text{as} \quad \nu \to 0\,.$$

Now let $\tilde{w}^\nu = \nu\tilde{u}^\nu_t$, then $\tilde{w}^\nu \in C^-_{\eta,L^2(D)}$ and

$$\tilde{w}^\nu_t(t,\omega) = A^\nu(t,\omega)\tilde{w}^\nu(t,\omega) + \nu B^\nu(t,\omega), \quad \tilde{w}^\nu(0) = \nu\tilde{u}^\nu_t(0),$$

with

$$\begin{aligned} A^\nu(t,\omega) &= \Delta + Df(\tilde{u}^\nu(t,\omega) + z^{*\nu}(t,\omega)),\\ B^\nu(t,\omega) &= Df(\tilde{u}^\nu(t,\omega) + z^{*\nu}(t,\omega))z^{*\nu}_t(t,\omega). \end{aligned}$$

Under the assumptions of Theorem 12.12, $\tilde{w}^\nu \in C^-_{\eta,L^2(D)}$ is equivalent to the statement that $\tilde{w}^\nu$ has the following form for $t \le 0$,

$$\begin{aligned} \tilde{w}^\nu(t,\omega) = S^\nu(t,\omega)P_N\tilde{w}^\nu(0) &+ \nu\int_0^t S^\nu(t-s,\omega)P_N B^\nu(s,\omega)\,ds\\ &+ \nu\int_{-\infty}^t S^\nu(t-s,\psi^\nu\omega)(I-P_N)B^\nu(s,\omega)\,ds \end{aligned}$$

where

$$S^\nu(t,\omega) = \exp\left\{\int_0^t A^\nu(s,\omega)\,ds\right\}.$$

Notice that $\tilde{u}^\nu(t,\omega)$ lies on the random invariant manifold $\widetilde{\mathcal{M}}_{L^2(D)}(\psi^\nu\theta_t\omega)$, and $z^{*\nu}$ is stationary, then by the assumption $(\mathbf{H}_{\text{lip}})$, $S^\nu(t,\omega)$ is nonuniformly pseudo-hyperbolic [4], and that $B^\nu(t,\omega)$ is tempered and locally integrable in t. Then by a similar discussion to that for random evolutionary equation [4], we also can construct a random invariant manifold. Then, by the same discussion above for the estimates $\nu\|e^{-\eta t}\tilde{u}^\nu_t(t,\omega)\|_{L^2(D)}$ and the estimate for $z^{*\nu}$ [22, Appendix], we have the estimate

$$\nu\|e^{-\eta t}\tilde{w}^\nu_t(t,\omega)\|_{L^2(D)} \to 0 \quad \text{almost surely for any } t \le 0\,, \quad \text{as} \quad \nu \to 0\,,$$

which completes the proof. □

Now we establish the following theorem on the relation between $\overline{\mathcal{M}}^{\nu}_E(\omega)$ and $\widetilde{\mathcal{M}}_{ER}(\psi^\nu\omega)$.

Theorem 12.13. *Suppose $f \in C^1(L^2(D), L^2(D))$ and $N > 0$ large enough. Then for any $R > 0$,*

$$\lim_{\nu\to 0} \mathrm{dist}_E \left(\widetilde{\mathcal{M}}_{ER}(\psi^\nu\omega), \overline{\mathcal{M}}^{\nu}_E(\omega)\right) = 0\,.$$

Proof. We adapt the discussion for the deterministic case [10].

Let any element $(\tilde{u}_0, \tilde{u}^\nu_t(0, \tilde{u}_0)) \in \widetilde{M}_{ER}(\psi^\nu\omega)$ with $\tilde{u}^\nu$ satisfying equation (5.19) with initial condition $\tilde{u}^\nu(0) = \tilde{u}_0$. Define

$$\tilde{U}^\nu(t) = \left(\tilde{u}^\nu(t), \frac{1}{2}\tilde{u}^\nu(t) + \nu\tilde{u}^\nu_t(t)\right),$$

then $\tilde{U}^\nu = (\tilde{u}^\nu, \tilde{v}^\nu)$ satisfies

$$\dot{\tilde{U}}^\nu(t,\omega) = C\tilde{U}^\nu(t,\omega) + \begin{bmatrix} 0 \\ f(\tilde{u}^\nu(t,\omega) + z^{*\nu}(t,\omega)) \end{bmatrix} + \begin{bmatrix} 0 \\ \nu\tilde{u}^\nu_{tt}(t,\omega) \end{bmatrix}.$$

Let $\bar{U}^\nu \in \overline{\mathcal{M}}^{\nu}_E(\omega)$ be a solution of equation (5.5) and $0 < \nu < \nu_0$. Let

$$\hat{U}^\nu(t,\omega) = \tilde{U}^\nu(t,\omega) - \bar{U}^\nu(t,\omega).$$

Hence, $\hat{U}^\nu(t,\omega)$ satisfies

$$\begin{aligned}\dot{\hat{U}}^\nu(t,\omega) &= C\hat{U}^\nu(t,\omega) + \begin{bmatrix} 0 \\ \nu\tilde{u}^\nu_{tt}(t,\omega) \end{bmatrix} \\ &\quad + \begin{bmatrix} 0 \\ f(\tilde{u}^\nu(t,\omega) + z^{*\nu}(t,\omega)) - f(\bar{u}^\nu(t,\omega) + z^{*\nu}(t,\omega)) \end{bmatrix}.\end{aligned}$$

Notice that by the Lipschitz property of f and by Lemma 12.3,

$$\Delta\tilde{u}^\nu(t,\omega) = \tilde{u}^\nu_t(t,\omega) - f(\tilde{u}^\nu(t,\omega) + z^{*\nu}(t,\omega)) \in C^-_{\eta,L^2(D)}\,.$$

Then by the interpolation between $H^2(D)$ and $L^2(D)$,

$$\tilde{u}^\nu \in C^-_{\eta,H^1_0}\,.$$

By Lemma 12.3 we have $\hat{U}^\nu \in C^-_{\eta,E}$, then by the construction of solution in $C^-_{\eta,E}$,

$$\begin{aligned}\hat{U}^\nu(t) &= e^{Ct}P\hat{U}^\nu(0) \\ &\quad + \int_0^t e^{C(t-s)}P\left\{\begin{bmatrix} 0 \\ f(\tilde{u}^\nu + z^{*\nu}) - f(\bar{u}^\nu + z^{*\nu}) \end{bmatrix} + \begin{bmatrix} 0 \\ \nu\tilde{u}^\nu_{tt} \end{bmatrix}\right\} ds \\ &\quad + \int_{-\infty}^t e^{C(t-s)}Q\left\{\begin{bmatrix} 0 \\ f(\tilde{u}^\nu + z^{*\nu}) - f(\bar{u}^\nu + z^{*\nu}) \end{bmatrix} + \begin{bmatrix} 0 \\ \nu\tilde{u}^\nu_{tt} \end{bmatrix}\right\} ds.\end{aligned}$$

Since $PE = E_1$ is of finite dimension, we can choose $\bar{u}(0)$ and $\bar{u}_t(0)$ such that $P\hat{U}(0) = 0$. Hence,

$$e^{-\eta t}\|\hat{U}^\nu(t)\|_E$$

$$
\begin{aligned}
&\le e^{-\eta t}\int_t^0 e^{\lambda_N^+(t-s)}\left\{\left\|\begin{bmatrix}0\\ f(\tilde{u}^\nu+z^{*\nu})-f(\bar{u}^\nu+z^{*\nu})\end{bmatrix}\right\|_E+\left\|\begin{bmatrix}0\\ \nu\tilde{u}_{tt}^\nu\end{bmatrix}\right\|_E\right\}ds\\
&\quad+e^{-\eta t}\int_{-\infty}^t e^{\lambda_{N+1}^+(t-s)}\left\{\left\|\begin{bmatrix}0\\ f(\tilde{u}^\nu+z^{*\nu})-f(\bar{u}^\nu+z^{*\nu})\end{bmatrix}\right\|_E+\left\|\begin{bmatrix}0\\ \nu\tilde{u}_{tt}^\nu\end{bmatrix}\right\|_E\right\}ds\\
&\le 3L_f\int_t^0 e^{(\lambda_N^+-\eta)(t-s)}\|\hat{U}^\nu\|_{C_{\eta,E}^-}\,ds+3L_f\int_{-\infty}^t e^{(\lambda_{N+1}^+-\eta)(t-s)}\|\hat{U}^\nu\|_{C_{\eta,E}^-}\,ds\\
&\quad+\nu e^{-\eta t}\int_t^0 e^{\lambda_N^+(t-s)}\|\tilde{u}_{tt}^\nu\|\,ds+\nu e^{-\eta t}\int_{-\infty}^t e^{\lambda_{N+1}^+(t-s)}\|\tilde{u}_{tt}^\nu\|\,ds.
\end{aligned}
$$

Then by Lemma 12.3, for N large enough we have $\|\hat{U}^\nu\|_{C_{\eta,E}^-}\to 0$, as $\nu\to 0$. Hence, $\|\hat{U}^\nu(0)\|\to 0$, as $\nu\to 0$. The proof is complete. □

Next we show the approximation of the random dynamics u^ν. For this we define the random sets

$$\overline{\mathcal{M}}^\nu_{L^2(D)}(\omega)=\left\{\bar{u}:\bar{U}=(\bar{u},\bar{v})\in\overline{\mathcal{M}}^\nu_E(\omega)\quad\text{for some }\bar{v}\in L^2(D)\right\}$$

and

$$\overline{\mathcal{M}}^\nu_{L^2(D),R}(\omega)=\left\{\bar{u}\in\overline{\mathcal{M}}^\nu_{L^2(D)}(\omega):\bar{U}=(\bar{u},\bar{v})=\xi+h^\nu(\xi,\omega),\ \|\xi\|_{L^2(D)}\le R\right\}.$$

We next prove that for small parameter $\nu>0$, $\overline{\mathcal{M}}^\nu_{L^2(D),R}(\omega)$ is approximated by $\widetilde{\mathcal{M}}_{L^2(D)}(\psi^\nu\omega)$.

Theorem 12.14. *Suppose $f\in C^1(L^2(D),L^2(D))$ and $N>0$ large enough. Then for any $R>0$,*

$$\lim_{\nu\to 0}\operatorname{dist}_{L^2(D)}\left(\overline{\mathcal{M}}^\nu_{L^2(D),R}(\omega),\widetilde{\mathcal{M}}_{L^2(D)}(\psi^\nu\omega)\right)=0\,.$$

Proof. The proof is similar to that of the deterministic case. Let $\bar{U}_0=(\bar{u}_0,\bar{v}_0)$ with $\bar{u}_0\in\overline{\mathcal{M}}^\nu_{L^2(D),R}(\omega)$ and $\bar{U}^\nu(t,\omega)=(\bar{u}^\nu(t,\omega),\bar{v}^\nu(t,\omega))$ be the solution of (5.5) with $\bar{U}^\nu(0)=\bar{U}(0)$. By the invariance of $\overline{\mathcal{M}}^\nu_E(\omega)$,

$$\bar{u}_t^\nu=A\bar{u}^\nu+f(\bar{u}^\nu+z^{*\nu})-\nu\bar{u}_{tt}^\nu\,.$$

Let $\tilde{u}^\nu(t,\omega)$ be the solution of equation (5.19) on the random inertial manifold $\widetilde{\mathcal{M}}_{L^2(D)}(\psi^\nu\omega)$ with $\tilde{u}^\nu(0)=\tilde{u}_0\in L^2(D)$. Thus $\hat{u}^\nu(t,\omega)=\tilde{u}^\nu(t,\omega)-\bar{u}^\nu(t,\omega)$ satisfies

$$
\begin{aligned}
\hat{u}_t^\nu(t,\omega)&=A\hat{u}^\nu(t,\omega)+f(\tilde{u}^\nu(t,\omega)+z^{*\nu}(t,\omega))\\
&\quad-f(\bar{u}^\nu(t,\omega)+z^{*\nu}(t,\omega))+\nu\bar{u}_{tt}^\nu(t,\theta_{-t}\omega).
\end{aligned}
$$

Notice that $\tilde{u}^\nu(t,\omega)\in C^-_{\eta,L^2(D)}$ and $\bar{U}^\nu\in\overline{\mathcal{M}}^\nu_E(\omega)$, we have $\hat{u}^\nu\in C^-_{\eta,L^2(D)}$. Then,

$$
\begin{aligned}
\hat{u}^\nu(t,\omega)=e^{At}P_N\hat{u}^\nu(0)+\int_0^t e^{A(t-s)}P_N[&f(\tilde{u}^\nu(s,\omega)+z^{*\nu}(s,\omega))\\
&-f(\bar{u}^\nu(s,\omega)+z^{*\nu}(s,\omega))+\nu\bar{u}_{tt}^\nu(s,\omega)]\,ds
\end{aligned}
$$

$$+\int_{-\infty}^{t} e^{A(t-s)}(I-P_N)[f(\tilde{u}^\nu(s,\omega)+z^{*\nu}(s,\omega))$$
$$-f(\bar{u}^\nu(s,\omega)+z^{*\nu}(s,\omega))+\nu\bar{u}^\nu_{tt}(s,\omega)]\,ds\,.$$

We need an estimate on $\nu\bar{u}^\nu_{tt}$. Suppose we have the following expansion in the orthonormal basis of the eigenfunctions e_k of A, $\bar{u}^\nu_t=\sum_{k=1}^{\infty}\bar{u}^\nu_{t,k}e_k$. Then by integration by parts,

$$\begin{aligned}
&\sup_{t\le 0} e^{-\eta t}\left\|\int_0^t e^{A(t-s)}P_N\bar{u}^\nu_{tt}(s)\,ds\right\|_{L^2(D)}\\
&=\sup_{t\le 0} e^{-\eta t}\left\|P_N\bar{u}^\nu_t(t)-e^{At}P_N\bar{u}^\nu_t(0)-\sum_{k=1}^{N}k^2\int_0^t e^{-k^2(t-s)}\bar{u}^\nu_{t,k}(s)e_k\,ds\right\|_{L^2(D)}\\
&\le \sup_{t\le 0} e^{-\eta t}\|P_N\bar{u}^\nu_t(t)\|_{L^2(D)}+\sup_{t\le 0} e^{-(\eta+N^2)t}\|P_N\bar{u}^\nu_t(0)\|_{L^2(D)}\\
&\quad+\sup_{t\le 0} e^{-\eta t}\left\|\sum_{k=1}^{N}k^2\int_0^t e^{-k^2(t-s)}\bar{u}^\nu_{t,k}(s)e_k\,ds\right\|_{L^2(D)}.
\end{aligned}$$

For the last term in the above equation, we consider its square as

$$\begin{aligned}
&\sup_{t\le 0} e^{-2\eta t}\left\|\sum_{k=1}^{N}k^2\int_0^t e^{-k^2(t-s)}\bar{u}^\nu_{t,k}(s)e_k\,ds\right\|^2_{L^2(D)}\\
&=\sup_{t\le 0}\sum_{k=1}^{N}\left[k^2\int_0^t e^{-k^2(t-s)}e^{-\eta(t-s)}e^{-\eta s}\|\bar{u}^\nu_{t,k}(s)\|_{L^2(D)}\,ds\right]^2\\
&\le \sup_{t\le 0}\sum_{k=1}^{N}\left[k^2\int_0^t e^{-k^2(t-s)}e^{-\eta(t-s)}\,ds\right]^2\|\bar{u}^\nu_{t,k}\|^2_{C_{\eta,\mathbb{R}}}\\
&\le \|P_N\bar{u}^\nu_t\|^2_{C^-_{\eta,L^2(D)}},
\end{aligned}$$

where we use that $\eta<-N^2$. Then

$$\sup_{t\le 0} e^{-\eta t}\left\|\int_0^t e^{A(t-s)}P_N\bar{u}^\nu_{tt}(s)\,ds\right\|_{L^2(D)}\le 3\|P_N\bar{u}^\nu_t\|_{C^-_{\eta,L^2(D)}}. \tag{5.22}$$

For the higher modes of $\bar{u}^\nu_t$, similarly we have

$$\begin{aligned}
&\sup_{t\le 0} e^{-\eta t}\left\|\int_{-\infty}^t e^{A(t-s)}(I-P_N)\bar{u}^\nu_{tt}(s)\,ds\right\|_{L^2(D)}\\
&\le 3\sup_{t\le 0} e^{-\eta t}\|(I-P_N)\bar{u}^\nu_t(t)\|_{L^2(D)}\\
&\le 3\|(I-P_N)\bar{u}^\nu_t\|_{C^-_{\eta,L^2(D)}}.
\end{aligned} \tag{5.23}$$

Since $\bar{U}^\nu_t=(\bar{u}^\nu_t,\bar{v}^\nu_t)\in C^-_{\eta,E}$, and by the same discussion as that for Theorem 12.13, we have

$$|\hat{u}^\nu|_{C^-_{\eta,L^2(D)}}\to 0,\quad\text{as}\quad \nu\to 0\,. \tag{5.24}$$

This completes the proof. □

Remark 12.4. As stationary solutions lie on a random invariant manifold, the above approximation for a random invariant manifold implies that the distribution of stationary solutions to both SWE (1.5)–(1.6) and SHE (1.7) coincide with each other. This coincidence was also shown by Cerrai and Freidlin [6] under certain conditions.

References

[1] Arnold, L. (1998). Random Dynamical Systems, (Springer–Verlag, New York/Berlin).

[2] Barbu, V. and Da Prato, G. (2002). The stochastic nonlinear damped wave equation, Appl. Math. Optim., **46(2–3)**, 125–141.

[3] Billingsley, P. (1999). Convergence of Probability Measures 2nd, (John Wiley & Sons, New York).

[4] Caraballo, T., Duan, J., Lu, K. and Schmalfuß, B. (2010). Invariant manifolds for random and stochastic partial differential equations, Adv. Nonlinear Stud., **10(1)**, 23–52.

[5] Caraballo, T., Langa, J. A. and Robinson, J. C. (1998). Upper semicontinuity of attractors for small random perturbations of dynamical systems, Communications in Partial Differential Equation, **23**, 1557–1581.

[6] Cerrai, S. and Freidlin, M. (2006). On the Smoluchowski–Kramers approximation for a system with an infinite number of degrees of freedom, Prob. Th. and Relat. Fields, **135**, 363–394.

[7] Cheban, D. (2004). Global attractors of nonautonomous dissipative dynamical systems, Interdisciplinary Mathematical Sciences, **1**, (World Scientific).

[8] Chow, P. (2002). Stochastic wave equation with polynomial nonlinearity, Ann. Appl. Probab., **12(1)**, 361–381.

[9] Chow, P., Kohler W. and Papanicolaou, G. (1981). Multiple Scattering and Waves in Random Media, (North-Holland, Amsterdam).

[10] Chow, S. and Lu, K. (1988). Invariant manifolds for flows in Banach spaces, J. Diff. Equa., **74**, 285–317.

[11] Crauel, H. (2008). Measure attractors and Markov attractors, Dynamical Systems, **23**, 75–107.

[12] Crauel, H. and Flandoli, F. (1994). Attractor for random dynamical systems, Probability Theory and Related Fields, **100**, 365–393.

[13] Crauel, H., Debussche, A. and Flandoli, F. (1997). Random attractors, J. of Dynamics and Differential Equations, **9**, 307–341.

[14] Duan, J., Lu, K. and Schmalfuß, B. (2003). Invariant manifolds for stochastic partial differential equations, Annals Probability, **31(4)**, 2109–2135.

[15] Hale, J. K. (1988). Asymptotic Behavior of Dissipative Systems, (Amer. Math. Soc. Providence, RI).

[16] Kornev, A. (1999). On a criterion of the complete continuity of an attractor with respect to a parameter for a certain class of semidynamical systems, Dikl. Akad. Nank Rus., **369(5)**, 597–599.

[17] Da Prato, G. and Zabczyk, J. (1992). Stochastic Equations in Infinite Dimensions, (Cambridge University Press).

[18] Da Prato, G. and Zabczyk, J. (1996). Ergodicity for Infinite Dimensional Systems, (Cambridge University Press).

[19] Liu, Z. (2010). Stochastic inertial manifolds for damped wave equations, Stoch. Dyn., **10(2)**, 211–230.

[20] Lu, K. and Schmalfuß, B. (2007). Invariant manifolds for stochastic wave equations, J. Diff. Equa., **236(2)**, 460–492.

[21] Lv, Y. and Wang, W. (2008). Limit dynamics for stochastic wave equations, J. Diff. Equa., **244**, 1–23.

[22] Lv, Y., Wang, W. and Roberts, A. J. (2014). Approximation of the random inertial manifold of singularly perturbed stochastic wave equations, Stoch. Dyna., **14(2)**, 1350018 (21 pages).

[23] Mora, X. (1987). Finite-dimensional attracting invariant manifolds for damped semilinear wave equations, Contributions to Nonlinear Partial Differential Equations **II** (Paris), 172–183. Pitman Res. Notes Math. Ser., **155**, (Longman Sci. Tech., Harlow).

[24] Wang, W. and Lv, Y. (2010). Limit behavior of nonlinear stochastic wave equations with singular perturbation, Disc. and Cont. Dyna. Syst., **13(1)**, 175–193.

[25] Pazy, A. (1983). Semigroups of Linear Operators and Applications to Partial Differential Equations, Applied Mathematical Sciences, Vol. 44, (Springer–Verlag, New York).

[26] Brzeniak, Z., Ondrejt, M. and Seidler, J. (2015). Invariant measures for stochastic nonlinear beam and wave equations, J. Differential Equations, **260(5)**, 1–23.

[27] Sanz-Sol, M. and Quer-Sardanyons, L. (2004). A stochastic wave equation in dimension 3: smoothness of the law, Bernoulli, **10(1)**, 165–186.

[28] Rokhlin, V. A. (1962). On the fundamental ideas of measure theory, Tran. Amer. Math. Soc. **1(10)**, 1–54.

[29] Schmalfuß, B. (1992). Backward cocycles and attractors of stochastic differential equations, In International Seminar on Applied Mathematics-Nonlinear Dynamics: Attractor Approximation and Global Behaviour, Reitmann, V., Riedrich, T. and Koksch, N. (Eds.), 185–192.

[30] Schmalfuß, B. (1999). Measure attractors and random attractors for stochastic partial differential euqations, Stoch. Anal. and Appl., **17**, 1075–1101.

[31] Sell, G. R. and You, Y. (2002). Dynamics of Evolutionary Equations, (Springer, New York).

[32] Temam, R. (1997). Infinite-Dimensional Dynamical Systems in Mechanics and Physics 2nd edition, (Springer–Verlag, New York).

Lightning Source UK Ltd.
Milton Keynes UK
UKHW051004151119
353519UK00005B/109/P

9 789811 200342